Coastal and Estuarine Studies

Series Editors:
Malcolm J. Bowman   Christopher N.K. Mooers

Coastal and Estuarine Studies

53

D. G. Aubrey and C.T. Friedrichs (Eds.)

Buoyancy Effects on Coastal and Estuarine Dynamics

American Geophysical Union

Washington, DC

Series Editors

Malcolm J. Bowman
Marine Sciences Research Center, State University of New York
Stony Brook, N.Y. 11794, USA

Christopher N.K. Mooers
Division of Applied Marine Physics
RSMAS/University of Miami
4600 Rickenbacker Cswy.
Miami, FL 33149-1098, USA

Editors

David G. Aubrey
Department of Geology and Geophysics
Woods Hole Oceanographic Institution
360 Woods Hole Road, MS 22
Woods Hole, MA 021543-1541

Carl T. Friedrichs
Virginia Institute of Marine Science
School of Marine Science
College of William & Mary
Gloucester Point, VA 23062-1346

Library of Congress Cataloging-in-Publication Data

Buoyancy effects on coastal and estuarine dynamics / D.G. Aubrey and
   C.T. Friedrichs (eds.).
         p.   cm. -- (Coastal and estuarine studies, ISSN 0733-9565 ; 53)
         Articles based on presentations at the 7th International Biennial
Conference on Physics of Estuaries and Coastal Seas, held at the
Woods Hole Oceanographic Institution, Nov. 28-30, 1995.
         Includes bibliographical references and index.
         ISBN 0-87590-267-7
         1. Estuarine oceanography--Congresses. 2. Ocean circulation--
Congresses. 3. Coasts--Congresses. I. Aubrey, David G.
II. Friedrichs, C.T. (Carl T.) III. International Conference
on Physics of Estuaries and Coastal Seas (7th : 1995 : Woods Hole
Oceanographic Institution) IV Series.
GC96.5.B86   1996   96-38037
551.46'09--dc21   CIP

ISSN 0733-9569
ISBN 0-87590-267-7

CONTENTS

PREFACE

Coastal water quality, flooding, estuarine habitat diversity, and distribution of coastal organisms depend in part on the dynamics of the coastal water column. Particularly within coastal embayments and estuaries, areas within the influence of freshwater from surface and ground water sources, the water column may be stratified by temperature and/or salinity. Resulting density gradients affect the behavior of the water column, including mixing and transport processes. Understanding physical processes associated with buoyancy in the coastal oceans is a requisite first step towards understanding the effects of buoyancy on coastal processes, including geological, biological and geochemical aspects.

This volume presents 23 papers addressing various aspects of buoyancy in the coastal oceans, including plumes, tidal interaction with buoyancy, shelf dynamics and mixing processes, and estuarine dynamics of buoyancy. The interwoven common thread amongst these articles is how buoyancy processes affect the density stratification and dynamics of shallow coastal flows.

The articles in this volume are based on presentations made at the 7th International Biennial Conference on Physics of Estuaries and Coastal Seas (PECS), held at the Woods Hole Oceanographic Institution (WHOI) November 28-30, 1995. The conference, subtitled "Buoyancy Effects on Coastal Dynamics," was attended by nearly 100 participants with broad international participation. Previous PECS meetings, held in Hamburg, Miami, Qingdao, Monterey, Gregynog, and Margaret River, have resulted in focused volumes published within the Coastal and Estuarine Studies series. PECS was created to foster interaction amongst international scientists and engineers specializing in coastal circulation and mixing processes, two disciplines which have approached the topic from different perspectives.

A second volume also has arisen from the 7th International PECS Conference, focusing on sediment dynamics in estuaries and coastal seas. Being published simultaneously with the present volume as a Special Issue of the Journal of Coastal Research, the six papers discuss related topics of sediment transport processes within these coastal environments.

I wish to thank the many people who made this conference and this volume possible. First, the International Steering Committee graciously accepted Woods Hole as the venue for the 7th International Conference. I would like to thank this committee (Charitha Pattiaratchi of the Centre for Water Research, The University of Western Australia; Ralph Cheng of the U.S. Geological Survey; David Prandle of the Proudman Oceanographic Laboratory; J. Dronkers of the National Institute for Coastal and Marine Management of The Netherlands; and Jacobus van de Kreeke of the University of Miami) for allowing us to showcase our wonderful spot of the globe. The local organizing committee of W. R. Geyer, R. P. Signell, and B. W. Tripp helped to assure a balanced technical content and strong poster session. Pamela Barrows of WHOI performed yeoman's chores for this conference: she helped organized the meeting, produced the abstracts volume, implemented the meeting, and followed through on the scientific publications by typing, proofing, and organizing this final volume. My sincerest gratitude to Pam, without whom neither the meeting nor this volume would have emerged.

In addition to Pam, numerous others from WHOI contributed to the success of this conference, including Linda Lotto, Tim Silva, Jeannine Pires and Jack Cook, and others too numerous to be named. Funding for the conference was provided in part by the Office of Naval Research, Department of the Navy (under grant N00014-94-1-1055); the NOAA National Sea Grant College Program Office, Department of Commerce, under grants NA90AA-D-SG480 and NA46RGO470, Woods Hole Oceanographic Institution projects A/S-27-PD and A/S-30-PD; the U.S. Geological Survey through its cooperative program with WHOI; the Woods Hole Oceanographic Institution's Coastal Research Center; and WHOI, which provided the meeting venue.

Finally, I wish to thank the reviewers for their valuable work and the authors themselves, as we struggled with formats and figures to the very end.

David G. Aubrey

1

Introduction

C. T. Friedrichs and D.G. Aubrey

Part I: The Importance of Buoyancy in Estuarine and Coastal Environments

The shores and resources of estuarine and coastal waters have attracted human settlement for millennia. Though not obvious to most of these "settlers," buoyancy dynamics contribute to the attraction of these estuarine and coastal waters. The combination of intermittent stratification with sufficient shallowness for frequent nutrient delivery by complete vertical mixing favors a convenient, productive fishery. Strong density gradients in coastal waters usually indicate a connection between rivers and the sea, often providing a natural highway for trade or a harbor for protection. And land bordering the coast may well have been a flood plain in recent geologic time, leaving soils well-suited for agriculture. As human population has mushroomed during the past two hundred years, environments dominated by buoyancy dynamics have also been among those most severely altered. Strong buoyancy gradients often imply restricted water exchange or restricted mixing, indicating an environment susceptible to over-enrichment by agricultural runoff or sewage, or one easily contaminated by industrial waste. Restricted exchange also favors ecological isolation and areas particularly sensitive to over-fishing or to introduction of new species by human activities. Shallow regions whose dynamics respond most strongly to buoyancy input are often those most easily affected by dredging, diking or changes in relative sea level. In recent years, the competing interests of commerce, agriculture and industry versus sustainable fisheries, quality of life and tourism have come to a head and helped accelerate the impetus for study of coastal and estuarine buoyancy dynamics.

Study of estuarine and coastal environments during the last four decades has highlighted many physical processes influenced by buoyancy at first order. In the 1950's, the first process-oriented classification systems for estuaries (salt wedge, partially mixed, homogeneous) were based on buoyancy distribution, whereas in the 1960's dynamical models for estuarine circulation treated the along-channel density gradient as the dominant forcing. The important role of gravitational circulation in the transport of larvae, contaminants and sediment (e.g., the turbidity maximum) was recognized early on. About the same time, a separate but similarly simplified view of buoyant flows on the inner shelf developed based on the interaction of the coastal wall, the earth's rotation and idealized density gradients, while neglecting bottom friction and intra- and infra-tidal effects. In the 1970's and 1980's, it became increasingly apparent that the restrictive assumptions of two-dimensional circulation, idealized density fields, and simplistic tidal averaging represented the lowest order behavior of only a small fraction of environments strongly influenced by buoyancy input. Lines of research developed investigating topographic control and transverse circulation, the roles of spatially-varying density gradients and fronts, intra- and infra-tidal exchange and mixing processes, controls on tidally-averaged dispersion, and alternative sources and sinks of buoyancy, to name just a few. More subtle aspects of buoyancy dynamics were found to be highly applicable in both channelized estuaries and less confined coastal seas, such as intratidal mixing, rapidly

Buoyancy Effects on Coastal and Estuarine Dynamics
Coastal and Estuarine Studies Volume 53, Pages 1-6
Copyright 1996 by the American Geophysical Union

evolving fronts, differential advection and straining of the density field, and controls on dispersion by interacting spatial and temporal scales.

1. Prominent Themes

This volume represents a maturation of several lines of research beyond the classical regime of estuarine and coastal circulation driven by steady, idealized density gradients. Three themes which receive particular attention in this volume and cut across the traditional geographic distinctions between estuaries and coastal seas are: (i) topographic steering and hydraulic control of buoyant flows; (ii) dispersion and time-scales of shear and mixing; and (iii) estuarine and coastal systems influenced by negative sources of buoyancy.

1a) Topographic Steering and Hydraulic Control of Buoyant Flows

Many of the papers in this volume emphasize the role of irregular bathymetry in influencing coastal and estuarine circulation and exchange in a variety of dynamic settings. In deeper environments such as coastal seas (tens to hundreds of meters), conservation of vorticity under near geostrophy requires flow driven by buoyancy to closely follow irregular isobars, as is the case in Hannah and Loder's modeling of the Scotian Shelf and Gulf of Maine region. An extension of this simple lateral balance also explains why estuarine plumes follow the curves of the coastline with the shoreline on the right in the northern hemisphere (Blanton; Mavor and Huq; Visser) and suggests a preference for northern hemisphere mean flow along the right side of relatively deep estuaries. However, three papers in this volume which address intermediate depth, partially-mixed estuaries (Friedrichs and Hamrick; Valle-Levinson and O'Donnell; Wong), observe the influence of Coriolis on the lateral distribution of along-channel mean velocity to be minor. Rather, mean buoyancy-driven flow over channel-parallel shoals tends to be seaward, whether on the left or right side of the channel, whereas flow over deeper areas tends to be landward. Lateral segregation of mean flow in these systems appears to be explained largely by a laterally invariant baroclinic pressure gradient acting over a depth-varying cross-section.

Several articles address the importance of bathymetric irregularities in the form of rapid along-channel changes in channel depth or width via their role in hydraulic control of two-layer flow (Chadwick et al.; Cudaback and Jay; Kay et al.; Geyer and Nepf; Uncles and Stephens). Plunging intrusion fronts are observed on the flood tide near sudden increases in channel depth or width where the associated decrease in tidal velocity can no longer push back the more buoyant estuarine water found up-estuary (Uncles and Stephens). On the ebb, tidal outflow is observed to lift off the bottom near an expansion when the reduced inertia no longer prevents denser water downstream from nosing underneath (Chadwick et al.). If the local acceleration term becomes large in comparison to the convective term, however, the quasi-steady approximation begins to break down, and the dynamic effects of time-dependence on the hydraulic response must be considered (Cudaback and Jay). Significant salt transport due to tidal pumping may result if the hydraulic responses triggered by a constriction during flood and ebb are asymmetric (Geyer and Nepf; Kay et al.). Under high discharge conditions in the Hudson, for example, Geyer and Nepf observed strongly sheared conditions to persist on the ebb, causing the sharp pycnocline to deepen within a lateral constriction. During the flood, the hydraulic response was subdued due to reduced shear, and the tidally-averaged transport of salt due to this tidal pumping effect was the same order as that due to gravitational circulation.

1b) Dispersion and Time-Scales of Shear and Mixing in Buoyant Flows

A theme which appears in several articles is the relationship between time-scales for shear and mixing and their control on the magnitude of dispersion.   Buoyancy plays various important roles controlling shear and/or mixing in estuarine and coastal environments. Shear and mixing control dispersion, which then feeds back to determine the distribution of buoyancy. For classical circulation in partially-mixed tidal estuaries, the horizontal density gradient brought about by fresh water leads to vertical shear in the along-channel flow, whereas tides and bottom friction create (and stratification inhibits) vertical mixing perpendicular to the shear (Friedrichs and Hamrick; Park and Kuo). In a neat trade of roles in more strongly tidal estuaries, differential tidal advection creates cross-channel shear in the along-channel flow, and the resulting cross-channel density gradient can lead to cross-channel gravitational circulation which "mixes" across the shear (Dronkers; R. Smith; Valle-Levinson and O'Donnell; Uncles and Stephens). In coastal plumes and regions of freshwater influence, vertical shear in the cross-shelf current is provided by wind, the cross-shelf density gradient, and tidal interaction with stratification; by contrast, vertical mixing perpendicular to the shear is provided by the wind and tide (Blanton; Souza and Simpson; Visser).

In each of the above cases, the effective diffusivity in the direction parallel to the sheared flow is determined by the time-scale of the shear relative to the time scale of the mixing. When time scales for mixing are short relative to time scales for shear, shear dispersion becomes larger as mixing decreases (R. Smith; Visser). This property can be seen in the partially-mixed estuary model of Park and Kuo, for which spring-neap modulation of tidal mixing increases dispersion at neaps and decreases dispersion at springs. In the Rhine coastal plume, Visser identifies a similar pattern. Strong cross-shore winds create shear in the plume which alone should eventually spread out the plume in the cross-shore direction. However, strong winds cause strong vertical mixing over a time-scale much shorter than cross-shore advection. This means the cross-shore velocity has very little time to transport density horizontally before it is mixed vertically. R. Smith describes a similar ratio of time-scales in explaining the sensitivity of salt intrusions to the size of tidal estuaries. In small estuaries, where cross-channel mixing is fast, decreased freshwater input reduces cross-channel circulation and increases along-channel dispersion. Thus salt "diffuses" much farther up small estuaries under conditions of low river flow. In large estuaries, the time scale of cross-channel mixing is much longer than the tidal time-scale of cross-channel shear, so the interaction of cross-channel circulation and tidal shear isn't as important to longitudinal dispersion. Thus reduced freshwater input has less effect on the effective diffusion of salt.

1c) Estuarine and Coastal Systems Influenced by Sources of Negative Buoyancy

All the articles discussed under the two previous topics concern density gradients resulting from inputs of positive buoyancy into estuarine or coastal waters, and nearly all are due mainly to fresh water sources (as opposed to temperature). This is also the most commonly described case in the literature, perhaps because fresh water input dominates along the Atlantic coasts of Europe and North America. But several articles in this volume also address the quite different scenario of negative buoyancy inputs (Hearn & Largier; Largier et al.; Lin and Mehta; N. Smith; Yanagi et al.). Coastal embayments in hotter and/or drier climates are often influenced by removal of buoyancy by evaporation (Hearn & Largier; Largier et al.; N. Smith). Hearn & Largier and Largier et al. discuss estuaries in California which are characterized by longitudinal zones indicative of the locally dominant buoyancy source or sink: Beyond the immediate vicinity of the inlet, a "thermal" regime is entered wherein buoyancy input by heat dominates. Deeper in the basin, an evaporative increase in salinity dominates in the "hypersaline" region. Finally, an "estuarine" regime may be observed closest to the fresh water source.

Another source of negative buoyancy addressed in this volume is surface cooling (N. Smith; Yanagi et al.). N. Smith discusses a phenomena observed on the eastern margin of Great Bahama Bank, triggered by a combination of intense surface cooling and evaporation. Water on the shallow Bank soon becomes denser than the much deeper water to the east, and plunging density currents occur at the break in topography. Even fresh water can be a negative source of buoyancy if it is much colder than neighboring sea water. When this is the case, the usual density contrast associated with an embayment mouth may be largely subdued or reversed, and fronts where fresher water spreads under saltier water may develop (Yanagi et al.). Yet another source of negative buoyancy in estuaries and coastal seas is suspended sediment, which can contribute to vertical density gradients via near bottom turbidity layers (Yasuda et al.).

2. Organization of the Volume

This volume contains a subset of the papers presented at the 7th International Biennial Conference on the Physics of Estuaries and Coastal Seas, held in Woods Hole, Massachusetts, 28-30 November, 1994. Like previous PECS volumes published in the Coastal and Estuarine Studies Series, all papers included here were peer reviewed in order to maintain "journal standards." In addition, papers for this volume were selected on the basis of their contribution to the central theme of buoyancy dynamics. This represents a shift from past PECS volumes which have recently been one to two hundred pages longer and presented a less cohesive survey of physical processes. The hope is that this work represents more of a "text" on estuarine and coastal buoyancy dynamics within which complementary papers better foster a unified understanding. Despite the unifying themes discussed above, the contents of this volume follow a quasi geographic organization from buoyancy dynamics of plumes and coastal seas to estuarine exchange/lower estuary physics to buoyancy dynamics of the inner estuary. This choice has been made because several papers address more than one of the major themes highlighted in the previous sections and several others focus on other distinct aspects of buoyancy dynamics. Within each section, papers are organized according to common. physical processes and decreasing time scale to provide insight into the potential overlap of the dominant mechanisms considered.

Part II -- Buoyant Plumes and Buoyancy in Coastal Sas

These first six papers address buoyancy dynamics in coastal seas or on inner shelves outside the immediate vicinity of the river mouth or estuary. In the lead off paper, Hannah and Loder diagnostically model seasonal baroclinic circulation in the Scotian Shelf/Gulf of Maine and find circulation to be dominated by along-shelf flows with substantial seasonality and topographic steering by banks and basins. The next three papers examine the behavior of buoyant plumes as they move along the coast. Visser examines the persistence of the Rhine River plume over time scales of months, which Visser credits to vertical mixing by wind shutting down horizontal dispersion. Blanton examines estuarine-like circulation perpendicular to the coast within a lens of low density water observed along the southeastern U.S. inner shelf, and finds the circulation to be strongly modified by upwelling or downwelling favorable winds.

Mavor and Huq simulate the behavior coastal plumes using a rotating laboratory tank, and identify dimensionless parameters governing gravity current flow velocities and the behavior of instabilities. The last two papers address the interaction of tides and buoyancy in shelf seas. Sharples and Simpson examine the spring-neap advance of tidal mixing fronts in shelf seas and find lags of several days occur because spring tidal mixing must remove the buoyancy stored since the previous spring tide. Souza and Simpson investigate intratidal properties of Rhine's region of freshwater influence. They find semi-diurnal variability of stratification to be the result of interaction between the mean water column stability and tidal shear.

Part III -- Buoyancy, Salt Transport and Estuary-Shelf Exchange

Eight papers focus on mechanisms for buoyancy exchange near the mouth of an estuary or embayment. Yanagi, Guo and Ishimaru discuss the seasonal flow structure around a front at the mouth of Ise Bay, Japan, where surface cooling of relatively fresh bay water causes it to sink at the mouth of the bay. The remaining papers consider intratidal phenomena, with the first two specifically addressing the influence of shelf density structure. N. Smith discusses tidal exchange between Exuma Sound and the shallow Great Bahama Bank, where denser water from the Bank cascades down the narrow shelf of the Sound on the ebb and is replaced by entirely different Exuma Sound water on the flood. Next, Wong describes the effect of tidal motion on the salinity distribution in the lower-most Delaware Bay. Along with the role of cross-sectional variations in depth, Wong stresses the complex advection and mixing of distinct water masses within and outside the Bay. The next five contributions investigate hydraulic control on estuary-shelf exchange, with the final two emphasizing the ramifications of tidal pumping.

Cudaback and Jay apply a time-dependent hydraulic control model to tidal exchange at the mouth of the Columbia Estuary. Chadwick, Largier and Cheng document the role of thermal stratification in San Diego Bay leading to alternating sub- and super-critical flow at the mouth of the Bay. Along with variations in stratification correlated with the Richardson Number, Uncles and Stephens observe hydraulic control of plunging fronts near the mouth of the Tweed Estuary. Kay, Jay and Musiak perform salt transport calculations for the mouth of the Columbia and find landward transport of salt to be dominated by tidal advection of the tidal salinity field. Finally, Geyer and Nepf find a similar pattern holds along the Hudson River estuary.

Part IV -- Estuarine Dynamics and Buoyancy

Nine papers consider the dynamics of mixing and/or tidally-averaged flow within estuaries, which are presented here in approximate order of decreasing time-scale. Largier, Hearn and Chadwick describe a class of low inflow estuaries found in Mediterranean climates, where buoyancy fluxes during the dry season are dominated by air-water exchange. Under these conditions, positive buoyancy flux by heating nearly cancels negative flux due to evaporation, leading to very weak gravitational circulation. Next Hearn and Largier present a case study of a such an estuary as they document the response of Tomales Bay to historical changes in bathymetry. On seasonal time-scales, Schroeder, Wiseman, Pennock and Noble examine the relationships between river input, flushing time and salinity in Mobile Bay. The next four papers examine controls on tidally-averaged circulation and tidal mixing in partially to well-mixed coastal plain estuaries. Valle-Levinson and O'Donnell perform numerical experiments to investigate the tidally-averaged response of a central channel bordered by shoals and find mean inflow concentrated in the deep channel. Friedrichs and Hamrick find similar results based on analytical solutions and flow observations for a triangular cross-section of the James River estuary. Park and Kuo examine two opposing effects of vertical mixing over the spring-neap cycle in the Rappahannock Estuary. Mixing weakens circulation by enhancing vertical momentum exchange may potentially increase circulation by strengthening the longitudinal salinity gradient. Next, R. Smith discusses the critical relationship between the width of an estuary and the sensitivity of longitudinal dispersion to freshwater discharge. Lewis and Lewis describe vertical mixing events in detail over a tidal cycle in the Tees Estuary. Finally, Dronkers suggests that transverse gravitational circulation enhances erosion of sediment on channel bends and intertidal flats in Volkerak estuary and the Wadden Sea.

3. Conclusions

The 23 papers of this volume focus on the important coastal dynamics issue of buoyancy and its effects. By providing case studies as well as basic theoretical formulation of important processes in the coastal oceans, the topic has been introduced at a level we hope will be of benefit to the reader. Though not an inclusive treatment of buoyancy effects, which can be found in several extant works, this volume produces an update of recent thinking in this arena, and modern observational programs addressing these buoyancy effects.

2

Seasonal Variation of the Baroclinic Circulation in the Scotia Maine Region

Charles G. Hannah, John W. Loder and Daniel G. Wright

Abstract

Results from a diagnostic investigation of seasonal baroclinic circulation in the Scotian Shelf and Gulf of Maine region are presented. The investigation is based on the large historical temperature-salinity database for the region and uses a diagnostic finite-element circulation model to compute 3-d circulation fields. Steric height estimates and variations in potential energy along selected isobaths are used to interpret the circulation tendencies in the baroclinic pressure field and numerical model solutions. The transport streamfunction fields for six bi-monthly seasons show the spatial and temporal structure of the major features of the baroclinic circulation, such as the seasonally-varying flow on the western side of Cabot Strait which supplies the Nova Scotian Current off Halifax and the seasonal flow into the Gulf of Maine off southwestern Nova Scotia. Strong topographic steering is apparent, with anticyclonic baroclinic flow around the Banquereau-Sable-Emerald Bank complex and Georges Bank, and cyclonic flow around Emerald Basin and basins in the Gulf of Maine. Alternative choices of the horizontal correlation scales in the optimal interpolation procedure are considered for the winter and summer seasons in a second set of density fields on a refined mesh. These fields indicate that the large-scale features of the shelf circulation are robust, but there is some sensitivity to the inclusion of spatial anisotropy in the correlation scales, particularly along the upstream boundary. Collectively, the results indicate that the baroclinic circulation on the shelf in the Scotia-Maine region is strongly constrained by geometry: the pattern of shallow offshore banks and deep basins and channels, in conjunction with the location and timing of the sources and sinks of temperature and salinity.

1 Introduction

The low-frequency circulation in the Scotian Shelf-Gulf of Maine (henceforth Scotia-Maine) region (Fig. 1) is generally dominated by the along-shelf (equatorward) flow of relatively cold and fresh water, with substantial seasonality and topographic steering by submarine banks and basins [e.g. Smith and Schwing 1991; Brown and Irish 1992]. At depth, denser Slope Water moves on-shelf via several deep channels, providing an estuarine nature to the large-scale shelf circulation regime [e.g. Brooks 1985]. Most of this circulation has been attributed to density forcing, but the large-scale baroclinic circulation is neither well described nor well understood. Suggestions for important baroclinic dynamics have included a large-scale buoyancy-driven coastal current [Chapman and Beardsley 1989], buoyancy forcing from coastal freshwater discharge such as the Gulf of St. Lawrence outflow [Drinkwater et al. 1979, Brooks 1994], geostrophic adjustment to Slope Water intrusions [Brooks 1985],

Buoyancy Effects on Coastal and Estuarine Dynamics
Coastal and Estuarine Studies Volume 53, Pages 7-29
Published in 1996 by the American Geophysical Union

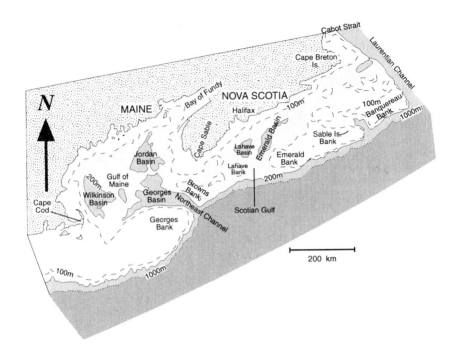

Figure 1. Location map for the Scotia-Maine region showing the extent of the model domain.

transport amplification through baroclinic current interactions with topography [e.g. Shaw and Csanady 1983], and seasonal intensification of bank-scale gyres by tidal-mixing fronts [Loder and Wright 1985].

Here we present results from a diagnostic investigation of seasonal baroclinic circulation in the Scotia-Maine region. Our overall goals are to obtain three-dimensional (3-d) climatological fields for the seasonal-mean circulation in the region, and to elucidate the underlying dynamics of the baroclinic flow component so that sensitivities to forcing variations can be assessed. The investigation is based on the large historical temperature-salinity database for the region, and uses the diagnostic finite-element circulation model of Lynch et al. [1992] and simple diagnostics for interpretation.

We start in Section 2 with a brief description of the database and methods used in the analysis. Our primary results are then presented in Section 3, using six climatological bi-monthly density fields estimated under the assumption that the horizontal covariance of density is isotropic over most of the shelf. (These fields were used by Naimie et al. [1994], Lynch et al. [1996] and Naimie [1996] in studies of composite seasonal-mean circulation on Georges Bank and in the Gulf of Maine.) In Section 4, we briefly discuss an exploration of the influence of horizontally-anisotropic density covariance associated with the local bottom slope, and we conclude with a brief discussion in Section 5.

2 Data and Methodology

Temperature and Salinity Database

The foundation of the study is a historical database of temperature and salinity observations, assembled at the Bedford Institute of Oceanography using data from national archives and recent cruises [also see Loder et al. 1996a]. Density profiles were computed from coincident temperature and salinity observations which passed standard quality control. The resulting database (records with temperature, salinity and density) that was used to compute the bi-monthly fields described in Section 3 comprised about 50,000 stations in the greater Scotia-Maine region, mostly scattered between 1950 and 1990. The density fields used in Section 4 were computed from an updated version of the database, comprising about 54,000 stations.

Optimal Linear Interpolation

Climatological-mean temperature, salinity and density fields for selected seasons were estimated from the databases using four-dimensional (4-d) optimal linear interpolation [Bretherton et al. 1976], with various choices of correlation scales in an assumed covariance function [eq. (1) in Naimie et al. 1994]. The optimal interpolation algorithm provides estimates of the mean fields at grid points in 4-d (x,y,z,t) space, from their nearest-neighbour data points based on separation distances scaled by the correlation scales. Only the 50 nearest neighbours were used in the interpolation, so the key feature of the correlation scales summarized below is the relative magnitudes in the four dimensions.

The horizontal grid points were the nodes of triangular finite-element meshes extending along-shelf from Cabot Strait to Long Island, and about 100 km offshore from the shelf-break to the region of Gulf Stream influences (e.g. meanders and rings), but not as far as the Stream's mean northern boundary (Fig. 1). The finite-element meshes allow variable grid size and hence increased resolution in target areas. Two meshes are used: the first, used in Section 3, is focused on the Gulf of Maine region [mesh g2s; see Fig. 6 of Naimie et al. 1994]; while the second, used in Section 4, has increased resolution over the shelf-break and on the Scotian Shelf [mesh gb3; see Fig. 1 of Greenberg et al. 1996]. Both meshes have realistic water depths shoreward of the 1000-m isobath, and a false bottom in the deep ocean which slopes gently to 1200 m at the offshore boundary. The vertical grid points for the estimated fields were levels of constant depth, at 10-m intervals from the surface to 60 m, increasing to 100-m intervals below 500 m [see Loder et al. [1996a] for more details]. The optimal interpolation procedure was also used to estimate the fields at levels below the sea floor, and values at the sea floor were obtained by linear interpolation in the vertical. The time grid point was taken as the mid-time of the particular period or season of interest, with data from all years used in estimating the present climatological fields.

The six bi-monthly fields used in Section 3 were estimated on the g2s mesh, with time grid points (correlation scales) of 1 February (45 d), 1 April (45 d), 1 June (30 d), 1 August (30 d), 1 October (30 d) and 1 December (45 d). The horizontal correlation scales were taken as isotropic over most of the shelf (H < 450 m where H is the water depth) with a value of 40 km for the 1 December, 1 February and 1 April fields, and 30 km for the others. However, in view of the pronounced influence of topography on the hydrographic structure on Georges Bank, anisotropic horizontal correlation scales were used in five subareas over the Bank's sides, with values of 60 or 100 km in the along-bank direction, and 20 km in the cross-bank direction. Anisotropic horizontal correlation scales were also used in the deep ocean where data density was generally sparser than on the shelf. For nodes with H > 1000 m, along-(cross-) shelf scales of 400 (60) km were specified with a uniform along-shelf direction of 65° T in order to smooth poorly-resolved structures in the fields, while an intermediate level of anisotropy was specified in a transition region over the continental slope. Finally, vertical correlation scales were specified to increase with depth below the surface, ranging from 15 m in the upper 60 m to 200 m at 1200 m. While the above strategy represents an ad hoc approach for objective analysis of spatially-anisotropic and inhomogeneous fields, it does crudely reflect hydrographic structure and data distribution in the region, and yields fields which appear to be good approximations to reality [e.g. Naimie et al. 1994].

In Section 4, we explore a more systematic domain-wide specification of spatial anisotropy in the horizontal correlation scales, using the 1 February and 1 August periods as representatives of winter and summer, respectively. While the horizontal mesh (gb3) and updated temperature-salinity database are also improvements, the principal difference in these fields is the shelf-wide parametric specification of the horizontal correlation scales in terms of the water depth and bathymetric gradient vector at each horizontal node [see Loder et al. [1996a] for details]. For nodes with H < 1000 m, increased correlation scales were parametrically specified in the along-isobath direction, taken as normal to the bathymetric gradient vector. Similarly, the cross-isobath correlation scales were specified to decrease from the flat-bottom shelf values (40 km for 1 Feb. and 30 km for 1 Aug.) to 15-25 km over the continental slope, and then increase again to approximate the deep-ocean cross-shelf scale of 60 km.

Steric Height

A diagnostic measure of the baroclinic circulation tendency in a 3-d density field is the 2-d steric height field — an estimate of the surface elevation under the simplifying assumption that the bottom geostrophic velocity is zero everywhere. Here, we follow the method suggested by Sheng and Thompson [1996] and estimate the steric height η relative to a reference depth H_0 taken as 1200 m, using

$$\eta = - \int_{-H_0}^{0} \varepsilon \, dz = - \int_{-H_0}^{-H} \varepsilon_b \, (z) dz - \int_{-H}^{0} \varepsilon(z) dz \qquad (1)$$

where z is the vertical coordinate (positive upward), $\varepsilon = (\rho - \rho_0)/\rho_0$ the nondimensional density anomaly, $\varepsilon_b(z)$ is the domain-average bottom density anomaly, ρ the density and ρ_0 the reference density. This computation is a 2-d generalization of Csanady [1979], and yields a unique but approximate value for the steric height at each location, through the use of the domain-average bottom density anomaly. The associated baroclinic circulation tendency is surface flow along steric height isolines with high steric values to the right.

Potential Energy

The circulation associated with horizontal density gradients in the presence of variable bottom topography is known to depend strongly on the along-isobath component of the density gradient, through vorticity dynamics that are referred to by some as the "Joint Effect of Baroclinicity and Relief" [JEBAR; e.g. Huthnance 1984] and by others as "pycnobathic" currents [Csanady 1985].

An estimate of the baroclinic transport (i.e. geostrophic transport with zero bottom velocity) across an isobath can be obtained from the along-isobath variation in the depth-integrated potential energy χ,

$$\chi = \frac{g}{\rho_0} \int_{-H}^{0} z\rho \, dz \qquad (2)$$

where g is gravity. Following Mertz and Wright [1992], the vorticity equation for depth-averaged, near-steady flow yields an expression for the total transport vector $\underline{U}$ (the vertical integral of the horizontal velocity):

$$\underline{U} \cdot \nabla(f/H) = \text{curl}_z \left[\frac{\tau_s - \tau_b - \rho_0 \nabla \chi}{\rho_0 H} \right] \qquad (3)$$

where f is the Coriolis parameter, τ_s is the surface stress, τ_b is the bottom stress, ∇ is the gradient operator, and curl_z is the vertical component of the curl operator. In the limit of negligible surface and bottom stress and constant Coriolis parameter, the shoalward directed component of the transport vector normal to an isobath is

$$U^{(n)} = -\frac{1}{f} \frac{\partial \chi}{\partial s} \qquad (4)$$

where s is the along-isobath coordinate, chosen such that deep water is to the left when facing in the direction of increasing s. Under these approximations, the net cross-isobath transport between two points s_1 and s_2 is

$$T = \int_{s_1}^{s_2} U^{(n)} ds = -\frac{1}{f} \left[\chi(s_2) - \chi(s_1) \right] \qquad (5)$$

where positive T implies transport from deep water to shallow. The same result is obtained by assuming that the cross-isobath flow is in geostrophic balance with the along-isobath pressure gradient and is zero at the bottom.

Equation 5 provides a simple quantitative diagnostic for the tendency of the baroclinic pressure field to force cross-isobath transport which, for the low-frequency circulation of interest here, must induce convergences or divergences in the net along-isobath transport between particular isobaths. Note that (5) only accounts for the cross-isobath transport due to the geostrophic flow (with zero velocity at the bottom); it does not account for transport in surface and bottom Ekman layers or for transport due to purely barotropic processes that have no expression in the density field. With these restrictions, the transport between s_2 and the coast is equal to the transport between s_1 and the coast plus the contribution from (5).

Diagnostic Circulation Model

To diagnose the total 3-d circulation associated with horizontal density gradients from the climatological fields, we use FUNDY5 [Naimie and Lynch 1993] — a version of the linear harmonic finite-element numerical model described by Lynch et al. [1992]. The model solves the linearized shallow water equations for a specified baroclinic pressure field and boundary conditions, with the hydrostatic and Boussinesq approximations, eddy viscosity closure in the vertical, and linearized bottom stress.

The model solutions for the fields in Sections 3 and 4 are obtained on the meshes g2s and gb3, respectively, which have slightly different depths and model parameters. The meshes have 21 unequally-spaced nodes in the vertical, with minimum spacing of 2.5 m at the surface and bottom to resolve the Ekman layers. Horizontally-varying bottom friction coefficients and eddy viscosity values were specified from solutions of the iterative nonlinear two-frequency version of the model that considers the M_2 tide and associated mean currents [Lynch and Naimie 1993]. For the g2s mesh, the friction representation is the same as that in the basic solution used by Ridderinkhof and Loder [1994], while for the gb3 mesh, the coefficients are taken from an analogous solution used in Greenberg et al. [1996].

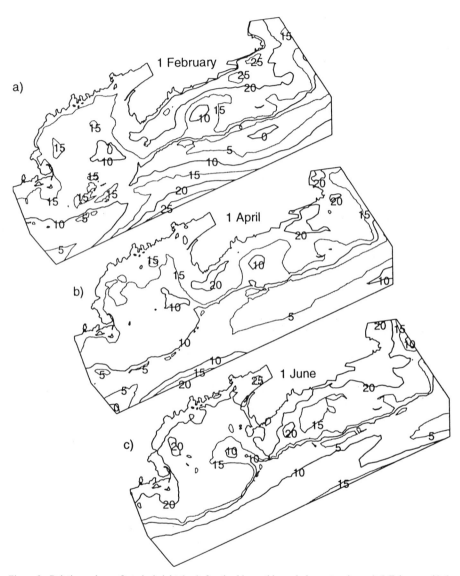

Figure 2. Relative values of steric height (cm) for the bi-monthly periods centered on a) 1 February, b) 1 April, and c) 1 June (after subtraction of the same constant for all fields). Note that the surface circulation tendency is anticyclonic around high steric values.

The baroclinic forcing in the model solutions has two components. First, the baroclinic pressure gradient fields (zero at the surface) are specified by computing horizontal density and then pressure gradients on the level surfaces of the optimal interpolation grid, and then interpolating vertically onto the finite element mesh. Second, steric elevations are specified on the upstream cross-shelf (Laurentian Channel) and offshore open boundaries. These elevations were estimated from the boundary density fields to provide no normal geostrophic flow at the sea floor, essentially ignoring the influences of purely barotropic flow from upstream or offshore. The forcing of the resulting "regional baroclinic" solutions [Loder et al. 1996a] is thus based entirely on the density field (however bottom geostrophic flows may occur as a result of the interior dynamics). A geostrophic outflow condition [Naimie and Lynch 1993] was used on the downstream (western) cross-shelf boundary, and no normal flow was specified on the various land and the Bay of Fundy boundaries, as in Naimie et al. [1994].

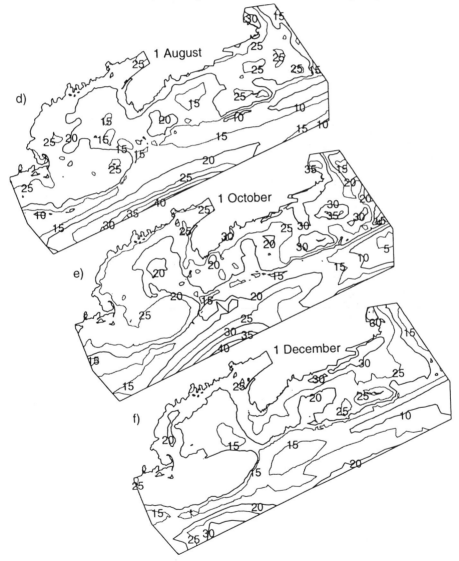

Figure 2. Relative values of steric height (cm) for the bi-monthly periods centered on d) 1 August, e) 1 October, and f) 1 December.

Each forcing and solution was represented with a low-frequency harmonic (period of 10^5 years) approximating a steady state. The model flow fields are displayed here with the (vertically-integrated) transport streamfunction [Lynch and Naimie 1993] to emphasize the large-scale features of the circulation.

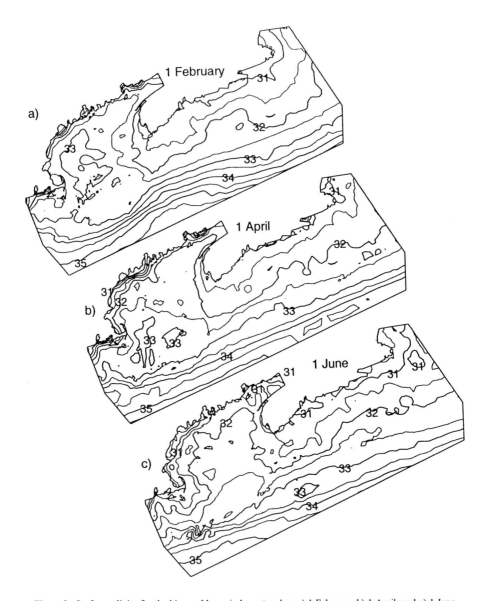

Figure 3. Surface salinity for the bi-monthly periods centered on a) 1 February, b) 1 April, and c) 1 June.

3 Seasonal Variation from Bi-Monthly Fields

In this section we discuss the seasonal variation of low-frequency circulation using the six bi-monthly fields estimated on the g2s mesh with horizontally-isotropic correlation scales on the shelf (except Georges Bank). We focus on the large-scale features of the steric height fields, the transport (streamfunction) fields from the diagnostic circulation model, and the potential energy variations along the 100 m and 200 m isobaths as indices for the inner and

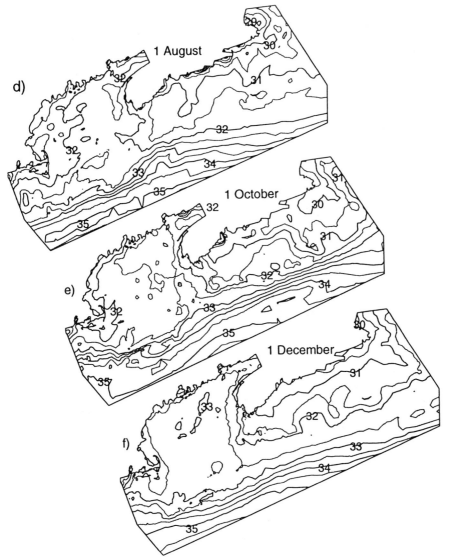

Figure 3. Surface salinity for the bi-monthly periods centered on d) 1 August, e) 1 October, and f) 1 December.

outer shelf, respectively. We emphasize from the outset that our focus is on the shelf and upper slope, and that the present treatment of the deep ocean is crude and intended to represent the qualitative features of the deep ocean and shelf-ocean connections (rather than being quantitatively realistic).

Since the circulation in the Scotia-Maine region is strongly influenced by the shelf morphology, we start with a brief introduction to the regional geography (Fig. 1). At the upstream end, Laurentian Channel is a major cross-shelf feature separating the Scotian Shelf from the southern Newfoundland Shelf and continuing into the Gulf of St. Lawrence through Cabot Strait – an important disruption in the coastline between Nova Scotia and Newfoundland. The coastline and inner shelf topography are relatively-uniform along the Scotian Shelf, while the Gulf of Maine provides a major coastal indentation between Cape Sable and Cape Cod. The dominant features of the shelf bathymetry are a series of relatively-deep basins forming an extended mid-shelf trough, and a series of shallow banks on the outer shelf with intervening channels connecting the shelf basins with the deep ocean. The major basins are Emerald and Lahave Basins on the Scotian Shelf, connected to the deep ocean by the Scotian Gulf between Emerald and Lahave Banks, and Georges, Jordan and Wilkinson Basins in the Gulf of Maine, connected to the ocean by Northeast Channel between Browns and Georges Banks.

Hydrography and Steric Height

The dominant features of the hydrography of the northwestern Atlantic Shelf (Labrador Shelf to Middle Atlantic Bight) are a shelf-wide prevalence of relatively fresh and cold water of primarily subpolar origin, and a strong annual temperature variation in the upper layers associated with air-sea interactions [e.g. Loder et al. 1996b]. In the Scotia-Maine region, the low-salinity shelf water is supplemented by discharge from the St. Lawrence River system flowing out through Cabot Strait and, to a lesser extent, by rivers discharging into the Gulf of Maine [Loder et al. 1996a]. The primary salinity source for the region is the large Slope Water mass which lies offshore between the shelf and the Gulf Stream [Csanady and Hamilton 1988], and which penetrates the shelf system at depth in the major channels [e.g. Brooks 1985]. Below the upper layers where there is significant summertime warming, the salinity and temperature contributions to density are partially compensating with salinity generally dominating.

The bi-monthly steric height fields (Fig. 2) contain a variety of persistent features that can be related to the salinity distributions and shelf morphology, as well as significant seasonal variations. The large-scale pattern in all six fields comprises high steric values along the coast and the offshore boundary of the domain considered, and a trough of low values over the continental slope. This pattern is qualitatively consistent with the conventional wisdom that the climatological mean circulation in the region involves generally along-shelf southwest-ward flow on the shelf and northeastward flow in the deep ocean near the Gulf Stream, with a cyclonic gyre in the intervening Slope Water [e.g. Csanady 1979]. Superimposed on this cross-shelf pattern are broad along-shelf variations comprising a steric decrease along the inner shelf downstream from Cabot Strait, and an apparent spatial meandering of the steric highs associated with the "Gulf Stream" influence towards the shelf off Georges Bank and the western Scotian Shelf. The latter contributes to a local steric maximum (saddle) in the slope trough, and hence separate cyclonic gyres off the Scotian Shelf and Middle Atlantic Bight. This pattern is qualitatively consistent with Csanady's [1979] findings that there are steric variations on scales of 200–400 km along the continental slope (rather than a dominant large-scale steric gradient of order 10^{-7}), but its reliability is uncertain in view of the present crude treatment of the deep ocean and data sparsity in the highly variable Gulf Stream region.

On the shelf, persistent patterns are also present in the steric fields on the scale of the major topographic variations, with cyclonic circulation tendencies over Emerald/Lahave Basins and the Gulf of Maine basins, and anticyclonic tendencies around the major bank complexes (Banquereau, Sable Island, Browns, Georges). As a result, the steric fields suggest that baroclinic circulation is important to cross-shelf transport and shelf-ocean exchange in areas such as Laurentian Channel, the Scotian Gulf and Northeast Channel where Slope Water enters the shelf basins. This Slope Water influence is illustrated clearly by the bottom salinity distributions [see Fig. 5 of Loder et al. 1996a] which suggest on-shelf penetration (at depth) in these channels, as well as to a lesser extent in the Gully to the east of Sable Island. In contrast, the surface salinity fields (Fig. 3) show only a weak influence of the bank-basin topographic structure, being dominated by the large-scale cross-shelf and along-shelf variations associated with coastal and upstream freshwater sources.

The surface salinity fields (Fig. 3) also illustrate the contribution of the seasonal hydrographic variation to baroclinic circulation tendencies (Fig. 2). In contrast to the limited seasonal variation in bottom salinity, the surface salinity has a shelf-wide seasonal variation with structure and phase related to the two primary freshwater sources: Cabot Strait outflow with minimum salinities in late summer/early fall, and Gulf of Maine river discharges which peak in spring. As a result, the annual salinity minimum of the shelf waters generally progresses from late summer/early fall on the eastern Scotian Shelf to late fall/early winter on the central/western Scotian Shelf to spring (summer) in the inner (outer) Gulf of Maine where local and upstream (advective) freshwater sources reinforce each other. Influences of the freshwater pulse on the seasonal variation in baroclinic circulation tendency can be seen off Halifax and Cape Sable in the 1 December and 1 February steric fields, and south of Cape Cod in the 1 June, 1 August and 1 October fields (Fig. 2). However, there are also important local contributions from the large upper-ocean temperature variations, such as in the Georges Bank region where spatially-variable water depths and vertical mixing rates can result in significant contributions to the local baroclinic circulation [e.g. Loder and Wright 1985]. In addition, the seasonal temperature variation contributes to a domain-wide seasonal variation in steric height which appears to be of secondary importance to the regional baroclinic circulation.

Circulation Model Results

The large-scale patterns in the bi-monthly transport streamfunction fields from the numerical model (Fig. 4) have many similarities with those in the steric height fields (Fig. 2), confirming the importance of the baroclinic circulation tendencies to the vertically-integrated circulation. The model solutions also provide quantitative estimates of the transports, including those arising from barotropic flows not reflected in the steric fields, as well as indications of areas where density data sparsity may be contributing to unrealistic flow structures. To aid our discussion of the transport fields, we also show the annual cycle in model transport across six key sections on the shelf (Figs. 5 and 6), and a comparison of model transports with observational estimates based on moored and hydrographic measurements from winter and summer (Table 1).

Table 1. Comparison of observed and model transports, in units of 10^6 m^3 s^{-1} (Sv), for sections shown in Fig. 5. The sources of the observed (Obs) transports are indicated in the footnotes, and the periods in the Months columns (using first letters of consecutive months). The model transports are shown for the 1 February (Winter) and 1 August (Summer) flow fields, from the solutions with shelf-wide isotropic (g2s) and anisotropic (gb3) horizontal correlation scales.

| Section | Transport | | | | | | | |
| | | Winter | | | | Summer | | |
	Obs	g2s	gb3	Months	Obs	g2s	gb3	Months
Halifax (to St5)[1]	1.0	0.9	1.1	DJFM	0.3	0.5	0.6	JAS
SW Nova Scotia[2]	0.3	0.5	0.4	JF	0.1	0.1	0.3	JA
GB S. Flank[3]	0.4	0.3	0.3	JFM	0.7	0.3	0.5	JAS
Nantucket Shoals[4]	0.3-0.4	0.3	0.3	all	0.3-0.4	0.3	0.4	all

[1]Anderson and Smith [1989]; Loder et al. [1996b]   [2]Smith [1983]
[3]Flagg et al [1982]   [4]Beardsley et al.[1985];   Ramp et al. [1988]

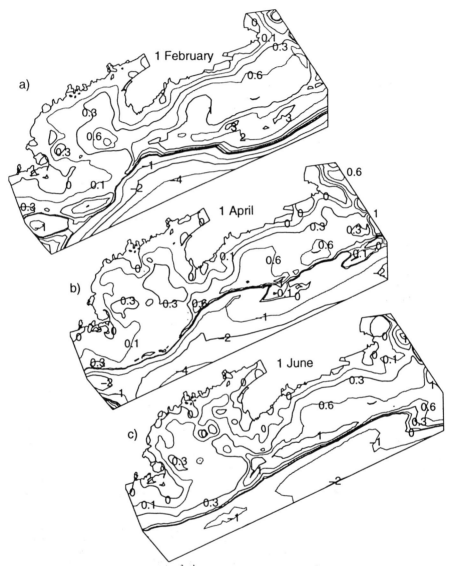

Figure 4. Transport streamfunction (106 m³ s⁻¹) for the bi-monthly periods centered on a) 1 February, b) 1 April, c) 1 June. Note that the coast corresponds to a zero streamline, so that the value on offshore streamlines indicates the integrated transport (in Sv) between that line and the coast. The flow is cyclonic (anticyclonic) around positive (negative) streamfunction values.

The model solutions show the broad-scale features of southwestward flow on the shelf, northeastward flow near the offshore boundary and cyclonic circulation over the continental slope, but with substantial variations with season. The along-shelf flow is strongest in the 1 December and 1 February solutions, with the winter transports generally in good agreement with the observational estimates (Table 1). The only exception to this is the southwest Nova

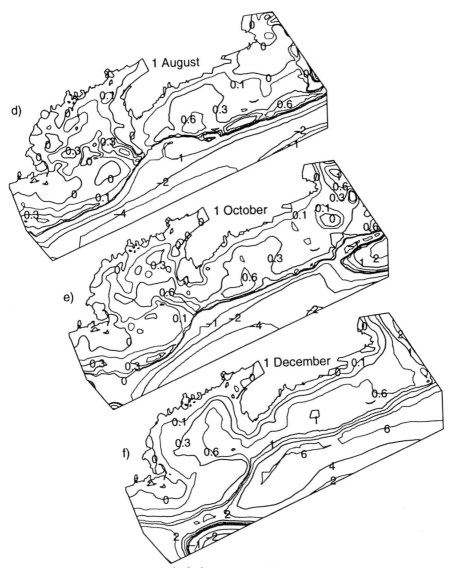

Figure 4. Transport streamfunction (10^6 m^3 s^{-1}) for the bi-monthly periods centered on d) 1 August, e) 1 October, and f) 1 December.

Scotia section where the transport estimates (both observed and modelled) are particularly sensitive to the section's offshore extent (into Northeast Channel) and where tidal rectification contributes to the observed currents [e.g. Han et al. 1996]. A new result is the model suggestion that about one-third of the approximately 1 Sv of transport on the inner Scotian Shelf off Halifax comes from and returns to the outer shelf, associated with the cyclonic circulation over Emerald Basin. In the Gulf of Maine, the winter throughflow from the Scotian

Shelf contributes to cyclonic circulation over the basins, with a peak transport exceeding 0.5 Sv over Georges Basin.

The 1 August and 1 October flow fields on the Scotian Shelf contrast those in winter, with the Halifax section transport reduced by about one-half and little transport past southwest Nova Scotia in summer. These changes are in approximate agreement with those in the observed transports. Associated with the reduced throughflow, there is increased structure in the shelf flow fields in spring/summer/fall. This is illustrated by the seasonal transport cycles for the four Gulf of Maine sections (Fig. 6b) which have different amplitudes and phases. The summer transport on the Nantucket Shoals section is the same as the winter value, in good agreement with the observations. However, the observed summer increase in transport on the southern flank of Georges Bank is not reproduced in the model, in part but not completely due to the neglect of tidal rectification [Naimie et al. 1994]. Considering the importance of tidal rectification off southwest Nova Scotia and on Georges Bank (see the model evaluations of seasonal composite circulation in Naimie et al. [1994] and Han et al. [1996]), the comparison with observed transports indicates that the major shelf transport features in the model solutions are generally realistic.

A persistent feature of the shelf fields is the transport from the inner shelf off Halifax to the outer shelf off southwest Nova Scotia. Much of this transport moves off the shelf and returns northeastward over the Scotian slope. This suggests that the relatively-narrow and shallow shelf off southwest Nova Scotia combines with the hydrographic structure to provide a bottle-neck in the larger-scale shelf flow regime.

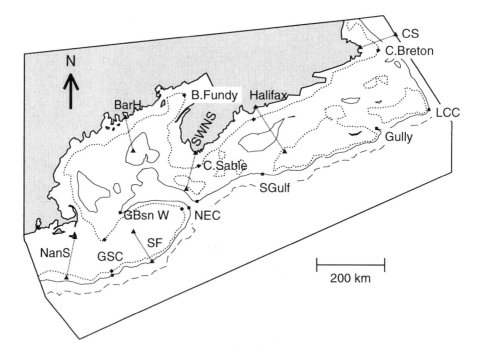

Figure 5. Location map for the transport sections and the reference positions for the potential energy diagnostic along the 100-m (dotted line) and 200-m (solid line) isobaths. The abbreviated section names are: CS - Cabot Strait; SWNS - southwest Nova Scotia; BarH - Bar Harbour to Jordan Basin; SF - southern flank of Georges Bank; NanS - Nantucket Shoals.

The spatial extents of the major deep-ocean flow features — the northeastward flow originating in the southwest, and the cyclonic gyres over the continental slope — show considerable changes with season. In most seasons the northeastward flow comes close to the slope over much of the domain (particularly near Georges Bank), resulting in a narrow cyclonic shear zone over the slope. The cyclonic gyres are largest in the 1 December and 1 February solutions, with a suggestion that the gyre off the Scotian Shelf expands rapidly in fall and contracts in winter. However, the reliability of the deep-ocean flow patterns remains unclear.

While the steric height fields, the diagnostic model solutions and various observations combine to provide confidence in the large-scale shelf circulation features described above, there are three specific areas of concern with the model flow fields. First, there are small-scale circulation features in some locations (with unrealistic associated velocities) which may be artifacts of unresolved or erroneous density features such as might arise from aliasing temporal variations in the density database into spatial variations. As a result, for example, it is unclear whether some of the interruptions of the streamlines along the shelf-breaks of Georges and Browns Banks (e.g. Fig. 4a,c,d,e) are reliable. Second, the throughflow in the solutions is strongly dependent on the upstream and offshore (for the deep-ocean) boundary conditions. In particular, there are questions regarding the appropriate (to the real-ocean) transports across these boundaries (e.g. possible additional barotropic inflows from upstream that are not included in the present solutions), as well as the possibility that the present boundary conditions may result in erroneous boundary density features being propagated through the domain. And finally, there is the question of the reliability of the circulation over the continental slope and in the deep ocean, in view of the present treatment of these regions.

The above three issues are illustrated by contrasting the 1 October and 1 December solutions (Fig. 4e,f). The 1 December solution has the greatest transport (over 1 Sv) through the shelf domain, as well as the largest and strongest cyclonic gyre off the Scotian Shelf. These

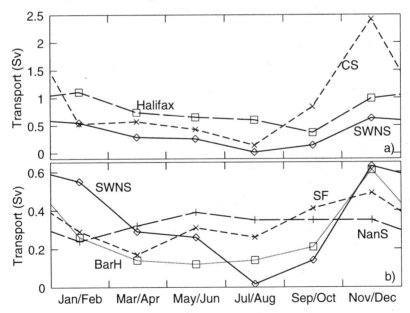

Figure 6. Annual cycle in transport across key sections in the g2s solutions for the a) Scotian Shelf and b) Gulf of Maine. See Fig. 5 for locations.

features can be traced to the relatively-smooth steric height variations in the upstream boundary region (Fig. 2f), involving abrupt changes across Cabot Strait and the shelf break and limited variability along Laurentian Channel. As indicated in Fig. 6, the large shelf throughflow (which apparently dominates over local baroclinic structure) can be partly attributed to the large (and perhaps unrealistic) transport across the Cabot Strait section. In contrast, the 1 October steric field has more small-scale structure in the Laurentian Channel region, and the flow field downstream has less throughflow and more structure.

In the next subsection, we use the potential energy diagnostic to further assess the reliability of the present shelf flow fields. Then, in Section 4, we present initial findings from the use of anisotropic correlation scales, in part motivated to smooth the density fields along isobaths (reducing artificial structures).

Potential Energy Diagnostic

Many of the flow features in the transport streamfunction fields (Fig. 4) can be interpreted by considering the variations in the potential energy χ along the 100- and 200-m isobaths. Recall from (5) that the cross-isobath baroclinic volume transport between any two points on an isobath is simply related to the difference in the potential energy between the two points. A potential energy difference of 25 $m^3 s^{-2}$ corresponds to 0.25 Sv of cross-isobath transport (for $f = 1.0 \times 10^{-4} s^{-1}$).

The locations of the 100- and 200-m isobaths on the g2s mesh, and of positions used in cross-referencing the potential energy variations are shown in Fig. 5. Note that the inshore 100-m isobath used in the following discussion remains within 100 km of the coast, except on Georges Bank and south of Cape Cod, while the offshore 200-m isobath follows the western edge of Laurentian Channel and the shelf break with an excursion into the Gulf of Maine around Georges Basin. The closed 100- and 200-m isobaths in the domain are not included in this analysis.

Figure 7a,b shows the potential energy variations along the 100-m isobath in the bi-monthly density fields, while those along the 200-m isobath are shown in Fig. 8a,b. In both cases there are persistent large-scale variations which are also present in the corresponding χ variations in the annual-mean fields (Figs.7c,8c). This persistence is consistent with the limited annual variation in hydrographic properties at depth in the region (the dominant contributor to χ), and with the persistent large-scale features of the steric height and transport fields (Figs. 2, 4).

Along the 100-m isobath, the potential energy generally decreases (larger negative numbers) with distance from Cabot Strait, as a result of the density (salinity) generally increasing downstream. This implies that there is a net shoreward baroclinic transport across the 100-m isobath over the Scotia-Maine region, which would require a corresponding increase in the along-shelf transport inside the 100-m isobath. Comparison (Fig. 7d) of the transport streamfunction values along the 100-m isobath obtained from the diagnostic model solution using the annual-mean density field with the transports expected from the annual-mean potential energy variation (using (5)) shows very good overall agreement, suggesting an associated along-shelf transport amplification of about 0.2 Sv. This provides a firm indication that the persistent throughflow on the inner shelf in the model solutions is a dynamically-robust feature associated with the large-scale density field. There are also seasonal and smaller-scale variations of χ along the 100-m isobath which have corresponding variations in the model flow fields, confirming that many of the model flow features are direct reflections of the density field (and probably, but not necessarily, the real ocean). For example, the potential energy decrease between the inner Scotian Shelf and inner Gulf of Maine is generally less in spring-summer than fall-winter (Fig. 7a,b), due to the increased importance of local runoff into

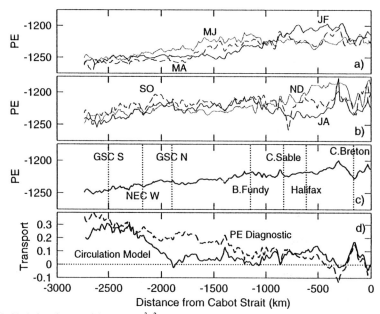

Figure 7. Variations in potential energy (m³s⁻²) along the 100-m isobath for (a,b) each of the six bi-monthly g2s fields and c) the annual-mean density field. The implied cross-isobath transport for the annual-mean case (dashed) is compared with the value of the transport streamfunction from the circulation model (solid) along the 100-m isobath in d). The reference positions in c) are identified in Fig. 5.

the Gulf in spring-summer (Fig. 3). This change in the hydrographic structure contributes to the reduced inflow from the Scotian Shelf (Fig. 4).

The χ variation along the 200-m isobath is dominated by an intriguing large-scale structure (Fig. 8a,b,c). The potential energy decreases (associated with density increases) from Cabot Strait to the Scotian Gulf, and then increases (density decreases) until the western end of Georges Basin, and finally decreases again around Georges Bank. Relative to the bottom flow, this implies that water moves onto the Scotian Shelf (from offshore) between Cabot Strait and the Scotian Gulf, and there is a corresponding (baroclinic) transport loss to deeper water between the Scotian Gulf and western Georges Basin. Finally, further cross-isobath (baroclinic) transport into shallow water is implied as the 200-m isobath goes around Georges Bank. The remarkable year-round persistence of this structure, together with the corresponding persistent offshore turning of the model transport streamlines in the western Scotian Shelf-Northeast Channel region (Fig. 4), indicates that the along-shelf transport bottle-neck off southwest Nova Scotia is associated with a robust feature of the large-scale hydrographic (and baroclinic pressure) fields. One possibility for the origin of this feature is the increased tidal (vertical) mixing in the region, resulting in reduced stratification and hence increased potential energy. This feature is further illustrated by the comparison (Fig. 8d) of the transport streamfunction values from the annual-mean model solution with those expected from the annual-mean χ variation, showing similar large-scale variations.

At this point we end our detailed discussion of the g2s bi-monthly fields with the conclusion that the various diagnostics considered here, together with Naimie et al.'s [1994] detailed current evaluation for the Georges Bank region, indicate that most of the large-scale shelf structures in the hydrographic and current fields are robust and realistic, although there are many outstanding issues, particularly regarding local structures.

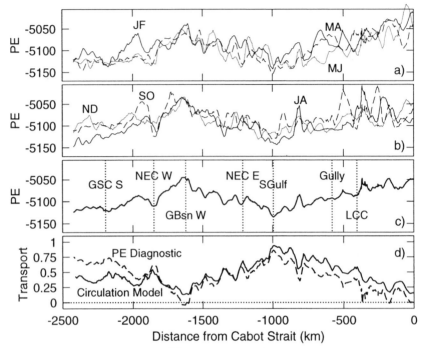

Figure 8. Variations in potential energy (m^3s^{-2}) along the 200-m isobath for (a,b) each of the six bi-monthly g2s fields and c) the annual-mean density field. The implied cross-isobath transport for the annual-mean case (dashed) is compared with the value of the transport streamfunction from the circulation model (solid) along the 200-m isobath in d). The reference positions in c) are identified in Fig. 5.

4 Sensitivity Indications for Winter and Summer

In this section we briefly describe alternative steric and transport fields for each of winter (1 February) and summer (1 August). The specific refinements are shelf-wide anisotropy of the optimal interpolation covariance scales, an updated historical database, and improved grid resolution primarily on the Scotian Shelf. The emphasis is on the new fields as indicators of the general robustness and sensitivity of the g2s fields presented in Section 3 (rather than a detailed sensitivity investigation).

The steric height fields for both seasons (Fig. 9) show strong similarities on all scales to the corresponding fields in Fig. 2. The primary difference over the shelf, present in both seasons, is a weak intensification of the basic circulation tendencies in the refined fields; i.e. increased throughflow, and cyclonic (anticyclonic) flow around basins (banks). This appears to be a result of the domain-wide use of anisotropic correlation scales in the density field estimation. A second notable difference in the 1 August field is the modified spatial structure in the Gulf Stream influence zone along the offshore boundary, associated with additional data for this region.

The numerical model transports associated with the refined fields (Fig. 10, Table 1) also show strong similarities to those in the corresponding seasons in Fig. 4, as well as some notable differences. On the shelf the primary difference, present in both seasons but greatest in the 1 August field, is increased flow through the domain and fewer local interruptions of the streamlines, for example, along the shelf-break off Browns and Georges Banks. While some of

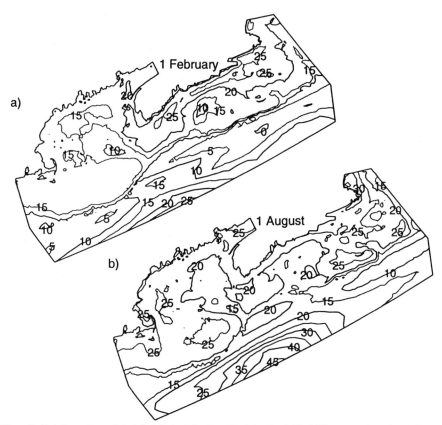

Figure 9. Relative values of steric height (cm) for the refined density fields (gb3), centered on a) 1 February and b) 1 August.

this arises from the increased tendency for along-isobath flow in the local density fields (e.g. Fig. 9), the dominant factor in both cases appears to be increased transports across the Cabot Strait boundary and onto the eastern Scotian Shelf associated with sharper steric gradients along the Cabot Strait boundary and smoother steric fields along the Laurentian Channel boundary. This illustrates both the general importance of the upstream boundary conditions in diagnostic model solutions, and the significance of the density field along the upstream boundary in the present boundary-condition strategy. Thus, the proper representation of along-isobath density gradients, as explored here with the anisotropic correlation scales, can affect the circulation both locally and downstream (in the sense of shelf-wave propagation) through either boundary conditions or baroclinic interactions with topography [Huthnance 1984; Csanady 1985].

A second notable difference between the transports associated with the refined fields (Fig. 10) and those in Fig. 4 is the increased size and strength of the cyclonic gyre offshore from the Scotian Shelf. The differences in the winter solutions are due primarily to differences in the boundary forcing and those in the summer to differences in the interior mass field over the continental slope.

These refined fields point to a general robustness of the large-scale baroclinic structures described in Section 3, particularly on the continental shelf where the density of historical

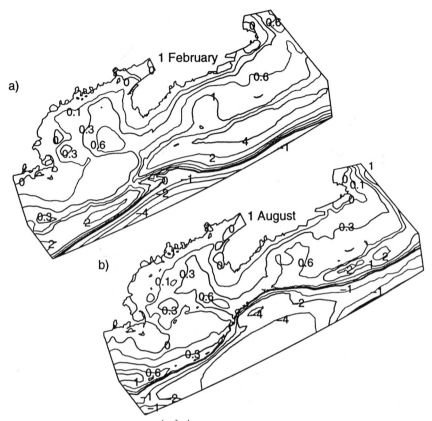

Figure 10. Transport streamfunction (10^6 m^3 s^{-1}) fields for the refined (gb3) fields, centered on a) 1 February and b) 1 August.

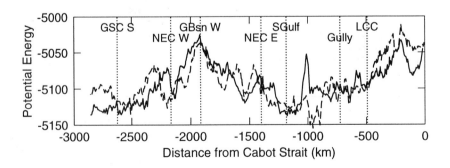

Figure 11. Variations in potential energy (m^{3}s^{-2}) along the 200-m isobath for the refined (gb3) fields for 1 February (dashed curve) and 1 August (solid curve).

hydrographic data is relatively high (compared to the deep ocean). The potential energy distribution in the refined fields, illustrated in Fig. 11 for the 200-m isobath, provides further confidence that there are persistent large-scale hydrographic structures over the Scotia-Maine shelf and slope regions that play critical roles in the regional circulation. In addition, there are numerous smaller-scale (e.g. topographic) structures in the steric and potential energy fields which are potentially both of real importance to the local circulation and useful diagnostic indices of the adequacy of the historical density database.

5 Discussion

These results indicate that a broad quantitative description of, and dynamical insights into, the elusive low-frequency baroclinic circulation on continental shelves can be obtained from historical hydrographic databases (where available) and appropriate diagnostics. For the Scotia-Maine region, the historical data coverage appears to be sufficient to provide a realistic diagnostic model computation of the large-scale circulation field and its seasonal variation, and the topographic-scale circulation field in many areas. This result represents a spatial extension of Naimie et al.'s [1994] detailed study of circulation on Georges Bank. Nevertheless, the regional hydrography remains sparsely sampled in general (considering its rich temporal variability), and there are local areas with inadequate data coverage for a reliable circulation diagnosis [also see Loder et al. 1996a].

The specific diagnostics explored here all provide promise as a complementary suite of analysis tools. The inclusion of topographic anisotropy in the optimal linear interpolation procedure, at a relatively weak level over most of the shelf in this initial exploration, generally reduces the influence of small-scale density structures, many of which may be artifacts of data sparsity. The steric height and potential energy fields, expected to be indicative of the surface and vertically-integrated circulation tendencies, show general consistency in the cases examined here, both with each other and the numerical model transport fields. Such direct diagnostics on the density fields thus show potential for the identification of baroclinic flow features, evaluation of density field reliability, and interpretation of numerical circulation models [also see Sheng and Thompson 1996]. Finally, the harmonic diagnostic circulation model allows evaluation of the 3-d circulation associated with various forcings, on realistic geometry, with spatially-variable (3-d) friction and with minimal computational demand. In addition to providing a valuable initial evaluation of the baroclinic circulation field, these tools can also be valuable in the application of more sophisticated prognostic numerical models [e.g. Lynch et al. 1996, Naimie 1996], for both initial and boundary conditions, and interpretative analyses.

Acknowledgements. We are grateful to the many individuals who have contributed to the database and our diagnoses. We extend special thanks to: Ken Drinkwater and Roger Pettipas for their lead role in the historical database development; Mary Jo Graça for computing the density and steric fields; Ian He, Gerry Boudreau and Ross Hendry for providing the revised optimal interpolation programs; Dan Lynch and Chris Naimie for providing the numerical model and related algorithms; Dave Greenberg and Yingshuo Shen for providing the gb3 grid and friction coefficients; and Brian Petrie, Jinyu Sheng, Peter Smith and Keith Thompson for helpful discussions. This is contribution 34 of the U.S. GLOBEC program, funded jointly by NOAA and NSF. This work is also funded by the (Canadian) Panel on Energy, Research and Development.

References

Anderson, C., and P.C. Smith, Oceanographic observations on the Scotian Shelf during CASP, *Atmos.-Ocean*, *27*, 130-156, 1989.

Beardsley, R.C., D.C. Chapman, K.H. Brink, S.R. Ramp, and R. Schiltz, The Nantucket Shoals Experiment (NSFE79): Part I. A basic description of current and temperature variability, *J. Phys. Oceanogr., 15,* 713-748, 1985.

Bretherton, F.P., R.E. Davis, and C.B. Fandry, A technique for objective analysis and design of oceanographic experiments applied to MODE-73, *Deep-Sea Res. I, 23,* 559-582, 1976.

Brooks, D.A., Vernal circulation in the Gulf of Maine, *J. Geophys. Res., 90,* 4687-4705, 1985.

Brooks, D.A., A model study of the buoyancy-driven circulation in the Gulf of Maine, *J. Phys. Oceanogr., 24,* 2387-2412, 1994.

Brown, W.S., and J.D. Irish, The annual evolution of geostrophic flow in the Gulf of Maine: 1986-1987, *J. Phys. Oceanogr., 22,* 445-473, 1992.

Chapman, D.C., and R.C. Beardsley, On the origin of shelf water in the Middle Atlantic Bight, *J. Phys. Oceanogr., 19,* 384–391, 1989.

Csanady, G.T., The pressure field along the western margin of the North Atlantic, *J. Geophys. Res., 84,* 4905-4915, 1979.

Csanady, G.T., "Pycnobathic" currents over the upper continental slope, *J. Phys. Oceanogr., 15,* 306-315, 1985.

Csanady, G.T., and P. Hamilton, Circulation of slopewater, *Cont. Shelf Res., 8,* 565-624, 1988.

Drinkwater, K., B. Petrie, and W.H. Sutcliffe, Seasonal geostrophic volume transports along the Scotian Shelf, *Estuar. Coast. Mar. Sci., 9,* 17-27, 1979.

Flagg, C.N., B.A. Magnell, D. Frye, J.J. Cura, S.E. McDowell, and R.I. Scarlet, Interpretation of the physical oceanography of Georges Bank, EG&G Environmental Consultants, Waltham MA. Final Report prepared for the US Dept. of Interior, Bureau of Land Management, 1982.

Greenberg, D.A., J.W. Loder, Y. Shen, D.R. Lynch, and C.E. Naimie, Spatial and temporal structure of the barotropic response of the Scotian Shelf and Gulf of Maine to surface wind stress, *J. Geophys. Res.,* submitted, 1996.

Han, G., C.G. Hannah, P.C. Smith, and J.W. Loder Seasonal variation of the three-dimensional mean circulation over the Scotian Shelf, *J. Geophys. Res.,* submitted, 1996.

Huthnance, J.M., Slope currents and "JEBAR", *J. Phys. Oceanogr., 14,* 795-810, 1984.

Loder, J.W., G. Han, C.G. Hannah, D.A. Greenberg, and P.C. Smith, Hydrography and baroclinic circulation in the Scotian Shelf region: winter vs summer, *Can. J. Fish. Aquat. Sci. Suppl.,* in press, 1996a.

Loder, J.W., B.D. Petrie, and G. Gawarkiewicz, The coastal ocean off northeastern North America: a large scale view, *The Sea,* submitted, 1996b.

Loder, J.W., and D.G. Wright, Tidal rectification and frontal circulation on the sides of Georges Bank, *J. Mar. Res., 43,* 581-604, 1985.

Lynch, D.R., and C.E. Naimie, The M_2 tide and its residual on the outer banks of the Gulf of Maine, *J. Phys. Oceanogr., 23,* 2222-2253, 1993.

Lynch, D.R., F.E. Werner, D.A. Greenberg, and J.W. Loder, Diagnostic model for baroclinic, wind-driven and tidal circulation in shallow seas, *Cont. Shelf Res., 12,* 37-64, 1992.

Lynch, D.R., J.T.C. Ip, C.E. Naimie, and F.E. Werner, Comprehensive coastal circulation model with application to the Gulf of Maine, *Cont. Shelf Res., 16,* 875-906 1996.

Mertz, G., and D.G. Wright, Interpretations of the JEBAR term, *J. Phys. Oceanogr., 22,* 301-305, 1992.

Naimie, C.E., Georges Bank residual circulation during weak and strong stratification periods - Prognostic numerical model results, *J. Geophys. Res., 101,* 6469-6486, 1996

Naimie, C.E., J.W. Loder, and D.R. Lynch, Seasonal variation of the three-dimensional residual circulation on Georges Bank, *J. Geophys. Res., 99,* 15,967-15,989, 1994.

Naimie, C.E., and D.R. Lynch, FUNDY5 Users' Manual Numerical Methods Laboratory, Dartmouth College, NH., 40 pp, 1993.

Ramp, S.R., W.S. Brown, and R.C. Beardsley, The Nantucket Shoals Flux Experiment: 3. The alongshelf transport of volume, heat, salt, and nitrogen, *J. Geophys. Res., 93,* 14,039-14,054, 1988.

Ridderinkhof, H., and J.W. Loder, Lagrangian characterization of circulation over submarine banks with application to the outer Gulf of Maine, *J. Phys. Oceanogr., 24,* 1184-1200, 1994.

Shaw, P.-T., and G.T. Csanady, Self-advection of density perturbations on a sloping continental shelf, *J. Phys. Oceanogr., 13,* 769-782, 1983.

Sheng, J., and K.R. Thompson, A robust method for diagnosing regional shelf circulation from density profiles, *J. Geophys. Res.*, in press, 1996.

Smith, P.C., The mean and seasonal circulation off southwest Nova Scotia, *J. Phys. Oceanogr.*, *13*, 1034-1054, 1983.

Smith, P.C., and F.B. Schwing, Mean circulation and variability on the eastern Canadian continental shelf, *Cont. Shelf Res.*, *11*, 977-1012, 1991.

3

Shear Dispersion in a Wind and Density Driven Plume

Andre W. Visser

Abstract

The lateral spreading induced by shear dispersion in a river plume is examined. In the absence of wind, the self-induced shear associated with the cross plume density structure promotes a lateral spreading characterised by horizontal diffusivity proportional to the square of the cross plume density gradient. Examining the time scales for this process indicates that the width of the plume goes as $t^{1/6}$ where t is the "age" of the plume water. Under normal circumstances for the Rhine plume, it appears that wind induced circulation dominates shear dispersion suggesting a much more rapid ($t^{1/2}$) spreading rate. However, there are two effects which moderate the wind efficiency in inducing lateral spreading: (a) convective overturn since wind induced circulation may promote unstable water column conditions, and (b) wind mixing which increases vertical mixing rates thus reducing effective horizontal diffusivity. Of these wind mixing appears to be a powerful agent in diminishing shear dispersion with increasing wind speed.

Introduction

Spreading and mixing of freshwater derived from river discharges often have a considerable impact on the dynamics and environmental conditions found in coastal seas. For instance, riverine water often carries elevated sediment and nutrient loads. The fate of freshwater, how fast it spreads and the lateral extent over which its influence is felt is of interest to various concerns. The dynamics of ROFI's; regions of freshwater influence, have received considerable attention over recent years. The dynamics of such regions are distinct from those of the adjacent shelf sea. Pressure gradients associated with density distributions drive circulations which are modified by friction, mixing and planetary rotation, and are often composed of

- an along shore buoyancy current and
- a cross shore "estuarine" with fresher (less dense) water transported seaward at the surface and saltier (denser) water landward near the bottom.

Furthermore, such regions are susceptible to periodic stratification [Simpson et al 1993; Simpson and Souza, 1994] which may serve to modify barotropic motion [e.g. tides, Visser et al 1994].

It is remarkable that river plumes remain as distinct as they do. Despite the highly energetic and turbulent environments in which they are often found, river plumes maintain their integrity, remain coastally trapped, and contribute to local dynamic conditions over long distances from the river mouth. Figure 1 for instance, shows the ensemble averages of surface salinity for the Rhine plume. While there is an initial period of rapid dispersion close to the river mouth, the plume subsequently follows the Dutch coast all the way into the German Bight, a distance of several hundred kilometres, with apparently very little seaward dispersion. The time period over which this along coast transport takes place (i.e. the "age" of the plume) is many months, and one would

Buoyancy Effects on Coastal and Estuarine Dynamics
Coastal and Estuarine Studies Volume 53, Pages 31-45
Copyright 1996 by the American Geophysical Union

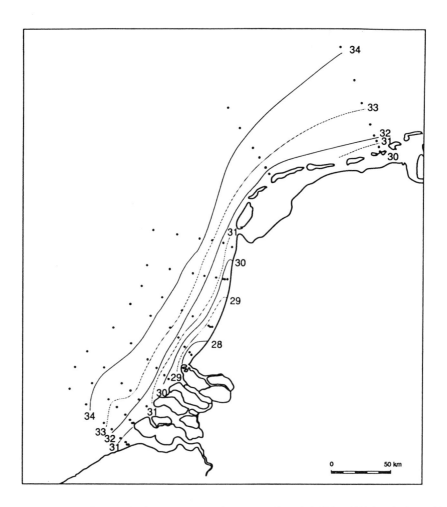

Figure 1. Average surface salinity along the Netherlands coast over the period 1973 to 1983 from hydrographic surveys conducted every 2 weeks; 170 surveys in all (from Visser et al, 1991).

expect that the plume water would have experienced a number of strong mixing events (storms, spring tides etc.). Nonetheless, the plume appears to maintain a constant width despite these events.

One feature of ROFI's which is noteworthy here is the prevalence of vertically sheared horizontal flow. These appear not only in the low frequency currents (density induced: [e.g. van der Giessen et al, 1990], wind induced [e.g. de Ruijter et al, 1992]), but also at tidal frequencies [Visser et al, 1994; Souza and Simpson, 1994]. This observation suggests that shear dispersion, the combined effects of vertical mixing with horizontal shears may prove to be a dominant factor contributing to the lateral dispersion (and hence dimensions) of a ROFI.

The principal of shear dispersion [cf. Bowden, 1965] proceeds on the premise that the processes controlling the advection / diffusion of density ρ (indeed any passive tracer) may be described by two distinct processes. Writing the water column depth as H and the vertical diffusivity as K_z, these processes may be summarised as:

- A balance between vertical mixing and horizontal advection by a vertically sheared flow (u',v') which maintains the steady state of the local density profile. This balance can be formalised as

$$u'\frac{\partial}{\partial x}\langle\rho\rangle + v'\frac{\partial}{\partial y}\langle\rho\rangle - K_z\frac{\partial^2\rho'}{\partial z^2} = 0 \tag{1}$$

with boundary conditions $K_z\,\partial\rho'/\partial z=0$ at $z=0$ and $z=-H$: no density flux through the surface or bottom. We use $\langle\ldots\rangle$ to indicate the vertical average of a quantity, and a prime its vertically dependent component. For instance $\rho = \rho' + \langle\rho\rangle$, $\langle\rho\rangle$ being the vertical integral of the density field, and ρ' being the vertically dependent departure from this.

- A vertically integrated density flux due to correlations of the velocity shear and density profile, gradients of which change the overall density of a water column as it moves.

$$\frac{d}{dt}\langle\rho\rangle + \frac{\partial}{\partial x}\langle u'\rho'\rangle + \frac{\partial}{\partial y}\langle v'\rho'\rangle = 0 \tag{2}$$

where $d/dt = \partial/\partial t + \mathbf{u}\cdot\nabla$ is the substantive derivative. This flux can be written as

$$\mathbf{F} = -\langle\mathbf{u}'\rho'\rangle. \tag{3}$$

Since rigorous treatments deriving these equations may be found elsewhere, [e.g. Turner, 1953; Bowden, 1965; Fisher et al, 1979], we will not reproduce them here. However, it is instructive to note that these two processes may be thought of as operating on separate time scales. In particular, there is a fast time scale associated with vertical mixing (H^2/K_z). A typical value for the Rhine ROFI along the Netherlands coast is of the order of hours to a few days [Van Alphen et al, 1987]. Thus, in the absence of other re-stratifying processes stratification would be rapidly eroded. A non-uniform density profile can be maintained by the advection of a mean density gradient by a vertically sheared flow, Eq. 1. In comparison, there is a slower time scale over which the large scale horizontal distribution varies, Eq. 2. As already noted, this may be of the order of weeks for the Rhine.

Suppose we have a density gradient predominantly in the across shore, y, direction. Using the short hand notation $\partial\langle\rho\rangle/\partial y=\langle\rho\rangle_y$, integrating Eq. 1 twice vertically gives

$$\rho'(z) = \rho'(0) + \frac{\langle\rho\rangle_y}{K_z}\int_z^0\int_z^0 v'\,dz dz \tag{4}$$

It is the interaction of this density profile with a vertically sheared flow that promotes an horizontal flux of density. In particular, Eq. 3 gives

$$F = K_H^*\langle\rho\rangle_y = -\langle v'\rho'\rangle = -\frac{\langle\rho\rangle_y}{HK_z}\int_{-H}^0 v'\left\{\int_z^0\int_z^0 v'\,dz dz\right\}dz \tag{5}$$

or an effective horizontal diffusivity given by

$$K_H^* = \frac{1}{HK_z}\int_{-H}^0\left\{\int_z^0 v'\,dz\right\}^2 dz \tag{6}$$

Assuming that both diffusivity K_z and eddy viscosity N_z are vertically uniform, it is relatively straight forward to determine the vertical velocity structure (u',v') for a density plume under the

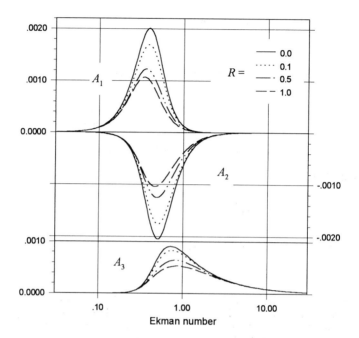

Figure 2. The coefficients, A_1, A_2 and A_3 plotted as functions of the Ekman number E, and various values of the normalised bottom stress number R.

influence of surface wind stress, $\tau(0) = (\tau_x, \tau_y)$, and linear bottom stress parameterized as $\tau(-H) = k\mathbf{u}(-H)$. Substituting the form of this solution (*cf.* appendix) into Eq. 6 gives:

$$K_H^* = \frac{2H^2}{fK_zN_z}\left\{\left[\tau_x - \frac{gN_z}{f\rho_0}\langle\rho\rangle_y\right]^2 A_1 + \tau_y\left[\tau_x - \frac{gN_z}{f\rho_0}\langle\rho\rangle_y\right]A_2 + \left[\tau_y\right]^2 A_3\right\} \qquad (7)$$

where $A_{1,2,3}$ are functions dependent on the Ekman number $E = \{2 N_z/f\}^{1/2}/H$, and $R = k/(fH)$ the ratio of the bottom spin down and the Coriolis frequency where k is the linear bottom friction velocity. These functions are discussed further in the appendix and are plotted in Figure 2 for a range of E and R. Note that substituting Eq. 7 into Eq. 5 would yield an expression very similar in form to that derived by Stommel and Leetmaa [1972]. The only qualitative difference being that here we allow for a finite bottom stress. A few points might be noted directly from Eq. 7.

- Firstly, as is often the case for shear dispersion processes, the effective horizontal diffusivity decreases for strong vertical mixing. As vertical mixing becomes very rapid, any differential advection by a vertically sheared horizontal velocity has very little time to transport density laterally before it is mixed vertically.
- In the absence of wind, the density flux is proportional to the cube of the density gradient. That is

$$F = \left\{\frac{gH}{f\rho_0}\right\}^2 \frac{2N_z}{fK_z}A_1(E,R)\langle\rho\rangle_y^3 \qquad (8)$$

This is a "self-induced" dispersion process arising through shear set up by the density gradient itself. As the density gradient weakens, both the sheared velocity and thus diffusivity become weaker (quadratic contribution) as well as the gradient upon which it acts (additional linear contribution).

- Examining the coefficients $A_{1,2,3}$ we see that the effective horizontal dispersion falls off for both large ($E>4$) and small ($E<0.03$) Ekman numbers. In the former case, this is because the water column is strongly coupled vertically so that very little shear develops. In the latter case, while strong shears can exist, these become confined to narrow layers at the surface and bottom so that the volume (and hence density) fluxes these shears may drive decrease. We see also that the coefficient associated with the across shore wind stress, A_1, has a peak skewed to lower Ekman numbers ($E<0.3$) compared to that associated with the along shore wind stress, A_3, which skews to higher Ekman numbers ($E>0.6$). This is because the Ekman transport at low Ekman numbers is more nearly at right angles to the wind stress so that an along shore wind induces an across shore velocity shear. On the other hand, as the Ekman number increases, Ekman transport aligns with the wind so that an across shore wind is more effective in producing an across shore shear.

Before proceeding, it may be instructive to examine parameter values typical for the Rhine plume. The linear bottom friction velocity k may be evaluated by equating it to the more commonly used quadratic stress law. That is

$$k \approx C_D |U(t)| \approx 8 C_D U_0 / (3\pi) \qquad (9)$$

where U_0 is the tidal current amplitude (range from 0.4 m/s for neap tides to 1.2 m/s for spring tides), and C_D is the drag coefficient appropriate for a vertically averaged velocity (0.0028). Taking a water depth H of 20 m, this gives

$$R = \frac{k}{fH} = \begin{cases} 03 & \text{neap tide} \\ 0.7 & \text{spring tide} \end{cases} \qquad (10)$$

We see here the importance of maintaining a finite bottom stress since the bottom spin down frequency may approach the Coriolis frequency.

Following Csanady [1976], an estimate of the bulk eddy viscosity may be given by:

$$N_z = \begin{cases} u^* H / 20 & H < H_i \\ u^{*2} / 200 f & H > H_i \end{cases} \quad \text{where} \quad \begin{array}{l} H_i = u^* / 10 f \\ u^* = \sqrt{C_D} |U(t)| \end{array} \qquad (11)$$

This gives a range of $N_z \approx [1.3 \times 10^{-2}, 7 \times 10^{-2}]$ $m^2 s^{-1}$ for neap and spring tides respectively. Ekman numbers associated with these range over $E \approx [0.7, 1.8]$. That is, Ekman numbers of order unity are characteristic of the Rhine ROFI. The associated vertical mixing time scale for momentum, $T_{u,v} = H^2 / N_z$, ranges from 1 to 10 hours. While the time scale for density mixing need not be identical, it should be of the same order thus confirming our fast time scale supposition in Eq. 2.

We can now evaluate the relative contributions of density and wind driven shear in effecting lateral dispersion in the Rhine ROFI. Considering only along shore wind stress, it may be readily shown from Eq. 7 that the two effects become comparable when

$$\tau_x = \frac{\rho_a}{\rho_0} C_D w_x^2 = \frac{g N_z}{f \rho_0} \langle \rho \rangle_y = E^2 \frac{g H^2}{2 \rho_0} \langle \rho \rangle_y \qquad (12)$$

where $\rho_a = 1.22$ kg/m^3 is the density of air, and $\rho_0 = 1028$ kg/m^3 is the density of sea water. The right hand side of this balance contains what might be interpreted as the density gradient induced stress. In particular we can write this as:

$$\tau_\rho = \frac{gH^2}{2\rho_0}\langle\rho\rangle_y \approx \frac{gH^2}{2\rho_0}\beta\frac{\Delta S}{\Delta Y} \tag{13}$$

where $\beta\approx0.8$ $kg/m^3/psu$ is the ratio of density to salinity. We set $\Delta S\approx6$ psu as the salinity difference found over a $\Delta Y\approx20$ km cross shore distance within the Rhine Plume. While this gradient is sharper than that evidenced in the averaged example shown in Figure 1, it is typical for a particular realisation. Substituting these values gives $\tau_\rho\approx4.6\times10^{-4}$ m^2/s^2. An Ekman number $E=1$ gives an equivalence along shore wind velocity of $|w_x|_{eq}=12m/s$ (Beaufort scale 6) which is moderately strong. This equivalence wind velocity increases linearly with Ekman number so that at spring tides it approaches storm conditions. For any particular condition, when the wind speed is less than $|w_x|_{eq}$, density driven shears will tend to dominate where as for wind speeds greater than $|w_x|_{eq}$, wind driven shears will dominate.

As already noted however, for Ekman numbers of order unity, it is across shore rather than along shore wind which is most effective in producing cross shore vertical shear. For instance at $E=1$, $R=0.5$, the coefficients in Eq.7 are $(A_1,A_2,A_3) \approx (1,-6,8)\times10^{-4}$. That is A_j, controlling the cross shore wind effect is nearly an order of magnitude greater than A_1, which controls the along shore wind effect. We might quantify this effect more rigorously by examining when the cross shore surface velocity $v'(z=0)=0$. From Eqs. a.6–a.10 we can write this as when

$$\tau_y + \left\{\tau_x - E^2\tau_\rho\right\}g_4/g_3 = 0 \tag{14}$$

This condition for $\tau_y=0$ has already been considered above. For $\tau_x=0$, the equivalence across shore wind velocity $|w_y|_{eq}$ depends on the ratio g_4/g_3. It appears (and may be shown although not readily so) that over the range of Ekman numbers of importance here, $g_4/g_3 \propto E^{-2}/4$ so that the surface velocity becomes zero when $\tau_y=\tau_\rho/4$. This gives $|w_y|_{eq}=6$ m/s (Beaufort scale 4) which is little more than a moderate breeze. Furthermore, $|w_y|_{eq}$ is independent of the Ekman number.

Given these estimates and the observation that winds along the Netherlands coast, both along and across shore often exceed 10 to 15 m/s, one might expect that wind driven shears almost always dominate the lateral dispersion process.

Lateral Spreading Rate

Before returning to wind effects, we examine the spreading rates expected for the self-induced shear dispersion component. Consider a low density, coastally trapped plume with off shore dimension Y, and with a uniform cross shore density variation $\rho(y)=\rho_c + y\{\rho_0-\rho_c\}/Y$ where ρ_c is the density at the coast and ρ_0 is the density of ocean (North sea) water, Figure 3. Typically, such a density gradient will drive an along shore current and an across shore estuarine circulation cell. This across shore sheared flow, interacting with rapid vertical mixing, produces a landward flux of sea water, increasing the density at the coast and the width of the plume. In the absence of wind stress, according to Eq. 8, this flux is proportional to the cube of the density gradient,

$$F= \frac{2A_1}{f}\left\{\frac{gH}{f\rho_0}\right\}^2\frac{N_z}{K_z}\frac{(\rho_0-\rho_c(t))^3}{Y(t)^3} = C\frac{(\rho_0-\rho_c(t))^3}{Y(t)^3} \tag{15}$$

Assuming there are no low density sources and sinks along the coast, and moving along shore with the mean velocity $\langle u\rangle$, then the total amount of fresh water within the moving cross section must remain constant over time.

$$\left(\rho_0-\rho_c(t)\right)\cdot Y(t) = \left(\rho_0-\rho_c(0)\right)\cdot Y(0) = P \tag{16}$$

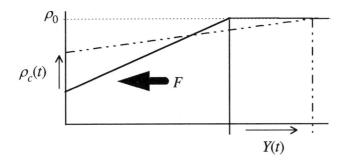

Figure 3. Spreading of a density plume with linear cross shore density gradient: $\rho(y) = \rho_0 - y(\rho_0 - \rho_c) / Y$ where Y is its cross shore dimension, ρ_0 the density of the adjacent shelf sea, and ρ_c the density at the coast. Shear dispersion effects will drive a landward density flux F which will (1) increase the plume dimension $Y(t)$, and (2) increase the density at the coast $\rho_c(t)$.

Thus a coast ward flux of dense water, equivalent to a seaward propagation of the outer plume boundary may be written as:

$$F(t) = \left(\rho_0 - \rho_c(t)\right)\frac{dY}{dt} = C P^3 / Y(t)^6 \tag{17}$$

Rearranging and integrating this gives:

$$Y(t) = [6 \, C P^2]^{1/6} \cdot t^{1/6} \tag{18}$$

This result shows the same behaviour as the similarity solution found by Smith [1982] and represents a very slow spreading rate which is in quantitative agreement with observations. For example, if the plume has achieved a certain off shore dimension Y_1 after 1 day, then it will take somewhat longer than 2 months for the plume to double its width to 2 Y_1. The mean along shore velocity may be given by:

$$\langle u \rangle = D \langle \rho \rangle_y = dX/dt = D P Y(t)^{-2} \tag{19}$$

where X is the position of the moving cross section considered above, and D is a coefficient depending on E and R. Solving Eq. 18 and 19 gives an expression for the cross shore width as a function of along shore distance

$$Y(X) = [2^{-1/6} \, 3^{-2/3} \, D^{-1/4} \, P^{1/4} \, C^{-5/12}] \cdot X^{1/4} \tag{20}$$

In this view, the plume also has a slow spreading characteristic. For instance, if the plume dimension at 10 km form the mouth is Y_1, then the plume will double its width to 2 Y_1 at a distance of 160 *km* from the mouth.

The calculations of the previous section seemed to indicate that under frequently encountered conditions along the Netherlands coast, wind driven effects tend to dominate. If this is true, then the cross shore density flux $F \propto \langle \rho \rangle_y$ rather than its cube. Applying the same arguments as above, this would yield a spreading rate $Y(t) \propto t^{1/2}$ or in terms of along shore distance $Y(X) \propto exp(X)$. Clearly this is a much faster spreading rate than is evident in nature. In the following section, we consider in more detail the effect of wind and how these considerations may be use to improve this simple shear dispersion model.

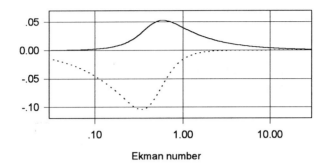

Figure 4. The coefficients $\langle h_1 \rangle$ and $\langle h_2 \rangle$ plotted as functions of E. $R = 0.5$

Wind Versus Density Driven Shears

Comparison of the density and wind driven effects in promoting lateral dispersion seem to suggest that wind is dominant for all but relatively calm periods of time. However, there are two overlooked dynamic effects which tend to moderate the wind's efficiency. These are (1) convective overturn since wind has a directional influence on the vertical shear and may in certain instances produce an unstable water column, and (2) wind mixing which tends to increase vertical exchange processes with increasing wind velocity. Both of these effects have tend to increase vertical mixing and thus decrease horizontal dispersion.

Convective Overturn

A strong wind blowing along the Netherlands coast out of the south west to north east quadrant will induce an Ekman transport with an onshore surface component. This shear, under normal conditions, will drive salty North sea water over fresher Rhine water producing an unstable water column susceptible to convective overturn. The condition when this happens can be deduced from Eq. 4 which gives the surface to bottom density difference:

$$\Delta \rho' = \frac{\langle \rho \rangle_y}{K_z} \left(\int_z^0 v' \, dz \right) = \frac{H^3 E \langle \rho \rangle_y}{N_z K_z} \{ \langle h_1 \rangle \tau_y + \langle h_1 \rangle (\tau_x - E^2 \tau_\rho) \} \tag{21}$$

where $\langle h_1 \rangle$ and $\langle h_2 \rangle$ are the vertical averages of the functions $h_1(z)$, $h_2(z)$ given in the appendix and plotted in Figure 4 for $R=0.5$. Examining the neutral stability criterion, $\Delta \rho'=0$, and noting that over the range of E expected, the ratio $\langle h_2 \rangle / \langle h_1 \rangle \approx 3/(8E^2)$ we can write

stable water column: $E^2(8 \, \tau_y + 3 \, \tau_\rho) - 3 \, \tau_x > 0$
unstable water column: $E^2(8 \, \tau_y + 3 \, \tau_\rho) - 3 \, \tau_x < 0$

If the density of a fluid element exceeds that of the ambient sea by $\Delta \rho'$, then the time it would take for this element to fall from the surface $z=0$ to the bottom at $z=-H$ is $T_{conv} \approx (H \rho_0 / (g \Delta \, \rho'))^{1/2}$. Taking this as a time scale for additional vertical exchange under unstable conditions, i.e. $K_z^{conv} = H^2/T_{conv}$, we can write an associated effective vertical diffusivity as:

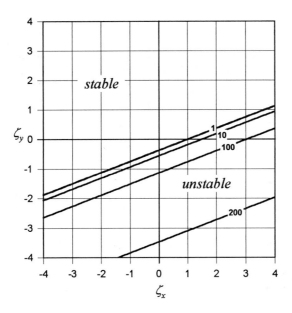

Figure 5. The increase in vertical exchange due to convective over turn for along (ζ_x) and across (ζ_y) shore wind stress.

$$K_z^* = K_z + K_z^{\text{conv}} \approx K_z \left\{ 1 + \frac{4\tau_\rho}{P_t^2 E^4 H^2 f^2} \left[\frac{\langle h_1 \rangle}{4EP_t} \left(3\zeta_x - 8E^2 \zeta_y - 3E^2 \right) \right]^{1/2} \right\} \quad (22)$$

where for convenience $(\zeta_x, \zeta_y) = (\tau_x, \tau_y)/\tau_\rho$ and P_t is the Prandlt number defined as $P_t = K_x/N_z$. In the absence of any evidence to the contrary, we set $P_t = 1$. Evaluating for typical conditions in the Rhine plume, ($E = 1, R = 0.5, \tau_\rho \approx 4.6 \times 10^{-4} \ m^2/s^2, H = 20$ m) we get

$$K_z^* \approx K_z \{1 + 40\{3 \ \zeta_x - 8 \ \zeta_y - 3\}^{1/2}\} \quad (23)$$

which are plotted in Figure 5. It is evident that even a small amount of instability radically increases the effective vertical mixing rate. With respect to horizontal dispersion, an unstable water column effectively halts lateral spreading.

Wind Mixing

Following e.g. Simpson [1981], the energetics of a water column under the influence of vertical mixing and buoyancy forcing B can be expressed as

$$\frac{d\phi}{dt} = \frac{dB}{dt} - \varepsilon_w \frac{\delta\rho_a C_D}{H}|\mathbf{w}|^3 - \varepsilon_t \frac{4}{3\pi} \frac{\rho_0 C_D}{H} U_0^3 \tag{30}$$

where $\phi = -\langle g z \rho' \rangle$ is the potential energy anomaly. The second and third terms on the right hand side represent vertical mixing due to wind (surface) and tidal (bottom) stresses respectively. $\mathbf{w}=(w_x, w_y)$ is the wind velocity and $\delta=0.03$ is the ratio of surface water to wind speed. The coefficients ε_w and ε_t are the wind and tidal mixing efficiencies.

Allowing for time dependence of ρ' in Eq. 1, the same form of expression may be derived in terms of vertical diffusivity:

$$\frac{d\phi}{dt} = \frac{g\langle\rho\rangle_y}{H}\langle v'z\rangle - \frac{gK_z}{H}\Delta\rho' \tag{31}$$

The first term on the right hand side of Eq. 31 represents the introduction of buoyancy by differential advection and is analogous to dB/dt in Eq. 30. The last term is the erosion of buoyancy by vertical mixing and may be equated to the wind and tidal mixing terms of Eq. 30. In particular

$$K_z g\Delta\rho' = C_D\left[\varepsilon_w \delta\rho_a|\mathbf{w}|^3 + \varepsilon_t 4/(3\pi)\rho_0 U_0^3\right] \tag{32}$$

While tidal mixing varies over the spring neap cycle, it is variations due to wind that we are concerned with here. We might write a wind dependant vertical diffusivity as:

$$K_z(\mathbf{w}) = K_z(0)\{1 + \gamma|\mathbf{w}|^3\} \tag{33}$$

where $\gamma=(3\pi\varepsilon_w\delta\rho_a)/(4\varepsilon_t\rho_0 U_0^3)$. Typical values of γ for the Rhine plume range form 4×10^{-4} to $1.1\times 10^{-2}s^3/m^3$ for spring and neap tide respectively and for mean tidal conditions $\gamma=3\times10^{-3}$ s^3/m^3. That is, under mean tidal conditions, a wind speed of 10 m/s (not untypical) increases vertical diffusivity by a factor of 4. This is in keeping with the observation *cf.* van Alphen et al [1987], that in the Netherlands coastal zone, wind induced mixing is more effective than tide induced mixing in affecting vertical exchange. The parameter γ depends on the ratio of the mixing efficiencies. In the above we choose values reported by Czitrom et al [1988]; ε_t=0.0028 andε_w=0.023. However we note the high degree of uncertainty in these values, particularly in applying them to different geographic locations. Nominally, we vary γ over the spring-neap cycle although a similar variation can be affected by these uncertainties.

Let us examine firstly the case for no along shore wind stress. Substituting Eq. 33 and a similar expression for $N_z(\mathbf{w})$ into Eq. 7 yields:

$$K_H^* = A_1 \tau_\rho^2 \frac{2H^2}{fK_z(0)N_z(0)}\left\{a\frac{|w_x|w_x}{1+\gamma|w_x|^3}-1\right\}^2 \tag{34}$$

where for convenience we define $a=(\rho_a C_D)/(\rho_0 \tau_\rho) \approx 1/200$ for the Rhine plume. As wind speed increases, horizontal dispersion increases but only to a point. After that increased vertical mixing tends to dominate over the wind induced shear and horizontal dispersion drops off. We illustrate this in Figure 6 showing $K_H^*(w_x)$ in units of $K_H^*(0)$, the purely density driven dispersion. Included is the trace for $\gamma=0$ when wind mixing effects are excluded. This show the rapid increase $K_H^*(w_x) \propto w_x^4$ suggested by simple shear dispersion considerations. We note that this has an unstable branch ($w_x>11$m/s) where horizontal dispersion essentially goes to zero. The plots for finite γ show quite different behaviour at large wind speed with $K_H^*(w_x) \propto K_H^*(0)$. In other words, the lateral spreading rate estimated in Eq. 18 is likely quite realistic despite the fact that we have ignored wind stress effects there. The effective horizontal diffusivity including wind effects vary at

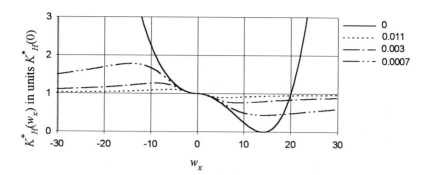

Figure 6. The effective horizontal diffusion coefficient including wind mixing effects as a function of along shore wind velocity, and given for various values of γ. The case where $\gamma = 0$ is for no wind mixing effect. The plots for $\gamma = (0.003, 0.011, 0.0007)$ are for mean, neap and spring tides respectively. Here there is no explicit wind mixing dependence in A_1, and $K_H(0)$ is the horizontal diffusivity for zero wind stress.

most by a factor of 2 form the zero wind estimate which will not effect the $t^{1/6}$ spreading rate. One last point to note is that for these selected values of γ, the water column never becomes unstable. This is because vertical mixing shuts down wind induced shear long before this shear can overwhelm the density (stabilising) driven flow.

Let us now consider an across shore wind stress since this is apparently more efficient in driving an across shore shear. Although the analytic proof is involved, it may be shown numerically that

$$A_1 = \langle h_2^2 \rangle / 4\alpha^2 \approx \langle h_2 \rangle^2 / 4\alpha^2$$
$$A_2 = 2\langle h_1 h_2 \rangle / 4\alpha^2 \approx 2\langle h_1 \rangle \langle h_2 \rangle / 4\alpha^2$$
$$A_3 = \langle h_1^2 \rangle / 4\alpha^2 \approx \langle h_1 \rangle^2 / 4\alpha^2$$

Further, since $\langle h_2 \rangle / \langle h_1 \rangle \approx -3/(8 E^2)$, for Ekman numbers of order unit, we can write

$$A_2 / A_1 \approx 2\langle h_1 \rangle / \langle h_2 \rangle \approx 16 E^2 / 3$$
$$A_3 / A_1 \approx \langle h_1 \rangle^2 / \langle h_2 \rangle^2 \approx (8 E^2 / 3)^2$$

Substituting into Eq. 7 yields

$$K_H(\mathbf{w}) \approx \frac{8}{f^3 H^2} \frac{N_z(\mathbf{w})}{K_z(\mathbf{w})} A_1 \left[\tau_\rho + \frac{8}{3}\tau_y \right]^2 \tag{34}$$

At first glance this might suggest that enhanced vertical mixing due to wind stirring has no effect since it appears equally in $K_z(\mathbf{w})$ and $N_z(\mathbf{w})$. However, it should be noted that A_1 is itself strongly dependent on vertical mixing. In particular, it might be noted that as E becomes large, A_1 falls off as E^{-4}. Since $N_z \propto E^2 \propto 1 + \gamma |\mathbf{w}|^3$, we can write

$$K_H^*(w_y) \approx K_H^*(0)[(1 + 2.67\ a|w_y|w_y)/(1 + \gamma|w_y|^3)]^2 \tag{35}$$

We plot this function in Figure 7. It appears that horizontal dispersion disappears for large wind speeds. This occurs sooner for smaller values of γ when tidal mixing contributes relatively less to vertical mixing (e.g. neap tides). All values of γ produce an unstable water column when the wind speed is less than $-8\,m/s$.

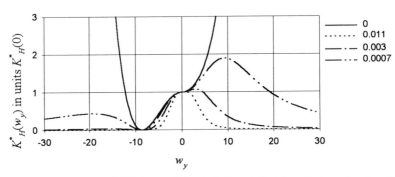

Figure 7. The effective horizontal diffusion coefficient including wind mixing effects as a function of across shore wind velocity, and given for various values of γ. The case where $\gamma = 0$ is for no wind mixing effect. The plots for $\gamma = (0.003, 0.011, 0.0007)$ are for mean, neap and spring tides respectively. Here an explicit wind mixing dependence is included in A_1 (in contrast to Fig (9)), and $K^*_H(0)$ is the horizontal diffusivity for zero wind stress.

Conclusions

It appears that the long term spreading characteristics of a density plume may be well represented by shear dispersion processes. Under the influence of self induced shear, this predicts a spreading rate proportional to time$^{1/6}$. This translates into an along shore increase of the plume width proportional to (distance along shore)$^{1/4}$. These qualitatively mirror the observed slow spreading characteristics of, for instance, the Rhine plume along the Netherlands coast.

With respect to wind driven effects, the two processes considered, namely convective over turn and wind mixing, appear to be powerful agents in moderating the effectiveness of wind induced shears in affecting horizontal dispersion. Of these wind mixing appears to be particularly effective. Strong winds, irrespective of their orientation, tend to shut down lateral dispersion. Here a type of "short circuiting" appears to be in effect. While increasing winds certainly increase shears, their effectiveness is rapidly diminished as vertical mixing erodes stratification faster than it is produced by differential advection.

These conclusions are not general and apply realistically only to the Rhine plume or river plumes with similar characteristics. In particular a river plume discharging into a mid-latitude shallow shelf sea with strong mixing through tides and winds. These essentially fix the Ekman number E and drag parameter R to be both of order unity. Further, the Rhine plume often exhibits only weak stratification. Thus the assumption of vertically uniform diffusivity and viscosity can be tenuously applied. River plumes discharging into deep shelf seas with little tidal energy would have different spreading characteristics, particularly in the susceptibility to wind effects.

Finally, we should perhaps list some of the limitations and overlooked effects which may influence these results:

- The shear dispersion process considered here ignores any explicit horizontal diffusion. Such additional horizontal exchange processes certainly exist, especially close to the river mouth where shear (Helmholtz) instabilities appear.
- We have ignored in this treatment, the effect of vertically sheared across shore flow at the tidal frequency. That such velocity shears exist and may be quite strong, particularly during stratified periods, has been demonstrated by Visser *et al* 1994, Souza and Simpson 1994, and Simpson and Souza 1994. Precisely what the effect of this is in affecting across shore spreading in yet to be determined.
- The assumption of vertical uniformity of N_z and K_z is a strong limitation to the present treatment. In particular, when the water column becomes strongly stratified, eddy motions in the vicinity of the pycnocline are suppressed. That is, vertical exchange by turbulent eddies

decreases as the water column stabilises. This process appears to be important in generating the across shore tidal shears mentioned above.

- The wind dependence of vertical diffusion Eq (31) shows a cubic dependence on wind speed. This is derived from energy and time scale considerations. If we extrapolate this argument to tidal currents we would get a similar cubic dependence of eddy viscosity on tidal current amplitude which would be in conflict with the parameterisation suggested by Csanady 1976, i.e. Eq (11). These two arguments could be brought closer into line if we suppose

$$N_z \propto K_z \propto \{1 + \gamma |w|^3\}^{1/2}$$

That is, the effect of increasing wind in generating vertical mixing goes as $w^{3/2}$ as opposed to the faster rate w^3 used in section 3.2. If we were to include this form of wind dependence, we would get a horizontal spreading rate at high wind speeds

$$K_H^* \propto a|w_x|^{-1} + b|w_y|$$

that is, falling off with increasing along shore winds, but increasing linearly with across shore winds. Here wind mixing still moderates dispersion but not as fast. It remains to be seen which parameterisation is more realistic and whether this would have a serious effect conclusions drawn here.

Acknowledgements. This work was conducted within PROFILE, a European Community funded project within MAST II, contract number MAS2-CT930054.

Appendix: Velocity Field of a Wind Driven Buoyancy Plume.

Suppose we have a density field $\rho(y)$ dominated by a strong vertically homogeneous cross shore (y direction) gradient. Allowing for the possibility of an across shore sea surface $\eta(y)$ slope set up in response to wind, and assuming a vertically uniform eddy viscosity N_z, then the steady state momentum balance may be written as:

$$v + \frac{1}{2\alpha^2} \frac{\partial^2 u}{\partial z^2} = 0 \tag{a.1}$$

$$u - \frac{1}{2\alpha^2} \frac{\partial^2 v}{\partial z^2} = -\frac{g}{f} \frac{\partial \eta}{\partial y} + z \frac{gH}{f\rho_0} \frac{\partial \langle \rho \rangle}{\partial y} \tag{a.2}$$

where $\alpha = \{fH^2 / (2 N_z)\}^{1/2} = E^{-1}$, E being the Ekman number. Eq (a.2) does not preclude the existence of a vertically dependent departure $\rho'(z)$ of the density field, only that this departure has a negligible cross shore variation. Boundary conditions appropriate for the momenta equations may be written:

$$\frac{\partial}{\partial z}(u,v) = \frac{H}{N_z}(\tau_x, \tau_y) \text{ at } z = 0 \tag{a.3}$$

$$\frac{\partial}{\partial z}(u,v) = \frac{kH}{N_z}(u,v) \text{ at } z = -h \tag{a.4}$$

where (τ_x, τ_y) are the wind surface wind stress components, and the bottom boundary condition matches internal stresses with a linear bottom stress parameterized as $k(u,v)$. The general form of the solutions is given by

$$u(z) = A_r \cos \alpha z \cosh \alpha z - A_i \sin \alpha z \sinh \alpha z +$$

$$B_r \cos \alpha z \sinh \alpha z - B_i \sin \alpha z \cosh \alpha z + z \frac{gH}{f\rho_0}\langle\rho\rangle_y - \frac{g}{f}\frac{\partial\eta}{\partial y} \tag{a.5}$$

$$v(z) = A_r \sin \alpha z \sinh \alpha z + A_i \cos \alpha z \cosh \alpha z +$$
$$B_r \sin \alpha z \cosh \alpha z + B_i \cos \alpha z \sinh \alpha z \tag{a.6}$$

The problem is closed, using the 4 boundary conditions and the additional constraint: $\langle v\rangle$=0; that there is no net transport of water in the on/off shore direction. In particular, from the surface boundary conditions we get:

$$B^+ = B_r + B_i = \frac{1}{\alpha}\frac{H}{N_z}\tau_y \tag{a.7}$$

$$B^- = B_r - B_i = \frac{1}{\alpha}\left\{\frac{H}{N_z}\tau_x - \frac{gH}{f\rho_0}\langle\rho\rangle_y\right\} \tag{a.8}$$

and from the y momentum bottom boundary condition and the integral constraint we get:

$$A_r = g_1 B^+ + g_2 B^- \quad\text{and}\quad A_i = g_3 B^+ + g_4 B^- \tag{a.9}$$

where

$$g_1 = [\alpha R(\cos 2\,\alpha - \cosh 2\alpha) - (\sin 2\alpha + \sinh 2\alpha)] / g_0$$
$$g_2 = (\cosh \alpha - \cos \alpha)[\alpha R(\cosh \alpha - \cos \alpha) + (\sin \alpha + \sinh \alpha)] / g_0$$
$$g_3 = [\alpha R(2 - \cos 2\alpha - \cosh 2\alpha) + (\sin 2\alpha - \sinh 2\alpha)] / g_0 \tag{a.10}$$
$$g_4 = (\sin \alpha - \sinh \alpha)[(\cosh \alpha - \cos \alpha) - \alpha R(\sin \alpha - \sinh \alpha)] / g_0$$
$$g_0 = 2[(\cos 2\,\alpha - \cosh 2\alpha) + \alpha R(\sin 2\alpha - \sinh 2\,\alpha)]$$

It may be readily show from Eq(a.6) that the vertically dependent cross shore velocity may be written in the form $v'(z) = B'f_1(z) + B'f_2(z)$ which leads to an evaluation of the integral in Eq(6) of the form

$$\int_{-1}^{0}\left[\int_z^0 v' dz\right]^2 dz = \frac{1}{4\alpha^2}\left\{\left[B^+\right]^2\langle h_1^2\rangle + 2\left[B^+B^-\right]\langle h_1 h_2\rangle + \left[B^-\right]^2\langle h_2^2\rangle\right\} \tag{a.11}$$

where

$$h_1 = (g_1 + g_3)\cosh \alpha z \sin \alpha z - (g_1 - g_3)\cos \alpha z \sinh \alpha z + \sin \alpha z \sinh \alpha z$$
$$h_2 = (g_2 + g_4)\cosh \alpha z \sin \alpha z - (g_2 - g_4)\cos \alpha z \sinh \alpha z + 1 - \cos \alpha z \cosh \alpha z$$

Integrals of these functions lead to the form functions $A_{1,2,3}$ (E,R) which determine in part the effective horizontal diffusivity induced by shear dispersion as expressed in Eq(7). In particular, $A_1 = \langle h_2^2\rangle/4\alpha^2$, $A_2 = 2\langle h_1 h_2\rangle/4\alpha^2$, and $A_3 = \langle h_1^2\rangle/4\alpha^2$.

References

Bowden, K. F., Horizontal mixing in the sea due to a shearing current. *J. Fluid Mech. 21*, 83-95, 1965.

Csanady, G. T., Mean circulation in shallow seas, *J. Geophys. Res., 81*, 5389-5399, 1976

Czitrom, S. P. R., G. Budéus and G. Krause, A tidal mixing front in an area influenced by land runoff, *Cont. Shelf Res.*, 8, 225-237, 1992.

de Ruijter, W. P. M., A. van der Giessen, and F. C. Groenendijk, Current and density structures in the Netherlands coastal zone, in *Dynamics and Exchanges in Estuaries and the Coastal Zone*, edited by D. Prandle, Coastal and Estuary Studies 40, AGU, Washington 529-550, 1992.

Fisher, H.B., E.J. List, R.C.Y. Koh, J. Imberger and N.H. Brooks, *Mixing in Inland and Coastal Waters* 481 pp., Academic Press, San Diego, 1979.

Giessen, A. van der, W.P.M. de Ruijter and J.C. Borst, Three dimensional current structure in the Dutch coastal zone, *Neth. J. Sea Res., 25*, 45-55, 1990.

Munchow, A., and R.W. Garvine, Buoyancy and wind forcing of a coastal current, *J. Mar. Res., 51*, 293-332, 1993.

Simpson, J.H., The shelf sea fronts: implications of their existence and behaviour, *Phil. Trans. R. Soc. Lond. A 302*, 531-546, 1981.

Simpson, J.H., W.G. Bos, F. Schirmer, A.J. Souza, T.P. Rippeth, S.E. Jones, and D. Hydes, Periodic stratification in the Rhine ROFI in the North Sea, *Oceanolog. Acta., 16*, 23-32, 1993.

Simpson, J.H., and A.J. Souza, Semi-diurnal switching of stratification in the Rhine ROFI, *J. Geophys. Res.,* , 1994.

Smith, R., Similarity solutions of a non-linear diffusion equation, *IMA J. Appl. Math., 28*, 149-160, 1982.

Souza, A.J., and J.H. Simpson, The modification of the tidal ellipses by stratification in the Rhine ROFI, *Cont. Shelf Res.,* , 1994.

Stommel, H., and A. Leetmaa, Circulation on the continental shelf, *Proc. Nat. Acad. Sci. USA, 69*, 3380-3384, 1972.

van Alphen, J.S.L.J., W.P.M. de Ruijter and J.C. Borst, Outflow and three dimensional spreading of Rhine river water in the Netherlands coastal zone, in *Physical Processes in Estuaries*, edited by J. Dronkers and W. van Leussen, Springer-Verlag, 70-92, 1988.

Visser, A.W., A.J. Souza, K. Hessner, and J.H. Simpson, The effect of stratification on tidal current profiles in a region of freshwater influence, *Oceanologica Acta, 17*, 369-381, 1994.

Visser, M., W.P.M. de Ruijter, and L. Postma, The distribution of suspended matter in the Dutch coastal zone, *Neth. J. Sea Res., 27*, 127-143, 1991.

4

Reinforcement of Gravitational Circulation by Wind

Jack Blanton

Abstract

Buoyancy inputs from riverine discharges form an estuarine-like frontal zone along coasts. The gravitational component of across-shelf circulation drives a near-surface flow seaward and a near-bottom flow shoreward. Using data from the southeastern U.S. continental shelf, the buoyancy field is examined under upwelling and downwelling favorable wind stress. Numerical simulations and field data show that prolonged upwelling-favorable winds reinforce the gravitational circulation by enhancing its offshore component. Larger-than-normal riverine discharges in 1993 showed that this reinforcement caused a lens of warm, low-salinity coastal water to become detached from the coastal front and be advected seaward. Reinforcement of this process enhances the loss of estuarine discharges from coastal areas. Climatological data show that this loss increases as the mean upwelling favorable wind stress for the year increases.

Introduction

This paper focuses on the reinforcement of the gravitational circulation component in coastal frontal zones on shallow continental shelves. This component is forced by the buoyancy of riverine and estuarine discharges along continental margins, discharges that form zones of low salinity water extending for distances of O(100 km) along the coast. Examples are found along the east coast of Central America [Murray and Young, 1985], the middle Atlantic shelf of the U.S, the Dutch and German coasts [Simpson et al., 1993], and the southern Atlantic shelf of the U.S. [Blanton, 1981].

The Atlantic U.S. continental shelf between the Pee Dee River in South Carolina south to Jacksonville, Florida has a multi-inlet coast line connecting low-lying coastal marshes to the ocean (Figure 1). Five of the ten rivers discharge freshwater at rates exceeding 100 m³/s, and all form plumes of low-salinity (low-density) water at the coast. Subsequent tidal and wind mixing blends the plumes into a band of low salinity that extends along the coast over a distance greater than 400 km. While we customarily define the inner shelf here to depths between 0 - 20 m, the seaward extent of the low-density band varies temporally and spatially and defines, in a dynamic sense, the inner shelf. In the absence of wind, a weak baroclinic coastal current flows southward within the low-density band at a speed less than 0.05 m s⁻¹ [Blanton et al., 1989].

The across-shelf density structure of many coastal regions resembles that of a partially mixed to well-mixed estuary. The density deficit of low-salinity water causes it to override ambient shelf water of higher density to form a coastal frontal zone (CFZ) that usually extends 20 - 30 km offshore. The pressure gradient from low-density water at the coast drives an across-shelf gravitational estuarine-like circulation that is offshore in surface layers and onshore near bottom. The strength of the gravitational component is proportional to the horizontal density gradient, which, in turn, is proportional to the magnitude of riverine discharge. The resulting

Buoyancy Effects on Coastal and Estuarine Dynamics
Coastal and Estuarine Studies Volume 53, Pages 47-58
Copyright 1996 by the American Geophysical Union

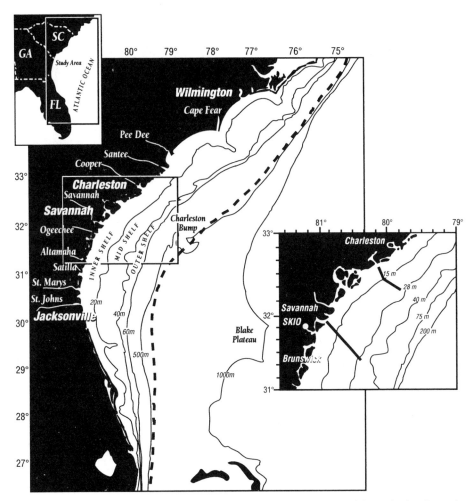

Figure 1. Map of continental shelf off the southeastern US. Inset on the right marks locations of oceanographic sections discussed in the text. River names are in italics.

across-shelf motion is enhanced in the coastal environments where the vertical Ekman number (ratio of frictional force per unit mass to Coriolis acceleration) is high [Garrett and Loder, 1981]. Model studies [Werner et al., 1993] and observations [Blanton, 1981; 1986] have shown that the strength of this across-shelf flow component is O(0.1 m/s).

This paper discusses the results of mass transfer through a CFZ due to, at least, three mechanisms: (1) gravitational circulation, (2) shear dispersion, and (3) wind-driven Ekman flux. The first two were compared by Garrett and Loder [1981] for low vertical Ekman number environments. The mass flux due to gravitational circulation through a coastal front with a mean horizontal density gradient is directly proportional to vertical eddy viscosity A_v. On the other hand, the mass flux due to shear dispersion is indirectly proportional to A_v. While Garrett and Loder [1981] suggested that the former mechanism was likely to be significantly greater than the latter for low Ekman numbers, the result was sensitive to the parameterization of A_v. Their relative importance in high vertical Ekman number environments is not presently understood.

A one-dimensional view of the role of A_v is adopted that assesses the effects of mechanical energy on vertical stratification induced by lateral buoyancy inputs. A_v "sets the stage" for the operation of wind stress on the gravitational circulation of the CFZ regardless of how the resulting stratification manages to integrate the effects of the other two processes in which A_v appears to play conflicting roles. (See Blanton et al., 1994 for an evaluation of shear dispersion by tidal currents in the CFZ off the southeastern U.S. coast.)

Background

The low-salinity band formed by estuarine discharges contains the main buoyancy source for the CFZ. Vertical density gradients exhibit a high degree of spatial and temporal variation due to (1) vertical mixing provided by tidal current friction at the sea bed, (2) surface friction provided by wind stress, and (3) the source strength of freshwater discharge and its spatial distribution.

The source strength is defined by the buoyancy flux, B, as follows:

$$B = \frac{g}{\rho} F \qquad (1)$$

where
F	=	freshwater flux (kg/m^2/s) = $\rho Q/LW$	
Q	=	freshwater discharge (m^3/s)	
L	=	shoreline length = 400 km	
W	=	mixing width = 20 km (offshore extent of CFZ)	
g	=	acceleration of gravity	
ρ	=	density of sea water	

B is directly proportional to freshwater discharge. We neglect sensible heat and evaporative fluxes which were shown to be small for the inner shelf of the southeastern U.S. coast relative to freshwater buoyancy sources [Blanton and Atkinson, 1983].

The significance of B can be quantified by calculating the power (V) required to vertically mix a given density deficit (ρ') at the surface relative to ambient shelf water to a given depth (H), or

$$V = \frac{H}{2}\rho'B \qquad (2)$$

This is approximately equivalent to calculating the work against the buoyancy force required to lower the center of gravity of a thin layer (thickness $\ll$ H) of surface water of density deficit ρ' to a position of (1/2) H. The power represented by V is supplied by wind stress, tidal currents, evaporative cooling and evaporation which leaves salt behind. The latter two processes are neglected for this region [Atkinson and Blanton, 1986].

Following van Aken [1986], tidal current power loss (P) to vertical mixing was estimated by:

$$P = \frac{4}{3\pi}\gamma\rho_o C_{dw} U_t^3 \qquad (3)$$

where $\gamma = 0.037$ is an efficiency factor, $C_{dw} = 0.005$ is a bottom drag coefficient, $\rho_o = 1000$ kg m^{-3} is sea water density, and U_t is the RMS amplitude of a sinusoidally varying vertically averaged tidal current.

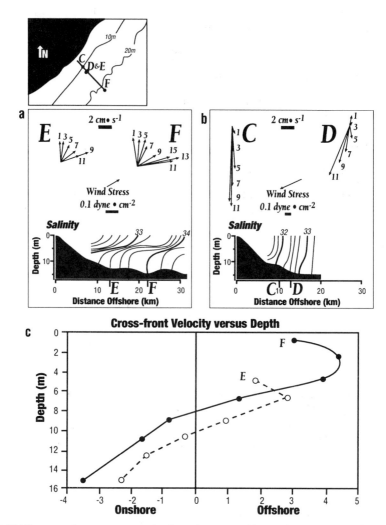

Figure 2. Tidally averaged current vectors at 1-m intervals at two positions within the CFZ. Numbers at the end of each vector designate distance above bottom. Upper diagram shows locations where current profiles were measured. (Data are from Blanton, 1986). (a) upwelling favorable wind; (b) downwelling favorable wind; (c) across- shelf velocity versus depth from data in panel 2a.

Similarly, wind mixing power loss (W) was determined by

$$W = \delta \rho_a^3 C_{da} U_w \qquad (4)$$

where delta = 0.001 is an efficiency factor, C_{da} = 0.002 is a wind drag coefficient, ρ_a = 1 is air density, and U_w is a suitably averaged wind speed.

Blanton and Atkinson [1983] determined that approximately 10^{-4} to 10^{-3} W m^{-2} is required to mix completely throughout depth observed ranges of total river discharges varying between 1000 m^3 s^{-1} in autumn to 8000 m^3 s^{-1} in spring. Estimates for tidal power dissipation at the bottom range from 0.3 to 2 x 10^{-4} Wm^{-2} off the Georgia coast. This is sufficient to vertically mix buoyancy due to heat additions but is too weak to vertically mix the buoyancy normally provided by freshwater discharge.   However, wind-induced surface stress can provide a significant increment of power over that of tidal currents.   Mixing power due to strong wind events can range between 1 and 3 x 10^{-3} W m^{-2} [Atkinson and Blanton, 1986], or an order of magnitude greater than that provided by tidal power.

Wind Effect on Coastal Gravitational Circulation

In addition to vertical mixing, wind can advect buoyancy to deeper/shallower water where, according to Eq. 2, the power required for complete vertical mixing is greater/smaller. Alongshelf wind stress produces an Ekman-induced across-shelf flow that combines with the gravitational circulation component. Upwelling favorable offshore Ekman transport spreads the front seaward, and there is an offshore component of flow above and within a strong pycnocline (Figure 2).   Currents are onshore below the pycnocline. Thus, upwelling reinforces the estuarine circulation component, and the across-shelf flow structure resembles the along-axis flow of a partially mixed estuary.

Onshore Ekman transport during downwelling-favorable wind (southward stress) advects the frontal zone shoreward and steepens the front. In contrast to the upwelling-favorable case, the downwelling-induced across-shelf flow opposes the gravitational component (Figure 2).

Examples of CFZs Under Contrasting Coastal Discharges

Discharges during summer 1992 and spring 1993 varied by a factor of four (Figure 3). Two examples of across-shelf density structure show the effects of this variation on buoyancy supply and the redistribution of buoyancy by upwelling-favorable winds.   The low buoyancy flux in summer 1992 in the presence of tidal mixing and low upwelling-favorable wind stress resulted in a relatively well-mixed CFZ out to 40 km from the coast (Figure 4a).   Contrast this with conditions in spring 1993 when the quadrupled freshwater discharge accompanied by predominantly upwelling-favorable winds stretched the CFZ seaward, and the water column extending 40 km offshore was vertically stratified (Figure 4b).

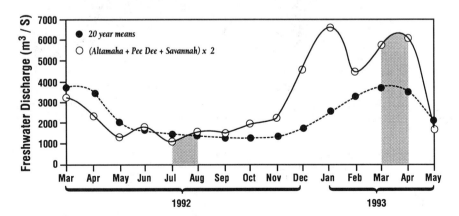

Figure 3.   Monthly freshwater discharge along the southeastern US. Values represent twice the sum of the discharges of the largest three rivers (Altamaha (GA), Savannah (GA/SC) and Pee Dee (SC). Dashed curve represents the 20-year mean. The two stippled areas mark the times of oceanographic section data shown in Figure 4.

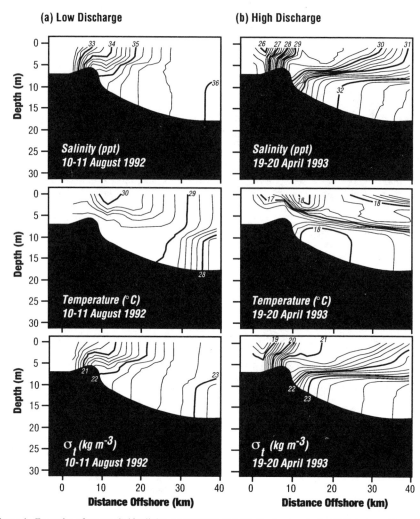

Figure 4. Examples of across-shelf salinity, temperature and density structure in the CFZ during a time of (a) low and (b) high river discharge. These sections were obtained along a line extending offshore of Savannah, GA (Figure 1). (From Nelson and Guarda, 1995).

The mixing energy requirements for these two seasons are illustrated in Table 1. The low discharge buoyancy can be completely mixed at H = 10 m by applying a mixing power of 0.2 milliwatts/m^2. In order to completely mix the high-discharge buoyancy at the same water depth, mixing power must increase by a factor of four since discharge is increased by this factor. For H = 20 m, the required power (V) is double that given in Table 1.

Note the strong vertical stratification present 40 km offshore where the mixing power by the wind was apparently insufficient to overcome the buoyancy imported from regions closer to shore by upwelling favorable winds. In other words, by reinforcing the ambient gravitational circulation, upwelling winds efficiently transport low-density water seaward where vertical mixing in deeper water is less effective which results in increased vertical stratification.

Across-shelf Advection of Buoyancy Under Upwelling Conditions

The effects of along shelf wind on gravitational circulation have been explored for the CFZ off the Georgia coast using a 2-D numerical model [Blanton et al., 1989]. This model showed that the continuous application of alongshelf upwelling favorable wind stress stretched the CFZ seaward (Figure 5). While wind and tidal currents efficiently mixed the shallow portions of the water column near the coast, the transport of buoyancy by the reinforced gravitational circulation increased vertical stratification farther offshore. Eventually, low-density water at the coast mixed with mid-shelf water coming shoreward underneath the stratified waters farther offshore. This mixture was advected seaward in the form of shallow lenses having a across-shelf scale of about 20 km. When winds reversed, the simulation showed that the gravitational circulation was essentially shut-down by onshore Ekman transport, and vertical stratification was destroyed within a day or so.

A series of 6 cross-sections off South Carolina (Figure 6) confirmed many of the aspects shown in Figure 5. Vector plots of wind stress are displayed with sub-tidal frequency currents measured near Station M2 along the sections (Figure 7). These data were obtained during the high runoff season in spring 1993 (Figure 3). The first three sections (Figure 6a) were completed under upwelling-favorable winds that persisted over a period of 5 days (Figure 7).

Survey 1 (Figure 6a) showed a CFZ that tilted downward and shoreward with 33 ppt water intersecting the bottom 6 km offshore at a depth of about 8 m. There was a shallow lens of low salinity and relatively warm water located shoreward of relatively strong vertical stratification.

One day later, under strengthening winds, Survey 2 showed a more flattened density structure and the disappearance of the shallow lens near the coast. Upwelling winds continued to increase in strength for the next four days then decelerated and reversed at the time of Survey 3. Note that measured currents had not yet reversed (Figure 7). Vertical stratification in Survey 3 increased even more farther offshore. High salinity water from the mid-shelf was apparently advected shoreward where the 34-ppt isohaline almost intersected the surface within 10 km of the coast. A shallow lens had reappeared near the coast perhaps due to the alongshelf advection of a low-salinity source directly downwind.

TABLE 1. Power required (V) to vertically mix a given amount of buoyancy for a water depth of 10 m. V = $(H/2)\,\rho'B$.

	Q (m^3/s)	F x 10^4 $(kg/m^2/s)$	B x 10^6 (m^2/s^3)	V (mw/m^2)	P (mw/m^2)	W (mw/m^2)
Low	1500	1.9	1.8	0.2	0.1	<0.1 breeze
High	6000	7.5	7.4	0.9	0.6	3 stormy

Q = Freshwater discharge between Charleston SC and Jacksonville FL
F = buoyancy source strength
B = buoyancy flux at surface
H = water depth = 10 m
ρ'= density deficit of low-density source = 25 kg/m^3

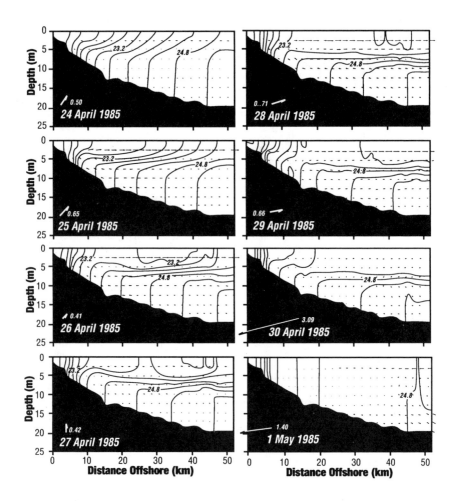

Figure 5. Two dimensional (across-shelf) numerical simulation of circulation and density structure in the CFZ off the southeastern US (from Blanton et al., 1989). All properties were uniform in the alongshelf direction. Actual winds were used to drive the model. The arrow in the lower left of each panel is the daily average wind stress vector applied to the cross-section.

Survey 4 (Figure 6b) was done about 12 hours later as downwelling winds set in. Measured currents were still weak but had reversed. The disappearance of strong stratification at the coastal boundary suggests that the currents there had reversed and were no longer importing water from a low-salinity source. Farther offshore, the strongly stratified water seen 20 km offshore in the previous survey actually extended at least 40 km offshore by the time the prolonged upwelling winds had reversed.

Surveys 5 and 6 (Figure 6b) were done during an interval of downwelling favorable winds. Measured currents had essentially stalled out during Survey 6 as upwelling favorable winds restarted. Note the reappearance of low-salinity water at the offshore edge of the section. Even though upwelling winds continued for an additional 3 days, the thin layer of

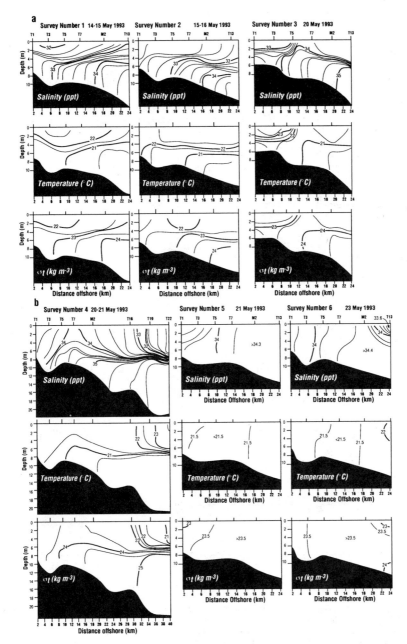

Figure 6. A series of six offshore surveys obtained along a line extending offshore of Charleston, SC (Figure 1) during a period immediately after high buoyancy discharge (Figure 3). Station labels are marked on the upper panels of each survey; (a) surveys 1-3; (b) surveys 4-6.

low-salinity stratified water so evident offshore in Surveys 3 and 4 never reappeared suggesting that the lens-like feature had either a finite alongshelf extent or was simply advected farther offshore.

Discussion

Cross-sections of hydrographic structure through a CFZ off the southeastern US continental shelf revealed that upwelling-favorable winds reinforce an estuarine-like circulation component that enhances the across-shelf transport of coastal water. The loss of coastal water appears to be in the form of shallow low-density lenses with a across-shelf scale width of about 20 km. The creation of lens of coastal water by upwelling favorable winds has also been seen off the coast of New Jersey (Munchow, personal communication).

These findings conform to results of numerical simulations published previously. For example, the lens seen in Surveys 3 and 4 is bounded on the shoreward side by the 34 ppt isohaline which marks a shoreward intrusion of high salinity water along the bottom. Vertical

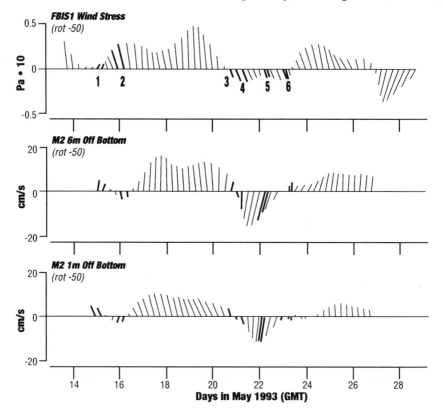

Figure 7. Vectors of wind stress and ocean currents (rotated 50 degrees clockwise to conform to shoreline orientation) throughout the time period of the offshore surveys in Figure 6. Vectors at the times of the surveys (Figure 6) are bold-faced and the survey numbers are marked on the wind stress plot. All data have been smoothed with a 40-hr low- pass filter. Wind stress data were obtained from NOAA Station FBIS1 on the coast near Charleston, SC. Current meter data were obtained from a mooring at Station M2 (see Figure 6 for location).

mixing at Station T7 actually brings high salinity water to the surface so that the across-shelf salinity gradient is initially positive then changes sign to negative offshore at T7. This is qualitatively similar to the density distribution simulated during persistent upwelling for 26 April (Figure 5) where one sees a similar intrusion of deeper higher density water approaching the surface 10 km offshore.

Spring and summer upwelling favorable winds provide favorable conditions for reinforcement of gravitational circulation. Blanton and Atkinson [1983] studied the rate of decrease of freshwater content (a proxy for buoyancy) for the inner shelf region of the SAB for eight spring seasons. The results suggest a correlation with the strength of seasonally averaged upwelling-favorable wind stress (Figure 8) which is consistent with a process that reinforces the gravitationally driven circulation of the inner shelf. While the tendency is for the CFZ to store the freshwater discharge, years when the average alongshelf wind stress is large are accompanied by relatively rapid decreases in freshwater content on the inner shelf. As a result, low-salinity water is historically found out to the 40-m isobath in April and May off the South Carolina coast [Atkinson et al., 1983].

The inference that Figure 8 provides information on the rate of freshwater export from the inner shelf can be used to estimate across-shelf transport during spring and summer when offshore transport is presumably maximized by upwelling-favorable winds. There is a flat area of the curve which was called the "threshold" loss rate by Blanton and Atkinson [1983] wherein it was suggested that tides could provide the mixing process to account for the threshold loss. Using an across-shelf diffusion coefficient of O (100) m^2 s^{-1} and a length scale of across-shelf freshwater gradient of 100 km [Blanton et al., 1994], the threshold rate amounts to about 9 x 10^{-4} m^3 s^{-1} per meter of coastline for an inner shelf 20 km wide. This amounts to about 370 m^3 s^{-1} for a 400-km coastline.

During the strong upwelling season of 1958, about 5 x 10^{-3} m^3 s^{-1} per meter of coastline was lost to the outer shelf, or almost 2000 m^3 s^{-1} for the entire length of 400 km. Thus, the export of coastal water to the outer shelf may be increased by an order of magnitude due to strong upwelling favorable wind and its reinforcement of the background gravitational circulation.

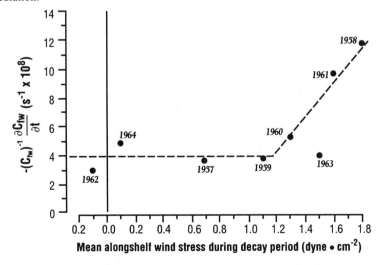

Figure 8. Rate of freshwater loss ($\partial C_{fw}/\partial t$) normalized by freshwater concentration at an inner shelf location off Savannah, GA (from Blanton and Atkinson, 1983). The decay period is defined as the period over which the maximum freshwater content decays to a minimum value (usually in June or July). The year for each spring season is denoted by each data point.

Acknowledgments. I wish to express my appreciation to several individuals. First my thanks go to Julie Amft who coordinated and implemented many aspects of the field work and data analyses. The able support of the captain and crew aboard Skidaway Institute's R/V *Blue Fin*, Jay Fripp (Capt), Raymond Sweatte and Chris Knight and Captain Paul Tucker of SC Marine Resources Research Institute's R/V *Anita* is also appreciated. Graphics were designed and carried out by Anna Boyette. I also thank Leo Oey and Francisco Werner who provided many stimulating discussions on coastal and estuarine processes. This work was sponsored by joint grants from the Georgia Sea Grant Program (Grant No. R/EA-15) to the Skidaway Institute of Oceanography and the South Carolina Sea Grant Consortium (Grant No. 93277) to the SC Marine Resource Research Institute. Further support is acknowledged from NOAA's Coastal Ocean Program (Grant No. NA16RG0481-01 to the Georgia Sea Grant College Program.)

References

Aken van, H.M., The onset of stratification in shelf seas due to differential advection in the presence of a salinity gradient, *Cont. Shelf Res.*, *5*, 475-485, 1986.

Atkinson, L.P., T.N. Lee, J.O. Blanton, and W.S. Chandler, Climatology of the southeastern United States continental shelf waters, *J. Geophys. Res.*, *88*, 4705-4718, 1983.

Atkinson, L.P., and J.O. Blanton, Processes that affect stratification in shelf waters, in *Baroclinic Processes on Continental Shelves, Coastal and Estuarine Sciences 3*, edited by C.N.K. Mooers, pp. 117-130, AGU, Washington, DC, 1986.

Blanton, J.O.., Ocean currents along a nearshore frontal zone on the continental shelf of the southeastern U.S., *J. Phys. Oceanogr.*, *11*, 1627-1637, 1981.

Blanton, J.O., Coastal frontal zones as barriers to offshore fluxes of contaminants, *Rapp. P. -v. Reun. Cons. int. Explor. Mer*, *186*, 18-30, 1986.

Blanton, J.O., and L.P. Atkinson, Transport and fate of river discharge on the continental shelf of the southeastern United States, *J. Geophys. Res.*, *88*, 4730-4738, 1983.

Blanton, J.O., L.-Y. Oey, J. Amft, and T.N. Lee, Advection of momentum and buoyancy in a coastal frontal zone, *J. Phys. Oceanogr.*, *19*, 98-115, 1989.

Blanton, J. O., F. Werner, C. Kim, L. Atkinson, T. Lee, and D. Savidge, Transport and fate of low-density water in a coastal frontal zone, *Cont. Shelf Res.*, *14*, 401-427, 1994.

Garrett, C.J.R., and J.W. Loder, Dynamical aspects of shallow sea fronts, *Phil. Trans. R. Society London A*, *302*, 563-581, 1981.

Murray, S. P., and M. Young, The nearshore current along a high-rainfall, trade-wind coast, Nicaragua, *Est. Coast Shelf Sci.*, *21*, 687-699, 1985.

Nelson, J.R., and S. Guarda, Particulate and dissolved spectral absorption on the continental shelf of the Southeastern United States, *Journal of Geophysical Research*, *100*, 8715-8732, 1995.

Simpson, J. H., W. G. Bos, F. Schirmer, A. J. Souza, T. P. Rippeth, S. E. Jones, and D. Hydes, Periodic stratification in the Rhine ROFI in the North Sea, *Oceanological Acta*, *16*, 23-32,1993.

Werner, F.E., J.O. Blanton, D.R. Lynch, and D.K. Savidge, A numerical study of the continental shelf circulation of the U.S. South Atlantic Bight during autumn of 1987, *Cont. Shelf Res.*, *13*, 971-997, 1993.

5

Propagation Velocities and Instability Development of a Coastal Current

Timothy P. Mavor and Pablo Huq

Abstract

The density difference between two water masses is often the driving force of a coastal current. Some characteristics of a buoyancy-driven coastal current are studied using a 1.2 m and a 2.4 m diameter turntable. Fresh water is released from a source along the vertical wall of a rotating, flat-bottomed basin filled with salt water. The source outflow flux Q, as well as the total water depth D, are varied. Non-dimensional nose velocities were found to decrease with time as $(fT)^{-1/2}$. Multiple instabilities are observed well behind the nose along the front separating the buoyant coastal current and the ambient shelf water. The position of the instabilities, as well as the time of occurrence, varied with changes in ratio of the buoyant layer depth to the total water depth, H/D.

Introduction

The presence of a density-driven coastal current is of great importance to nearshore physical and environmental processes. As a transport system for pollutants and waste, the current could threaten coastal environments by depositing the contaminants along the shoreline. Conversely, it could provide offshore transport and across-shelf mixing of these hazards. The coastal current may also transport nutrients upon which ecosystems depend. Hence, a better understanding of the dynamics of density driven coastal currents is necessary, and may be based on laboratory experiments that examine properties of such currents.

Buoyant coastal currents occur throughout the world. Estuaries, such as the Delaware Estuary, provide a source of low-density water which is discharged into higher-density sea water as a buoyant estuary plume [Münchow and Garvine, 1993]. When discharged from an estuary, the plume is turned, due to the Coriolis force. This is seen as a deflection of the plume to the right in the northern hemisphere. Subsequently, the low-density water approaches the coast and the buoyant plume forms a coastal current. Often, a well-defined boundary, or front, between the different water masses is observed. Though it is known that this front exhibits steep gradients in several fluid properties, the characteristics and dynamics of fronts are still largely unresolved.

Numerous coastal currents of various scales have been observed throughout the world. In addition to the Delaware Coastal Current, currents off the coasts of Algeria, Australia, Norway, New Jersey and Washington have been documented. Of particular interest is the occurrence of large eddy motions, which have been observed both in field and laboratory data. The dynamics and stability of the coastal current and the growth of instabilities will be addressed by laboratory experiments. The major goals are to investigate the nose velocity of the gravity current and the developing instabilities well behind the nose.

Buoyancy Effects on Coastal and Estuarine Dynamics
Coastal and Estuarine Studies Volume 53, Pages 59-69

Background

The dynamics of coastal currents have been researched by a variety of methods. The ability of laboratory experiments to simulate a buoyant coastal current is of great importance. Experiments undertaken on laboratory turntables have provided valuable information on fronts, meandering wave-like structures and other boundary current instabilities. Several different experimental designs have been employed in the past to study the various dynamical mechanisms often associated with coastal currents.

The structure of unstable waves has been of particular interest because of the waves' dramatic effect on the mean flow. The streamwise-uniform boundary currents produced by a sudden release of buoyant fluid along a wall were observed to be immediately unstable to long, slowly growing waves [Griffiths and Linden, 1982]. Two regimes were found. For wide currents, the dominant wave length was a constant multiple of the deformation radius. For narrow currents, however, the observed wavelengths were a constant multiple of the current width; the proximity of the wall was found to cause growing waves to be shorter than those of wide currents.

The transition from stable to unstable flow was observed in experiments with a continuous, slow flux of buoyant fluid from a line source along the wall [Griffiths and Linden, 1981]. In these experiments, the boundary current was stable until it reached a critical size, after which waves grew and transformed the flow into a very broad current containing eddies and large amplitude waves. The instability of the narrow current was attributed to the viscous dissipation of perturbations by Ekman-layer suction. This hypothesis was supported by observations of large wavelengths between the instabilities. For their point source experiments (which are more relevant to the findings discussed later), buoyant currents were found to become unstable at smaller Froude numbers than currents produced by a zonally independent (ring) source. It was postulated that the critical shear required for instability is exceeded at a smaller current width.

In a laboratory model of an upwelling front in a two-layer stratified system [Griffiths et al., 1982], the pycnocline sloped upwards toward a vertical coastline. It formed a surface front that was maintained by a geostrophic balance, but was unstable to wave-like disturbances. A conclusion was that a vertical coastline influenced large amplitude unstable waves, provided that the boundary was approximately three Rossby radii from the front (on the unstratified side).

Other laboratory results have shown that the coastal current is generally in geostrophic balance across-shelf [Vinger et al., 1981]. For supercritical flow (an initial velocity greater than the phase speed of long interfacial waves), the instabilities that developed along the front resembled Kelvin-Helmholtz billows, suggesting that the instabilities were barotropic. Subcritical flow was observed to separate from the wall. These observations led to the conclusion that the distinction between various modes of instability is connected to the flow's available energy.

Whitehead and Chapman [1986] found that a buoyant gravity current moving over a uniformly sloping bottom was wider and slower than a current next to a vertical wall. They also noted the movement of water ahead of the gravity current was consistent with frictionally damped barotropic shelf waves. This feature would contribute to the deceleration of the nose, along with the viscous drag and a finite volume source.

Garvine [1987] developed a layer model to study the steady state behavior of estuary plumes. A variety of plumes were investigated, in both rotating and non-rotating environments. The angle the estuary made with the coastline was also varied. For supercritical flow, the rotating plume displayed a boundary front that weakened soon after discharge. It formed a turning region, where Coriolis deflected the buoyant water towards shore. The buoyant water in the nearshore region hence forms a coastal current.

Other major studies on gravity currents in rotating and non-rotating environments include Benjamin [1968], who studied steady gravity currents in a variety of non-rotating settings, while Stern [1980] and Stern et al. [1982] discuss the rotating case. Principal findings were that the nose slows down and eventually stops, that the width of the bore is in agreement with theoretical models, and a that a fraction of the boundary transport is not carried by the nose but is deflected backwards. Griffiths and Hopfinger [1983] observed the velocity of the nose to decrease exponentially in time. It should also be noted that many experiments on gravity currents are done utilizing a dam-break technique, while our experiments utilize a steady, continuous point source.

Previous turntable experiments have also been performed to simulate dimensionless parameters of the Algerian Current [Chabert D'Hieres et al., 1991]. Data were collected regarding the wavelength and frequency of coastal current instabilities, along with the growth and development of instabilities. When the Burger number ($B_u = g'H/(fL)^2$) was decreased, the flow became less stable. This was noted by the presence of both cyclonic and anti-cyclonic eddies, which were similar in form to those also observed in the Algerian Current.

Experimental Procedure

The experiments are performed in Ocean Engineering Laboratory at University of Delaware. Two rotating turntables are available for our experiments. The smaller apparatus consists of a cylindrical flat-bottom basin, 1.2 m in diameter, mounted centrally on a drive unit. The basin can be filled with water to a maximum depth of 25 cm. A freshwater reservoir is secured 1.25 m above the turntable by metal framework connected directly to the turntable. The reservoir is connected to a flow meter, which regulates the flow of fresh water (density ρ_1) to the basin. The large turntable is a 2.4 m diameter flat-bottom basin, with a fresh water reservoir secured beneath the rotating turntable. The source water is pumped up through a flow meter, then into the basin.

The experimental procedure is similar for both apparatus. Figure 1 illustrates the experimental set-up used for the 1.2 m case. The salt water (density ρ_2) of depth D in the basin is spun up to the angular velocity ω. When fluid in the basin has reached solid body rotation, the fresh water is released through the flow meter. The source is positioned just below the free-surface interface, parallel to the vertical boundary. The cross-sectional shape of the source for experiments performed on the 1.2 m turntable was circular, with a diameter of 1 cm. The experiments performed on the 2.4 m turntable utilized a rectangular cross-section of dimensions 2.5 cm by 11 cm. In the data presented here, particular interest is focused on the nose of the current as it propagates around the basin, and the form of the front far from nose.

The angle of outflow is an important consideration in the study of front and shallow estuary plumes on a continental shelf. Flows out of estuaries at large angles to the shoreline will allow the right-hand streamline (in the northern hemisphere) to separate from the shore, possibly forming an interior front [Garvine, 1987]. The additional potential vorticity in flows leaving such estuaries may also be important. The alongshore orientation chosen allows investigation of instability development without this additional vorticity. Future experiments should give insight into the effect of the angle of outflow and also the effect on instabilities.

The effects of winds, tides and bathymetry on coastal current dynamics are often neglected in laboratory simulations of coastal currents. Wind stress can have a major effect on coastal currents. An upwelling favorable wind can contribute to movement of the buoyant plume offshore, whereas a downwelling favorable wind stabilizes the coastal current. Tidal action induces more mixing of the plume and ambient shelf water, and moves frontal boundaries where mixing is enhanced. Widely varied bottom topography may also be a factor. For this study, these effects are also ignored.

The reduced gravitational acceleration is $g' = g(\rho_2 - \rho_1)/\rho_1$, and has values of $O(0.1) \leq g' \leq O(10)$ cm s^{-2}. The Coriolis parameter $f = 2\omega$ is $O(0.1) \leq f \leq O(1)$ s^{-1}. D is varied from 5 to 25 cm, with the volumetric flow rate Q, monitored by the flow meter, set between $2.5 \leq Q \leq 10$ cc/s for the 1.2 m turntable. The larger turntable allows a parameter space approximately twice as large

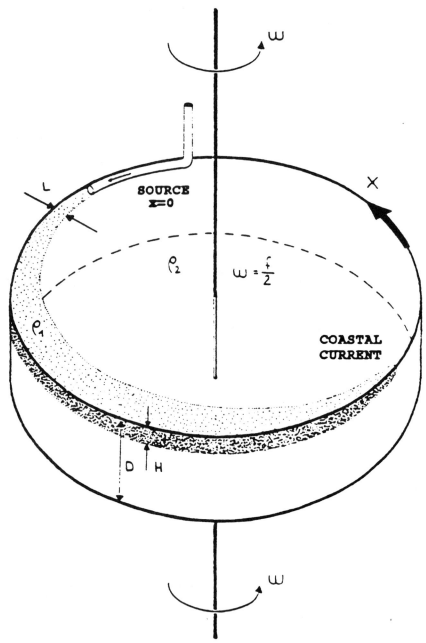

Figure 1. Sketch of 1.2 m diameter rotating turntable. Basin is filled with water of density ρ_2, while dyed water of density ρ_1 ($\rho_1 < \rho_2$) is released from the source.

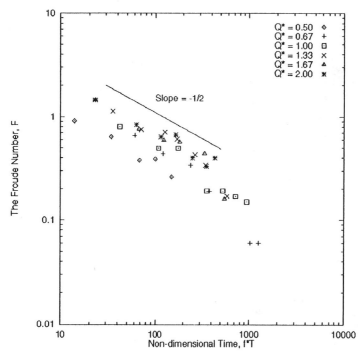

Figure 2. Non-dimensional plot of nose velocity versus time for various outflow fluxes. Nose velocity is non dimensionalized by $(g' H)^{1/2}$. All data from 1.2 m turntable experiments. H/D - 0.1, $g' = 9.8$ cm/s^2, $f = 2.1$ s^{-1}. One revolution corresponds to $fT = 12$.

for D, and a factor of eight times as large for Q. The depth H is taken as the depth of the outflow source. For all experiments, the bottom of the tank was flat.

These dimensional quantities yield several parameters which are useful in the scaling of our experiments. The *Internal Rossby deformation radius* is $R_d = (g'H)^{1/2}/f$, and will be used to non-dimensionalize length. Dimensionless time will be defined as dimensional time, T (sec), multiplied by the Coriolis parameter, f (sec^{-1}). Hence, one rotational period (or one day) corresponds to $fT \approx 12$. A depth ratio H/D is also considered, particularly in connection with instability development.

The convenient method to formulate a dimensionless velocity is the *Densimetric Froude number*, $F = U/(g'H)^{1/2}$. However, by taking advantage of the apparent semi-geostrophy of the system

$$u = -\frac{g'}{f}\frac{\partial h}{\partial y}$$

and the definition of volume transport

$$Q = \int_0^L hu\, dy$$

where L is the width of the current, $h(L) = 0$ and $h(0) \approx H$, results in $Q = g'H^2/2f$. The outflow flux Q will be non-dimensionalized by $g'H^2/f$, yielding a non-dimensional outflow flux denoted by Q^*. An expression for the gravity wave phase speed may also be taken from this, as $(g'H)^{1/2} = (2fQg')^{1/4} = (2fQ/H)^{1/2}$.

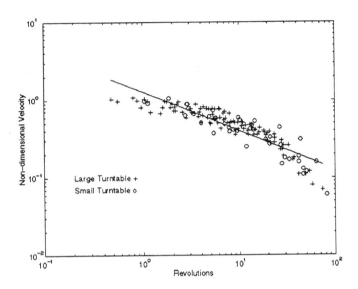

Figure 3. Non-dimensional plot of nose velocity versus time for experiments performed in various parameter spaces, as well as different turntables. Nose velocity is non dimensional by $(2fQg')^{1/4}$. One revolution corresponds to $fT \approx 12$. Correlation coefficient between data and best-fit line is 0.88.

A co-rotating VHS camcorder and a 35 mm motor-driven camera are used to record each experiment, with additional photography from a hand-held 35 mm camera. The flow is visualized by mixing fluorescent dye with the freshwater prior to the start of the experiment. Additionally, particle tracking determines flow-field velocities.

Results

The experiments involved observing two aspects of buoyant flow. The first was the effect of the source velocity on the coastal current. As this velocity is altered, the parameters of the flow are also changed, affecting both nose and frontal characteristics and thereby affecting the onset and growth of instabilities. The second feature under consideration by preliminary experiments was the effect of D, the total water depth, on the buoyant coastal current. Figure 2 shows nose velocities versus time for varied outflow volume fluxes. All the data is well approximated by a -1/2 slope for a non-dimensional time domain $fT = (10, 500)$. This time domain corresponds to between 1 and 40 revolutions. This slope contrasts to exponential decay of the nose velocity found by Griffiths and Hopfinger [1983] in dam-break experiments. As it takes the nose of the buoyant current approximately 100 revolutions to complete a trip around the perimeter of the tank, all data is collected before the nose passes the source. The velocities are non-dimensionalized by $(g'H)^{1/2}$, and the data appears to show that the non-dimensional velocity increases with non-dimensional outflow flux Q^*. Hence, when the velocities are non-dimensionalized by $(2fQg')^{1/4}$, a more concise collapsing of the data should occur. A compilation of a multitude of experiments where the outflow flux Q and Coriolis parameter f have been varied is shown in Figure 3 and does indeed show a collapse of the data with this particular scaling. No radiation of shelf waves was apparent from the leading edge of the front; this is in contrast to previous findings [Whitehead and Chapman, 1986] for a sloping bottom.

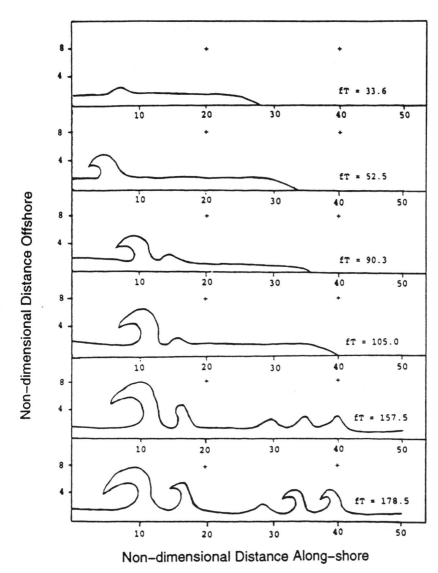

Non-dimensional Distance Along-shore

Figure 4. Schematic of experiment performed on 1.2 m turntable, utilizing non-dimensional distance and time. Distance is non-dimensionalized by R_d. Multiple instabilities arise for large values of fT. Note that circular geometry of turntable is represented as a straight coastline. $H/D = 0.1$, $Q^* = 0.5$, $g' = 9.8$ cm/s^2, $f = 2.1$ s^{-1}.

The occurrence of the first instability in a coastal current often was followed by a near-simultaneous occurrence of other instabilities. The first instability occurs nearest to the source, with the other instabilities occurring downstream. The amplitude A was measured from the edge of the mean current, and was non-dimensionalized by R_d, while λ/R_d was the non-dimensional wavelength between two instabilities, measured from crest to crest. Figure 4

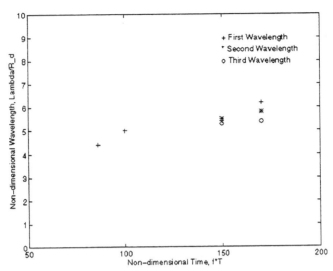

Figure 5. Non-dimensional plot of instability wavelength versus time for experiment represented in *Figure 4.* Wavelength is non-dimensionalized by the Internal Rossby Radius.

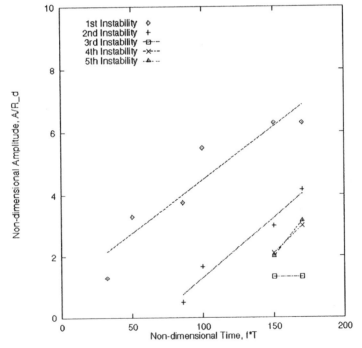

Figure 6. Non-dimensional plot of instability amplitude versus time for experiment represented in *Figure 4.* Best-fit lines have been plotted for each instability data set.

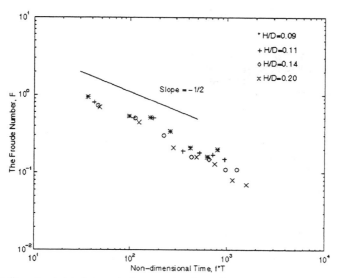

Figure 7. Non-dimensional plot of nose velocity versus time for various *H/D*. All data from 1.2 m turntable experiments, with nose velocity non-dimensionalized by $(g'H)^{1/2}$. $Q^* = 1.0$, $g' = 9.8$ cm/s², $f = 2.1$ s⁻¹.

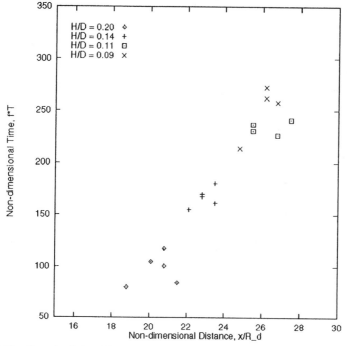

Figure 8. Non-dimensional plot of the time and position of the occurrence of first instability for various *H/D*. All experiments performed on 1.2 m turntable. $Q^* = 1.0$, $g' = 9.8$ cm/s², $f = 2.1$ s⁻¹.

shows a schematic of a particular experiment. Note that while the wavelengths remain approximately constant, the amplitudes of the instabilities do grow in time. The instabilities also appear to have small propagation velocities, with the largest of these being $F = O(0.1)$. Figure 5 shows that the wavelengths are approximately 5 R_d. The amplitude growth rates of the instabilities, as shown in Figure 6, increase linearly with respect to fT.

The second feature under consideration is the effect of the total water depth, D, on nose and frontal characteristics. A density-driven coastal current in relatively shallow waters may be affected by bottom friction forces. In these experiments, the net non-dimensional outflow flux Q^* is held constant, but the total water depth is changed. As can be seen in Figure 7, depth does play an important role in the evolution of the current with greater nose propagation velocities for larger depths. The plot of the Froude number versus fT for experiments with varied H/D shows that this data also exhibits a slope of -1/2. The variation in the Froude number apparent as H/D is varied suggests that if the ratio of $H/D \ll 1$, then bottom friction forces and the effect of horizontal movement in the bottom layer are negligible.

Several experiments were undertaken to study the effects of H/D on the occurrence of instabilities. Figure 8 presents the results of experiments with different depth ratios. The first instability occurred later and further downstream for experiments with a smaller depth ratio. This indicates that the instability process is affected by the value of H/D. The ratio of the top layer depth to the bottom layer depth had been found to be an important experimental parameter [Griffiths and Linden, 1981]. When this ratio is small, the instabilities were primarily barotropic; when the ratio was of $O(1)$, the instabilities were predominantly baroclinic.

Comparisons may be made with other simulations with regard to the instabilities occurring in our laboratory currents. In particular, the data of Chabert D'Hieres et al. [1991] provides a recent comparison with our experimental data, as the experimental set-ups are similar. They restrict themselves to H/D being very small, which we consider as a small part of the overall parameter space. Our data (Figure 8) for H/D of $O(0.1)$ is in agreement with their data concerning the location and time of appearance of the first instability.

Concluding Remarks

Our rotating turntable experiments have focused on several facets of buoyant coastal currents. We have found that non-dimensional nose velocities decrease with time as $(fT)^{-1/2}$. The values of the non-dimensional velocity are approximately unity after two periods of rotation $(fT \approx 25)$. Decreasing H/D over a series of experiments shows a systematic increase in nose velocities due to the negligible effects of bottom friction for small H/D. No shelf waves were evident ahead of the nose.

Multiple instabilities are observed along the front of the buoyant coastal current. Results show that the time fT and position x/R_d of occurrence of the first instability increase as H/D decreases. Wavelengths between instabilities remain constant $(\lambda/R_d \approx 5)$ with respect to time for the chosen experimental parameters. The propagation velocities of the instabilities are also small. However, instability amplitudes do vary with respect to time, displaying a growth rate that is approximately linear.

Acknowledgments. We would like to thank R. W. Garvine, K. -C. Wong, and others for helpful discussion of various aspects of this work, and to J. Kelly for his technical expertise on our experimental apparatus. We also wish to acknowledge the two anonymous reviewers for their helpful comments.

References

Benjamin, T. B., Gravity currents and related phenomena, *J. Fluid Mech., 31*, 209-248, 1968.

Chabert D'Hieres, G., H. Diddle, and D. Obaton, A laboratory study of surface boundary currents: application to the Algerian Current, *J. Geophys. Res., 96*, 12539-12548, 1991.

Garvine, R. W., Estuary plumes and fronts in shelf waters: A layer model, *J. Phys. Oceanogr., 17,* 1877-1896, 1987.

Griffiths, R. W., Gravity currents in rotating systems, *Ann. Rev. Fluid Mech., 18,* 59-89, 1986.

Griffiths, R. W., and E. J. Hopfinger, Gravity currents moving along a lateral boundary in a rotating fluid, *J. Fluid Mech., 134,* 357-399, 1983.

Griffiths, R. W., and P. F. Linden, Laboratory experiments on fronts. Part I: Density-driven boundary currents, *Geophys. Astrophys. Fluid Dyn., 19,* 159-187, 1982.

Griffiths, R. W., P. F. Linden, and F. Chia, Laboratory Experiments on fronts. Part II: The formation of cyclonic eddies at upwelling fronts, *Geophys. Astrophys. Fluid Dyn., 19,* 189-206, 1982.

Münchow, A., The formation of a buoyancy driven coastal current, Ph.D. thesis, University of Delaware, 1992.

Münchow, A. and R. W. Garvine, Buoyancy and wind forcing of a coastal current, *J. Mar. Res., 51,* 293-322, 1993.

Stern, M., Geostrophic fronts, bores, breaking and blocking waves, *J. Fluid Mech., 99,* 687-703, 1980.

Stern, M., J. A. Whitehead and B. -L. Hua, The intrusion of a density current along the coast of a rotating fluid, *J. Fluid Mech., 123,* 237-265, 1982.

Vinger, A., T. A. McClimans and S. Tryggestad, Laboratory observations of instabilities in straight coastal current, in *The Norwegian Coastal Current,* The University of Bergen, Norway, *Vol. 2,* 553-582, 1981.

Whitehead, J. A., and D. C. Chapman, Laboratory observations of a gravity current on a sloping bottom: the generation of shelf waves, *J. Fluid Mech., 172,* 373-399, 1986.

6

The Influence of the Springs-Neaps Cycle on the Position of Shelf Sea Fronts.

Jonathan Sharples and John H. Simpson

Abstract

A 1-dimensional numerical model, using a Mellor-Yamada level 2 turbulence closure scheme, is applied to the problem of the horizontal adjustment of shelf sea fronts resulting from the springs-neaps cycle of tidal mixing. The results are in general agreement with earlier prescriptive modelling, showing that inhibition of vertical mixing due to stratification is an important factor in determining the extent of frontal adjustment. The results also show that the horizontal extent of the adjustment can be significantly reduced in regions of highly cyclonic tidal currents, as the result of boundary-layer limitation of vertical turbulent exchange. The results are used to suggest that springs-neaps adjustment of shelf sea fronts could result in an increase in near-surface nutrient concentrations by approximately 2 mmol m^{-3}

Introduction

Shelf sea fronts mark the transition between thermally stratified and vertically mixed waters, and are a common feature of many shelf regions during late spring and summer. A number of hypotheses have been suggested to explain the generally steady positions of shelf sea fronts. Simpson and Hunter [1974] showed that a consideration of the energy balance between tidal mixing (acting to reduce stratification) and solar heating (generating stratification) implied that fronts should be found at a critical value of h/u^3, with h the water depth and u a measure of the tidal current amplitude. Simpson et al. [1978] and Loder and Greenberg [1986] have shown that surface wind-induced mixing also plays an important role in the control of frontal position, typically by raising the critical value of h/u^3 as a result of the extra turbulent kinetic energy input at the surface boundary. Limitation of the vertical extent of turbulent mixing by the thickness of the bottom boundary layer may also play a role in determining the position of a front [Garrett et al., 1978; Soulsby, 1983]. This thickness results from the interaction between the earth's rotation and the polarisation of the tidal current ellipse; in regions where boundary-layer limitation of turbulence was the dominant control, then the front would be expected to follow a critical value of h/u [Loder and Greenberg, 1986; Stigebrandt, 1988]. In Simpson and Sharples [1994] we demonstrated that, for typical fronts in temperate waters, control of frontal position is dominated by the balance between heating and stirring inputs. The influence of boundary-layer limitation of vertical turbulent exchange becomes important only for fronts in deep water (>100 metres), or for fronts in regions where there is a significant cyclonic component to the tidal currents.

Given the u^{-3} dependence of frontal position arising from consideration of the local energy balance, it is reasonable to expect a regular change in the geographical location of the energy balance as tidal currents are modulated by the springs-neaps cycle. Through the months of late

Buoyancy Effects on Coastal and Estuarine Dynamics
Coastal and Estuarine Studies Volume 53, Pages 71-82

spring and summer the position of shelf sea fronts has been observed to change very little [see, for example, Bowers and Simpson, 1987], typically by 10 - 20 kilometres. Most of this movement can be attributed to tidal advection of the front, but a small but significant part of the observed variation in frontal position (2 - 4 kilometres) can be correlated with the springs-neaps cycle with a lag of approximately 2 days [Simpson and Bowers, 1981]. More recently Bisagni and Sano [1993] have presented a detailed analysis of the Georges Bank frontal region observed by Advanced Very High Resolution Radiometer. Both within the frontal zone, and in the stratified and mixed regions either side of it, there was a significant negative correlation between the sea surface temperature residual about the seasonal mean, and the tidal current speed. A time lag between tidal current amplitude and the temperature residual of approximately 3 days was observed, with the temperature residual reaching its maximum negative value 3 days after springs, and maximum positive 3 days after neaps.

Simpson and Bowers [1981] investigated whether or not the observed response of shelf sea fronts to the springs-neaps cycle could be explained by a steady-state application of the critical h/u^3 criterion. They showed that the resulting predicted horizontal adjustment was far in excess of that typically observed, and attributed the over-estimate to the inapplicability of the steady-state assumption. Two physical processes were implicated in the breakdown of the steady-state energy balance over the springs-neaps cycle. First, the increasing tidal mixing as spring tide approaches not only has to counteract the buoyancy that is being input by the surface heating, but must also remove the buoyancy stored since the previous spring tide, so slowing the progress of the front into the stratified region. Second, as the front moves into stratified water, the thermocline will inhibit the efficiency of vertical mixing and so slow the progress of the front further.

Simpson and Bowers [1981] investigated these effects using a 1-dimensional numerical model in which the vertical water column structure was determined by the energetics of wind and tidal mixing, and solar heating. Including the effect of stored buoyancy reduced the predicted extent of springs-neaps frontal adjustment by approximately 50%, though this was still significantly higher than that typically observed. Stability-dependent mixing efficiencies were then introduced by multiplying both the wind and tidal mixing efficiencies by the factor

$$ F = \left(\frac{\phi_0}{\phi_0 - \phi} \right)^{\frac{1}{2}} \tag{1} $$

with ϕ the potential energy anomaly of the water column (a measure of the energy required to vertically mix a water column) and ϕ_0 a constant. It was found that the modelled extent of frontal adjustment was further reduced towards that observed by using a value for ϕ_0 of 5 J m^{-3} and specifying a lower limit to F of 0.25. Yanagi and Tamaru [1990] have successfully used the same model in an investigation of a tidal front in the Bungo Channel, Japan.

While this "prescriptive" method has been successful in demonstrating the effect of vertical stability, the use of mixing efficiencies related to stability via equation (1) remains arbitrary. Reducing the boundary-layer thickness (by, for instance, increasing the cyclonic component of the tidal currents) will also reduce the efficiency of tidal mixing, and so it is possible that a significant reduction in the thickness of the tidal boundary-layer could further alter the behaviour of a shelf sea front.

In this paper we deal with 3 main foci. First, we employ a commonly used turbulence parametrisation to model springs-neaps frontal adjustment with a more fundamental relationship between mixing efficiency and vertical stability (referred to as the T/C model). The results of this approach are compared to the earlier model of Simpson and Bowers [1981] (referred to as the prescriptive model). Second, the dynamically more fundamental nature of the T/C model allows investigation of the effect of boundary-layer limitation of tidal mixing, and the consequences for the amplitude of the springs-neaps frontal adjustment. Finally, the results of the T/C model are interpreted in terms of the potential for localized nutrient input and the enhancement of primary productivity in the vicinity of shelf sea fronts.

Method

Here we use a modelling method introduced in Simpson and Sharples [1994]. The core model is 1-dimensional (vertical). The total water column has a fixed depth of 80 metres, split into 20 grid cells of equal size. Integration is a simple, forward stepping explicit scheme with a timestep of 0.01 hours. The equations of motion in the x and y directions are:

$$\frac{\partial u}{\partial t} = -\sum_{i=1}^{n} A_{ix} g \cos(\omega_i t - \varphi_{ix}) + fv + \frac{\partial}{\partial z}\left(N_z \frac{\partial u}{\partial z}\right) \tag{2}$$

$$\frac{\partial v}{\partial t} = -\sum_{i=1}^{n} A_{iy} g \cos(\omega_i t - \varphi_{iy}) - fu + \frac{\partial}{\partial z}\left(N_z \frac{\partial v}{\partial z}\right) \tag{3}$$

where u and v are the x and y components of current velocity, g is gravitational acceleration, f the Coriolis parameter, and N_z the coefficient of eddy viscosity. $A_{ix, iy}$ and $\varphi_{ix, iy}$ are the amplitudes and phases of the tidally oscillating sea surface slopes in the x and y directions for the ith tidal constituent. The sum is taken over the number, n, of tidal constituents, each with frequency ω_i. For all the model runs presented, two tidal constituents are used (M2 and S2). Control of the polarisation of the tidal current ellipse is maintained by calculating the slope amplitudes from

$$A_{ix} = C_i(\omega_i + \lambda f); \quad A_{iy} = C_i(f + \lambda \omega_i) \tag{4}$$

with λ the polarisation of the current ellipse, ranging from $\lambda = -1$ for purely anticyclonic currents to $\lambda = 1$ for purely cyclonic currents. The slope phases are $\varphi_{ix} = 0$ and $\varphi_{iy} = \pi/2$. The factor C_i determines the strength of the tidal currents, and its value for the S2 tidal constituent is set at 30% of the M2 constituent. Surface and bottom boundary conditions are driven by surface wind stress, and quadratic bottom friction respectively. Solar irradiance is input at the surface, and a back radiation related to humidity, surface temperature, and wind speed output from the surface, following Edinger et al. [1968] and Simpson and Bowers [1984]. Seasonal variation of solar irradiance and surface winds is via the sinusoidal functions used by Simpson and Bowers [1984] and Simpson and Sharples [1994]. Vertical transfer of temperature, T, is controlled by

$$\frac{\partial T}{\partial t} = \frac{\partial}{\partial z}\left(K_z \frac{\partial T}{\partial z}\right) \tag{5}$$

with K_z the coefficient of vertical eddy diffusivity.

Vertical turbulent transfer of momentum and scalars, via N_z and K_z, is related to local water column stability at each level of the model grid by the Mellor-Yamada level 2 closure scheme via

$$N_z = S_M lq; \quad K_z = S_H lq \tag{6}$$

where S_M and S_H are stability functions related to the local Richardson number [see Mellor and Yamada, 1982, for a full description]. These stability functions represent the main control on the efficiency of vertical mixing (i.e. the magnitudes of N_z and K_z) by the local stability. The turbulent velocity scale, q, is calculated in the level 2 closure scheme by assuming a local equilibrium between the production of turbulent kinetic energy, work against buoyancy, and dissipation, so that

$$N_z\left[\left(\frac{\partial u}{\partial z}\right)^2 + \left(\frac{\partial v}{\partial z}\right)^2\right] + K_z\left(\frac{g}{\rho}\frac{\partial \rho}{\partial z}\right) = \frac{q^3}{B_1 l} \tag{7}$$

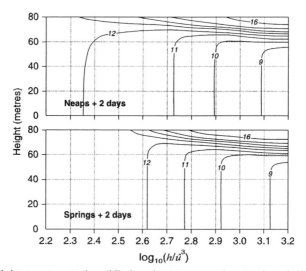

Figure 1. Modelled temperature sections (°C) through post-neaps and post-springs shelf sea fronts (T/C model, latitude=50°N, degenerate tidal ellipse).

with the density, ρ, calculated from temperature, and B_1 a laboratory-determined constant. We choose the profile of the turbulent lengthscale, l, to have the form

$$l(z) = \kappa z \left(1 - \frac{z}{h} \right)^2 \qquad (8)$$

with z the height above the seabed, h the total depth, and $\kappa = 0.41$ is von Karman's constant.

A frontal section was generated by running the model over a range of tidal current amplitudes, determined by the factor C_i of equation (4). For each tidal current amplitude the model was run for two full annual cycles, with data output daily during the second cycle. Running the model through an annual cycle implicitly includes the effect of stored buoyancy as a previously stratified water column is re-mixed by increasing tidal currents, while the effect of water column stability in reducing the efficiency of vertical mixing is inherent in the turbulence closure scheme. Boundary-layer determination of the extent of vertical turbulent exchange is also included within the T/C model.

A vertical slice normal to the front was then built up by combining the model results from different tidal intensities, with height above the seabed as the vertical axis and $\log_{10}(h/\hat{u}^3)$ as a geographically-fixed horizontal axis represented in terms of the tidal intensity. Here u is the depth averaged current speed and $\hat{u}^3$ is the average of u^3 over the springs-neaps cycle.

Results

Figure 1 illustrates the typical frontal sections produced by the T/C model over the springs-neaps cycle, generated by running the model with a degenerate (λ=0) tidal current ellipse. A change in position of the front is clearly evident, as is a general weakening of the horizontal thermal gradients near neap tides caused by the weaker horizontal gradient in tidal current amplitude. This "smoothing" of the frontal structure may be a result of the quasi-2 dimensional approach not including cross-frontal transfer (see discussion later), though it has been observed by, for instance, Yanagi and Tamaru [1990] and presents a problem in defining frontal position from the model output. In both Simpson and Bowers [1981] and Simpson and

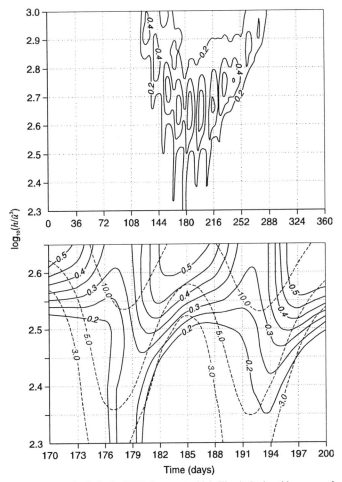

Figure 2. T/C model results (latitude=50°N, degenerate tidal ellipse). Spring tides occurred on day 183, neap tides on day 190. Day 0 is January 1st. (a) Seasonal variation of horizontal temperature gradient (in units of °C per 0.1 log$_{10}$ (h/$\bar{u}^3$). (b) Mid summer variation of horizontal temperature gradient (solid line, same units as (a)) and potential energy anomaly (dashed line, J m^{-3}).

Sharples [1994] the locus of the front was defined in terms of a critical value of the potential energy anomaly, ϕ_{crit}. An alternative descriptor of frontal position is the magnitude of the horizontal temperature gradient. Figure 2 illustrates the annual variation of the modelled horizontal temperature gradient, along with a more detailed picture of the gradient during mid summer with contours of $\phi = 3$, 5, and 10 J m^{-3} superimposed. The springs-neaps variation in horizontal temperature gradient and in the potential energy anomaly are clear. It is evident, however, that the amplitude of frontal excursion inferred from the modelled variation in ϕ is sensitive to the choice of the value for ϕ_{crit}. Similarly the amplitude of frontal excursion described by changes in the horizontal temperature gradient is dependent on the spatial scale represented by the y-axis of figure 2. Horizontal gradients in h/u^3 near shelf fronts, inferred from Pingree and Griffiths [1978] and Bowman and Esaias [1981] vary by almost an order of magnitude. Hence the choice of relevant temperature gradient contour (based on the thermal gradient resolution of, for instance, a satellite radiometer) is not obvious.

In Simpson and Sharples [1994] we used a value of $\phi_{crit}= 5$ J m^{-3} to define frontal position, based on an assessment of the cross-frontal variation of ϕ in allowing a clear distinction between mixed and stratified water columns. In figure 2 the close association between the $\phi = 5$ J m^{-3} contour and the position of maximum horizontal temperature gradient over most of the springs-neaps cycle, suggests such a critical value to be justified. For this present study we therefore retain this value for the critical potential energy anomaly defining the position of the front.

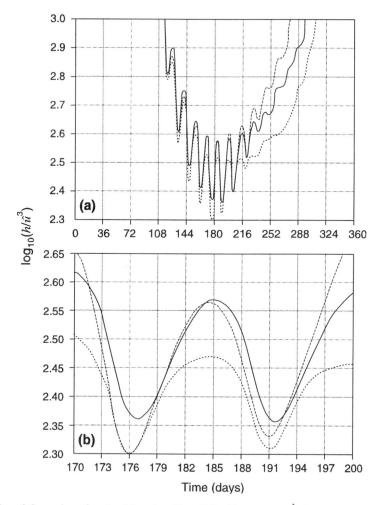

Figure 3. Comparison of predicted frontal positions (defined by $\phi = 5$ J m^{-3}) from the T/C model (solid line), the prescriptive model with constant mixing efficiencies (dashed line), and the prescriptive model with stability-dependent mixing efficiencies (dotted line). Spring tides occurred on day 183, neap tides on day 190. Day 0 is January 1st. (a) Annual variation of frontal position (b) Mid summer variation of frontal position.

Comparison with the prescriptive model

The prescriptive model of Simpson and Bowers [1981] was run under exactly the same conditions as the T/C model, with both constant mixing efficiencies and stability-dependent mixing efficiencies, the latter being calculated using equation (1). The results of the two versions of the prescriptive model are compared to those of the T/C model in figure 3. In all three model runs, spring tides occurred on day 183, and the following neap tide on day 190.

All results show a lag of 2 to 3 days of the adjustment of the front behind the springs-neaps cycle, in particular showing a slightly longer lag between springs and the adjustment to the highest $\log_{10}(h/\hat{u}^3)$ compared to that between neaps and the adjustment to lowest $\log_{10}(h/\hat{u}^3)$. The T/C model resulted in a phase lag consistently 1 day longer than either of the prescriptive models, though it is not clear if this is a real difference in prediction of the form of frontal adjustment, or simply due to a slight mismatch of the T/C and prescriptive springs-neaps cycles.

The amplitude of springs-neaps frontal adjustment illustrated in figure 3 is different for all three model runs. The prescriptive model with constant mixing efficiency resulted in the largest excursion, with the introduction of stability-dependent mixing efficiency causing a reduction of the excursion by approximately 40%, in agreement with the results of Simpson and Bowers [1981]. The T/C model predicted a frontal adjustment amplitude intermediate between the two prescriptive cases, suggesting that the stability-dependent mixing efficiency inherent in the level 2 turbulence closure scheme is not as effective in the inhibition of vertical mixing as the empirical relation of Simpson and Bowers [1981] used in the prescriptive model.

Variation of frontal adjustment with tidal current polarisation

The thickness of the tidal boundary-layer is proportional to $u_*/(\omega \pm f)$, with u_* the friction velocity, ω the tidal frequency, and f the Coriolis parameter [see, for instance, Soulsby, 1983]. The + and the − relate to cyclonic and anticyclonic tidal current ellipses respectively. At temperate latitudes this leads to an order of magnitude difference between the thicknesses of cyclonic and anticyclonic tidal boundary-layers, and implies significant boundary-layer limitation of vertical turbulent transfer for the case of a cyclonic tidal current ellipse.

An advantage of the T/C model over the prescriptive approach is that the dynamics inherently include this interaction between the earth's rotation and the polarisation of the tidal currents in determining the thickness of the bottom boundary-layer. In Simpson and Sharples [1994] we demonstrated that the boundary-layer limitation of turbulent mixing begins to be important as the tidal currents become dominantly cyclonic, ultimately leading to a significant reduction in the efficiency of tidal mixing. One result of this reduction in mixing efficiency is the manifestation of the front in a region of higher tidal currents (lower $\log_{10}(h/\hat{u}^3)$)) as, with an increased cyclonic component to the tides, the thickness of the boundary-layer can only be maintained by increasing u_*.

Such a marked reduction in the mixing efficiency as the polarisation of the tides changes should also affect the extent of springs-neaps frontal adjustment, in the same way that inhibition of vertical mixing by stability reduces the horizontal amplitude of frontal adjustment. Figure 4a shows the results of T/C model runs with tidal ellipse polarisations of λ=−1, 0, 0.5, and 1. The shift of the mean frontal position towards higher tidal energy regions as the proportion of cyclonic currents increases is again clear. The phase of the springs-neaps adjustment appears constant with ellipse polarisation, but there is a significant reduction in the amplitude of the excursion between polarisations of 0.5 and 1.0. This latter point is further illustrated in figure 4b, by showing the modelled excursion amplitude during a mid-summer springs-neaps cycle over the full range of tidal current polarisations. There is a gradual but small change in excursion amplitude from anticyclonic currents to λ=0.5, followed by a rapid decrease in amplitude towards zero between λ=0.5 and completely cyclonic currents.

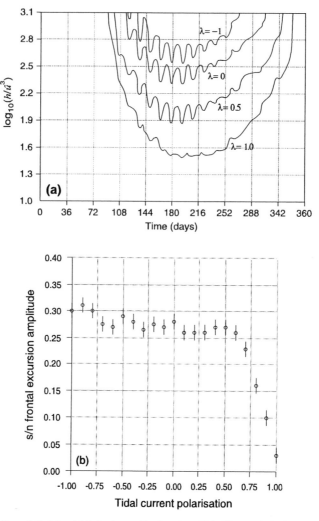

Figure 4.   T/C modelled frontal behaviour with changing tidal ellipse polarisation (latitude=50°N): (a) Annual variation of frontal position (defined by $\phi = 5$ J m-3) for tidal ellipse polarisations of $\lambda = -1$, 0, 0.5, and 1. (b) Variation of springs-neaps frontal excursion amplitude (units of $\log_{10}$ ($h/\hat{u}^3$) with tidal ellipse polarisation. The vertical lines represent the uncertainty arising from the model resolution along the $\log_{10}$ ($h/\hat{u}^3$) axis.

Summary and Conclusions

Two 1-dimensional numerical models have been compared in an investigation of shelf sea frontal adjustment in response to the springs-neaps tidal cycle, the first based on the prescriptive energetics model of Simpson and Bowers [1981] and the second on a turbulence closure model used in Simpson and Sharples [1994]. The prescriptive model was run with constant mixing efficiencies, and with mixing efficiencies inversely related to the stability of the water column. Results from a range of tidal mixing regimes were assembled to form cross-frontal vertical slices of shelf sea fronts as a method of synthesising a 2-dimensional frontal section with a 1-dimensional model.

The success of this quasi-2 dimensional approach rests on the applicability of the assumption that the position and structure of a shelf sea front are determined primarily by vertical processes. Away from significant horizontal density gradients this will be justifiable. However, Garrett and Loder [1981] have shown that significant cross-frontal transfer can be expected in the vicinity of shelf sea fronts, set up by internal friction. They noted that this cross-frontal transfer will result in changes in density of the same order as those expected from purely vertical transfer models, with the likely outcome being a sharpening of the frontal structure in the surface convergence region. The modelling approach used in this present work does not include cross-frontal transfer, and so the structure of the computed fronts is likely to differ from those produced by a fully 2 dimensional model. The basic results of this work are not expected to be affected; the exact prediction of the amplitude of springs-neaps frontal adjustment may change with a 2 dimensional model (and may in fact be easier to interpret in comparison with observations if the frontal structure is sharpened), but the response to the tidal boundary layer is expected to be similar.

All three model runs produced springs-neaps frontal adjustment with a time lag of between 2 and 3 days behind the springs-neaps tidal cycle, in agreement with observations [for example, Simpson and Bowers, 1981, Bisagni and Sano, 1993]. There was no significant difference between any of the model results in the extent of the lag, suggesting that the delay between the springs-neaps modulation of the tidal current and the adjustment of the front is not affected significantly by the reduction in mixing efficiency caused by vertical stability. Instead it is due to (i) thermal inertia as stratification progresses into the mixed water towards neap tides, and (ii) the need to re-mix stored buoyancy as the stratified region is eroded by the increasing tidal currents towards springs.

Stability-dependent vertical mixing efficiencies had a strong effect on the horizontal extent of springs-neaps frontal adjustment. As found by Simpson and Bowers [1981] the horizontal amplitude of the frontal adjustment predicted by the prescriptive model was considerably reduced when stability-dependent mixing efficiencies were introduced. The T/C model, using a less arbitrary relationship between stability and mixing efficiency, produced an amplitude intermediate between the results of the two prescriptive models, implying that the level 2 turbulence closure scheme reduced vertical mixing efficiencies in response to vertical stability to a lesser extent than the empirically-based method used by Simpson and Bowers [1981]. Comparison between the modelled and observed amplitudes of springs-neaps adjustment are hampered by the difficulty in relating modelled frontal parameters (potential energy anomaly or maximum horizontal thermal gradient) to those generally used in observational studies (i.e. thermal gradients from satellite imagery). However, Pingree and Griffiths [1978] showed gradients of h/u^3 in the vicinity of fronts on the NW European shelf of between 0.02 and 0.06 $\log_{10}(m^{-2} s^3)$ km^{-1}. Mid summer excursion amplitudes predicted by the change in the position of a critical value of the potential energy anomaly in the model results were 0.27, 0.17, and 0.22 $\log_{10}(m^{-2} s^3)$ for the prescriptive model with constant mixing efficiencies, the prescriptive model with stability-dependent mixing efficiencies, and the T/C model respectively. These suggest ranges of 5-13, 3-9, and 4-11 kilometres, compared to the observed range of 2-4 kilometres described by Simpson and Bowers [1981].

The more fundamental physical basis of the T/C model allowed investigation of the effects of changes to the tidal boundary-layer. Reducing the tidal boundary-layer thickness by increasing the proportion of cyclonically-rotating tides resulted in only a small reduction in the springs-neaps frontal adjustment between tidal current ellipse polarisations of $\lambda = -1$ and $\lambda = 0.5$. The extent of the adjustment then reduced rapidly as λ increased from 0.5 to 1.0, as the efficiency of vertical mixing was affected by the decreasing thickness of the bottom boundary-layer. The effect of this reduced boundary-layer is likely to be more pronounced in deeper water columns [Simpson and Sharples, 1994].

The movement of fronts in response to tidal cycles is not uncommon in coastal and shelf waters. Changes in frontal position can be either a broad, real horizontal movement of the entire structure in response to tidal flows, or an "apparent" movement of the front generated by changes in the intensity of vertical mixing. It has already been noted that shelf sea fronts are observed to change position under the influence of both of these processes. Similarly coastal fronts generated by freshwater inputs can show both types of behaviour; for instance the plume front outside the entrance to the Amazon river has been observed to be advected more than 20 km by tidal currents, but also exhibits strong variability driven by springs-neaps modulation of tidal mixing [Geyer, 1995]. Other examples of freshwater-induced coastal fronts responding to the springs-neaps cycle have been described by, for example, Nunes and Lennon [1987], Sharples and Simpson [1993], and Simpson et al. [1993]. In these latter three cases the springs-neaps cycle does not simply alter the position of the front, but instead determines whether or not the front exists through the interaction between the freshwater buoyancy input and the modulated tidal mixing. The low mixing around neap tides allows a vertically-mixed, relatively weak horizontal density gradient to collapse as a gravity current, resulting in the formation and propagation of a front; this has also been investigated in laboratory experiments by Simpson and Linden [1989].

An important aspect of a front that is responding to changes in vertical mixing, rather than horizontal advection, is that within the region of adjustment the same water mass (and its chemical and biological constituents) is undergoing a cycle of mixing and re-stratification. Of all the coastal fronts responding to tidal cycles this is particularly relevant to the region bounded by the springs-neaps adjustment of a shelf sea tidal mixing front. The surface water is periodically coupled to the deeper, nutrient-rich water, and so there is the potential for a fortnightly pulse of nutrient to the surface, which is then held within the photic zone during the following period of stratification. Pingree et al. [1975] and Loder and Platt [1985] have suggested this mechanism to be one possible explanation for the higher biomass associated with shelf fronts [see, for example, Le Fèvre and Grall, 1970; Pingree et al., 1977; Videau, 1987]. Such nutrient input, and subsequent pulses in phytoplankton growth, have been observed by Morin et al. [1985] at the Ushant tidal front. More recently Morin et al. [1993] have used satellite sea surface temperature images and measurements of surface nitrate concentration, in the vicinity of the Ushant front, to calculate a temperature-nitrate relation applicable to the springs-neaps adjustment region of the front.

It is possible to estimate the potential for extra primary production from the results of the T/C model, using the temperature-nitrate regression calculated by Morin et al. [1993] as -1.9 mmol N m^{-3} °C^{-1}. Figure 5 shows the cross-frontal surface temperature variation over the

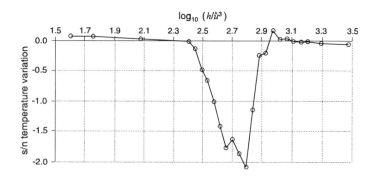

Figure 5.   T/C modelled surface temperature variation (°C) across a shelf sea front, over a springs-neaps cycle. Latitude=50°, $\lambda=-1$.

springs-neaps cycle from the T/C model run at a latitude of 50° and with anticyclonic tidal currents. Within the adjustment region the mean temperature change is approximately -1°C between neap and spring tides. If it is assumed that the nutrient replenishment within the transition region is likely to be low enough to be unaffected by the finite concentration of nutrient on the mixed side of the front, then this implies a mean increase in nutrient within the modelled transition region of 1.9 mmol N m^{-3}. Such a concentration is approaching the typical levels of nutrient observed within the mixed region, and so may, in some cases, be an over-estimate. Loder and Platt [1985] for instance, suggested a springs-neaps related nutrient enhancement of approximately 1 mmol N m^{-3}, assuming a concentration of 2 mmol N m^{-3} in the mixed region. By taking typical observed values for the chlorophyll:nitrogen ratio for phytoplankton, Simpson and Tett [1986] estimated a nitrate enhancement of 0.05 mmol N m^{-3} could support an extra biomass of between 0.05 and 0.25 mg chl m^{-3}. The results of the T/C model, therefore, suggest a potential additional biomass, supported by the springs-neaps frontal adjustment alone, of between 2 and 10 mg chl m^{-3}, though observation of these levels of chlorophyll would be affected by removal of biomass by grazing. Typical values of chlorophyll concentration at shelf sea fronts range between 1 and 20 mg chl m^{-3} [for instance Pingree et al., 1975; Savidge, 1976; Pingree et al., 1978; Bradford et al., 1986].

In summary, the T/C model resulted in computed springs-neaps frontal adjustment in broad agreement with observations. The amplitude of this adjustment was found to decrease rapidly in response to the decreasing thickness of the tidal boundary layer as the tidal current polarisation approached fully cyclonic. While springs-neaps frontal adjustment will not be responsible for all of the observed enhancement in biomass, the T/C model results suggest that, away from regions of highly cyclonic tidal currents, it can provide a significant amount of additional nutrient within the adjustment zone.

Acknowledgements. This work was funded by New Zealand's Foundation for Research in Science and Technology, contract number CO1304.

References

Bisagni, J.J., and M.H. Sano, Satellite observations of sea surface temperature variability on southern Georges Bank, Continental Shelf Research, 13, 1045-1064, 1993.

Bowers, D.G., and J.H., Simpson. Mean positions of tidal fronts in European shelf seas, Continental Shelf Research, 7, 35-44, 1987.

Bowman, M.J., and W.E. Esaias, Fronts, stratification, and mixing in Long Island and Block Island Sounds, Journal of Geophysical Research, 86, 4260-4264, 1981.

Bradford, J.M., P.P. Lapennas, R.A. Murtagh, F.H. Chang, and V. Wilkinson, Factors controlling summer phytoplankton production in greater Cook Strait, New Zealand, New Zealand Journal of Marine and Freshwater Research, 20, 253-279, 1986.

Edinger, J.E., D.W. Duttweiler, and J.C. Geyer, The response of water temperatures to meteorological conditions, Water Resource Research, 4, 1137-1145, 1968.

Garrett, C.J.R., J.R. Keeley, and D.A. Greenberg, Tidal mixing versus thermal stratification in the Bay of Fundy, Maine, Atmosphere-Ocean, 16, 403-443, 1978.

Garrett, C.J.R., and J.W.Loder, Dynamical aspects of shallow sea fronts, Philosophical Transactions of the Royal Society, London, A302, 563-581, 1981.

Geyer, W.R., Tide-induced mixing in the Amazon frontal zone, Journal of Geophysical Research, 100, 2341-2353, 1995.

Le Fèvre, J., and J.R. Grall, On the relationships of Noctiluca swarming off the western coast of Brittany with hydrological features and plankton characteristics of the environment, Journal of Experimental Marine Biology and Ecology, 4, 287-306, 1970.

Loder, J.W., and D.A. Greenberg, Predicted positions of tidal fronts in the Gulf of Maine region, Continental Shelf Research, 6, 397-414, 1986

Loder, J.W., and T. Platt. Physical controls on phytoplankton production at tidal fronts, in Proceedings, 19th European Marine Biology Symposium, edited by P.E. Gibbs, pp 3-21, Cambridge University Press, 1985.

Mellor, G.L., and T. Yamada, Development of a turbulence closure model for geophysical fluid problems, Reviews in Geophysics and Space Physics, 20, 851-875, 1982.

Morin, P., P. Le Corre, and J. Le Fèvre, Assimilation and regeneration of nutrients off the west coast of Brittany, Journal of the Marine Biological Association of the U.K., 65, 677-695, 1985.

Morin, P., M.V.M. Wafar, and P. Le Corre, Estimation of nitrate flux in a tidal front from satellite-derived temperature data, Journal of Geophysical Research, 98, 4689-4695, 1993.

Nunes, R.A., and G.W.Lennon, Episodic stratification and gravity currents in a marine environment of modulated turbulence, Journal of Geophysical Research, 92, 5465-5480, 1987.

Pingree, R.D., and D.K. Griffiths, Tidal fronts on the shelf seas around the British Isles, Journal of Geophysical Research, 83, 4615-4622, 1978.

Pingree, R.D., P.M. Holligan, and R.N. Head, Survival of dinoflagellate blooms in the western English Channel, Nature, 265, 266-269, 1977.

Pingree, R.D., P.M. Holligan, and G.T. Mardell, The effects of vertical stability on phytoplankton distributions in the summer on the northwest European Shelf, Deep Sea Research, 25, 1011-1028, 1978.

Pingree, R.D., P.R. Pugh, P.M. Holligan, and G.R. Foster, Summer phytoplankton blooms and red tides along tidal fronts in the approaches to the English Channel, Nature, 258, 672-677, 1975.

Savidge, G., A preliminary study of the distribution of chlorophyll a in the vicinity of fronts in the Celtic and western Irish Seas, Estuarine Coastal Marine Science, 4, 617-625, 1976.

Sharples, J., and J.H.Simpson, Periodic frontogenesis in a region of freshwater influence, Estuaries, 16, 74-82, 1993.

Simpson, J.E., and P.F.Linden, Frontogenesis in a fluid with horizontal density gradients, Journal of Fluid Mechanics, 202, 1-16, 1989.

Simpson, J.H., C.M. Allen, and N.C.G.Morris, Fronts on the continental shelf, Journal of Geophysical Research, 83, 4607-4614, 1978.

Simpson, J.H., W.G.Bos, F.Schirmer, A.J.Souza, T.P.Rippeth, S.E.Jones, and D.Hydes, Periodic stratification in the Rhine ROFI in the North Sea, Oceanologica Acta, 16, 23-32, 1993.

Simpson, J.H., and D.G. Bowers, Models of stratification and frontal movement in shelf seas, Deep Sea Research, 28A, 727-738, 1981.

Simpson, J.H., and D.G Bowers, The role of tidal stirring in controlling the seasonal heat cycle in shelf seas, Annales Geophysicae, 2, 411-416, 1984.

Simpson, J.H., and J.R. Hunter, Fronts in the Irish Sea, Nature, 250, 404-406, 1974.

Simpson, J.H., and J. Sharples, Does the earth's rotation influence the location of shelf sea fronts? Journal of Geophysical Research, 99, 3315-3319, 1994.

Simpson, J.H., and P.B. Tett, Island stirring effects on phytoplankton growth, in Lecture Notes on Coastal and Estuarine Studies, Vol. 17; Tidal Mixing and Plankton Dynamics. edited by M.J. Bowman, C.M. Yentsch, and W.T. Peterson, pp 41-76. Springer-Verlag, 1986.

Soulsby, R.L., The bottom boundary layer of the shelf sea, in Physical Oceanography of Coastal and Shelf Seas, edited by B. Johns, pp 189-266, Elsevier, New York, 1983.

Stigebrandt, A.,. A note on the locus of a shelf front, Tellus, 40A, 439-442, 1988.

Videau, C., Primary production and physiological state of phytoplankton at the Ushant tidal front (west coast of Brittany, France), Marine Ecology Progress Series, 35, 141-151, 1987.

Yanagi, T., and H. Tamaru, Temporal and spatial variations in a tidal front, Continental Shelf Research, 10, 615-627, 1990.

7

Interaction Between Mean Water Column Stability and Tidal Shear in the Production of Semi-diurnal Switching of Stratification in the Rhine ROFI

Alejandro J. Souza and John H. Simpson

Abstract

Observations of vertical structure of density and velocity profiles in the Rhine ROFI system in the North Sea show evidence of the interaction between the mean water column stability and tidal shear in the production of semi-diurnal variability of stratification. During periods of stratification the surface tidal ellipses change from degenerate to a more circular pattern, with the surface ellipse rotating clockwise and the bottom ellipse rotating anti-clockwise. These changes in ellipticity introduce a cross-shore velocity component which enhances the vertical shear in the tidal flow. When the enhanced cross-shore tidal shear interacts with the horizontal density gradients it results in large semi-diurnal variations in stratification. This hypothesis is tested using a 1-d point model; the results exhibit the same qualitative behaviour as the observations, with the amplitude of semi-diurnal oscillations in stratification of the same magnitude as observed and confirming the role of cross-shore tidal straining.

Introduction

The average input of 2200 m^3s^{-1} of freshwater into the North Sea by the Rhine represents a major source of buoyancy which, as discussed in a previous publication [Simpson et al., 1993], has an important implication in the water column structure and dynamics. In this publication, we discussed the water column structure with the aid of time series from moorings and survey observations. These observations have helped us to clarify the way in which the Rhine ROFI responds to changes in stirring from tides and winds. We have also observed strong semi-diurnal variability with an amplitude comparable to that of the mean stratification. This feature that has not yet been observed in other ROFIs, such as Liverpool Bay [Sharples and Simpson, 1994] or the Merbok estuary [Uncles et al., 1992].

The presence of stratification not only affects the non tidal components of currents, but it also modifies the tidal flow. As discussed by Visser et al. [1994] and Souza and Simpson [1996], the tidal ellipses tend to be more clockwise at the surface when stratification is present. This is due to the reduction in eddy viscosity at the pycnocline, which will then decouple the bottom and surface layer modifying the clockwise tidal Ekman layers.

In this work we try to combine observations of velocities and density structure to elucidate the way in which various processes involved interact to explain some of the complexity in the observed pattern of variability in structure, and in particular the occurrence of marked semi-diurnal variability in stratification. This will clarify the role of tidal straining

Buoyancy Effects on Coastal and Estuarine Dynamics
Coastal and Estuarine Studies Volume 53, Pages 83-96

as an important source of stratification in the region and will allow some conclusions about the origin of the cross-shore component of tidal shear.

Observational Techniques

Observations of water column stability and flow were carried out in the Rhine ROFI from the 2 to 17 September 1992 using an array of 5 moorings. The array formed a square of 13 by 13 km and had a central position at about 16 km off-shore from Noordwijk (Figure 1.a). Each mooring was equipped with four current meters at depths of approximately 1,10, 13, 16 meters; each of which were capable of measuring temperature and conductivity. The conventional current meters were supplemented by the use of 1 MHz ADCPs to give greater vertical resolution. The time series were complemented by continuous CTD spatial survey observations using the SEAROVER (Figure 1b) undulator which provided measurements of temperature salinity fluorescence and optical beam transmittance between the surface and 3m from the bottom with a horizontal resolution of about 300m.

Rhine ROFI Study Area
Mooring Positions and Survey Track

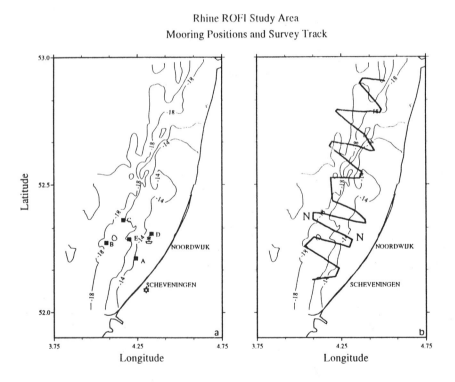

Figure 1. Rhine ROFI study area: (a) bathymetry in metres with mooring positions (■) and Noordwijk Tower; (b) SEAROVER tracks for survey in the period J254 21:40 to J256 18:33.

Observations of Stratification

The evolution of stratification $\Delta\rho$ at two of the mooring positions over the 1992 campaign is illustrated in Figure 2 together with the estimates of tidal stirring, based on the observed tidal velocity and wind stirring derived from wind observations at the Noordwijk tower, calculated using efficiencies from Simpson and Bowers [1981]. Strong bursts of wind stirring with energy inputs of up to 0.2 mWm^{-3} are seen to dominate at times over the tidal contribution. The influence of the combined stirring power inputs is reflected in the mean level of the stratification; low stratification at the start of density record follows a bout of strong wind mixing after which lower winds and neap tides allows the development of strong

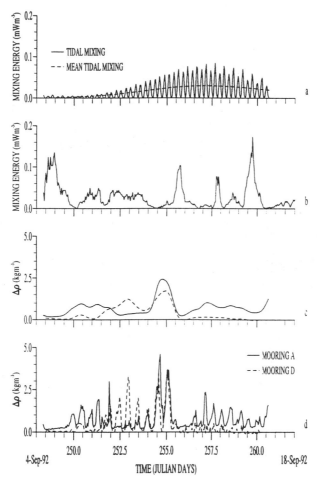

Figure 2. Time series of stirring power and stratification. (a) tidal stirring power: instantaneous value(dashed) and daily mean(continuous computed from current meter velocity at mooring A (~4m above bed). (b) wind stirring power calculated using wind data from the Noordwijk Tower. (c,d) stratification $\Delta\rho$, the density difference between 16m and 1m depth as daily mean (c) and instantaneous(d) values at mooring A (continuous line) and D (dashed). Mean water depth is 20m.

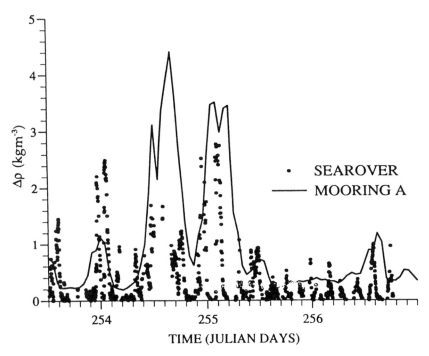

TIME (JULIAN DAYS)

Figure 3. Stratification time series shown as density difference Δρ: dots are calculated from SEAROVER data between depths of 12m and 1m, the continuous line is the density difference at mooring A data between 17m and 1m.

stratification at both moorings with values of the mean bottom to surface density difference Δρ ~2 kg m⁻³ on days 254-255. Immediately following this the combination of an episode of strong wind stirring augmented by the tidal stirring induces an almost complete vertical mixing.

Superimposed on the these changes in the average level of stratification is a very prominent semi-diurnal variation the amplitude of which is large in relation to the daily mean so that, at the time of minimum stratification, the system approaches complete homogeneity. This semi-diurnal signal is apparent at both moorings and shows similar phase and amplitude in both cases.

It is clear from the observations that, in the vicinity of the moorings, there is a significant evolution of the density field even on short time scales due to the marked semi-diurnal variation in column structure. If this pattern of variation is a general characteristic of the stratified region, it would seem that time-dependent changes will be interpreted as spatial structures in mapping exercises. In order to test the notion that there is a general semidiurnal oscillation under stratified conditions, we have combined the SEAROVER data with the mooring data for the same period in Figure 3. Surface to bottom density differences observed by the SEAROVER system when inside the ROFI (i.e. within 15 km of the coast) are plotted as a time series in parallel with Δρ from mooring A. There is a significant coherence between the envelope of the SEAROVER data and the mooring observations, with an R^2=0.65 with the SEAROVER data lag of approximately 2 hours. This strongly suggests that the semi-diurnal oscillation is a general characteristic of the ROFI. While the lag between fixed and moving sensors may be expected due to the change in the phase in tidal processes over the area.

Modification of the Tidal Ellipses by Stratification

In order to observe the daily changes in the vertical structure of the tidal currents, in response to the variations in water column stability, a moving average M_2 least square fit was carried out for each overlapping 25 hour periods of the data an with it we generate the respective tidal ellipse characteristics as suggested by Souza and Simpson [1996].

The stratification time series, plotted as the difference between bottom and surface density (Figure 4a), indicates three periods in which significant stratification was present at mooring A: (i) from day 250 to 252, (ii) between days 254 and 255 and (iii) from day 256 to day 259. In between these periods, almost complete mixing prevailed and there was very little vertical structure in ellipticity and ellipse orientation (Figure 4). The surface ellipses were almost degenerate ($\varepsilon \sim 0$) with increasing anticlockwise ellipticity at the bottom to values of the order of 0.1.

By contrast, during the periods of stratification, the surface ellipses acquired marked clockwise rotation with $\varepsilon \sim -0.2$ while in the lower layer the motion became anti-clockwise with $\varepsilon > +0.2$. The transition between the two regimes is well defined and occurs at a depth of ~8 which corresponds to the mean depth of the pycnocline observed in the SEAROVER sections.

Stratification was also observed to exert some influence on the ellipse orientation as indicated in Figure 4d. When the water column is mixed the surface and bottom ellipses are oriented in the same direction to within 5°, but when the water column stability develops there is a bottom surface orientation difference of up to 15° with the bottom maximum currents being deflected to the west relative to the surface currents.

To understand this contrasting behaviour under mixed and stratified conditions we need to examine the problem of tidal flow in two directions on a rotating frame (the Earth) under the influence of frictional stresses. The motion, in which the particle trajectories are generally elliptical in form, may usefully be decomposed into anti-clockwise and clockwise components. The characteristic boundary layer thickness for these two components is then given by [Prandle, 1982]:

$$\delta_- \simeq \left(\frac{2 N_z}{(\omega - f)} \right)^{\frac{1}{2}} \tag{1}$$

for clockwise motion and

$$\delta_+ \simeq \left(\frac{2 N_z}{(\omega + f)} \right)^{\frac{1}{2}} \tag{2}$$

for anti-clockwise motion, where ω is the frequency of the tidal constituent concerned, N_z is the eddy viscosity and f the Coriolis parameter. Outside these bottom boundary layers the tidal oscillation is not significantly influenced by frictional stresses and is depth independent.

At the latitude of the Rhine ROFI f~1.15 x 10^{-4} s^{-1} so that with $\omega = 1.41$ x 10^{-4} s^{-1}, we have:

$$\frac{\delta_+}{\delta_-} = \left(\frac{\omega - f}{\omega + f} \right)^{\frac{1}{2}} \simeq 0.3 \tag{3}$$

The very different scales for the clockwise and anticlockwise boundary layers accounts for the general tendency for rotation of the current ellipse to become more clockwise with increasing distance from the seabed. Figure 5a illustrates this for the case of equal clockwise

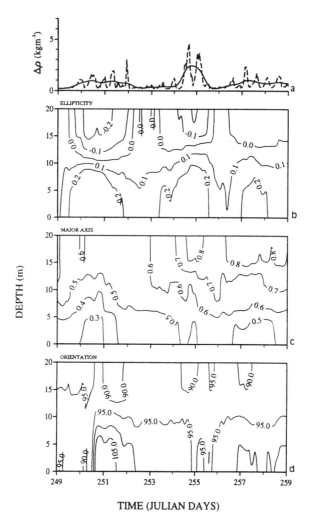

TIME (JULIAN DAYS)

Figure 4. Time series of ellipse characteristics and water column stratification for mooring A (water depth 20m). (a) The density difference between bottom and top current meters; the continuous line represent the mean stratification averaged over two tidal periods while the dashed line is the instantaneous value. (b) Ellipticity ε from bottom mounted ADCP data. (c) Semi-major axis of the ellipse in ms^{-1}. (d) Orientation of the ellipse major axis in degrees (positive anti-clockwise from the east).

and anti-clockwise components at the surface which produces a degenerate ellipse with rectilinear tidal flow. Near the bed the boundary layer structure reduces the clockwise component U_c relative to the anti-clockwise U_{ac} so that the resultant ellipse exhibits anti-clockwise rotation.

Consider now what happens when the water column becomes stratified while the tidal forcing remains the unchanged (Figure 5b). The effect of water column stability is to reduce the eddy viscosity N_z in the vicinity of the pycnocline with the consequence that the upper layer is largely de-coupled from frictional influence. The surface current ellipse thus acquires a

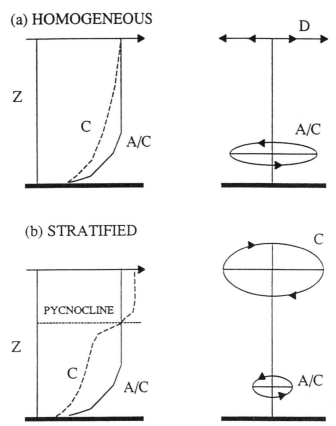

Figure 5. Schematic of the vertical profile of the clockwise and anti-clockwise rotary components and ellipse configuration. a) for the homogeneous case when N_z is constant. b) for the stratified case when N_z is reduced at the pycnocline.

clockwise rotation, while the near bottom current ellipse becomes more anti-clockwise due to a reduction in the clockwise component associated with the concentration of frictional influence below the pycnocline.

Interaction Between Mean Water Column Stability and Tidal Shear in the Production of Semi-Diurnal Switching of Stratification in the Rhine ROFI

Semi-diurnal variations in water-column stability of this kind have been previously identified in other ROFI and estuarine regimes [Bowden and Sharaf el Din, 1966; Uncles, 1992; Simpson et al., 1991] and attributed to the influence of tidal straining [Simpson et al,. 1990]. The amplitude of semi-diurnal variation in the Rhine ROFI is, however, considerably greater in relation to the mean amplitude than in previously observed cases. Reference to the 1990 observations show similar behaviour as is illustrated in Simpson et al. [1993] with even larger oscillations of salinity stratification (amplitude ~ 4 salinity units) than we observe in the 1992 measurements.

    The question then arises as to the role of tidal straining in the present case and whether or not other mechanisms are involved. As a first step in answering these questions we have investigated the relation between the density stratification and the relative cross-shore tidal displacement (Figure 6). This straining has been calculated by high pass filtering the ADCP data for the 4 day period. Prior to midday on day 255, there was a strong enhancement of the straining with a relative displacement of 7 km between current meters at depths of 4m and 16m. Following the onset of intense wind mixing on day 255, the tidal shear signal was greatly reduced. These changes in the straining signal are clearly reflected in the $\Delta\rho$ variation suggesting that straining is the primary process driving the semi-diurnal stratification cycle. *i.e.* for the maximum cross-shore gradient, on day 254 as calculated from mooring A and B, is: 0.4 kg m^{-3} per km, while the relative displacement, for the same time at mooring A, is $\Delta X = 7.6$ km, leading to a $\Delta\rho = 3.04$ kg m^{-3} which is very similar to the observed value (Figure 6a).
    This striking feature of the relatively large cross-shore straining which evidently operates during periods of stratification and which provides the primary drive for the semi-diurnal oscillation, appears to be associated with the change in the shape of the tidal ellipse which is brought about by the development of mean stratification as discussed by Souza and Simpson [1996]. The fact that there is a strong correlation between the cross-shore straining and the semi-diurnal variability in stratification, is strong evidence that the semi-diurnal stratification is mainly produced by this mechanism.

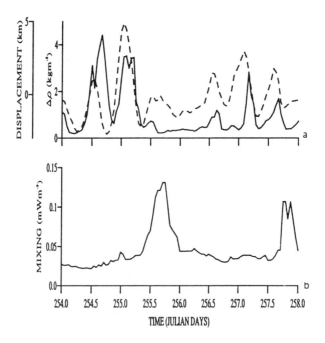

Figure 6. Tidal straining and periodic stratification: (a) time series of stratification (continuous line) and tidal displacement (dashed line); the displacement is calculated from the high-pass-filtered velocities from the top bin (3m deep) relative to the bottom bin (17m deep) from a 1.2 MHz ADCP at mooring A; positive displacement is off-shore; (b) estimates of the combined wind and stirring power for the same period.

Model

In order to combine all these mechanisms and test the above conceptual mechanism for the generation of the semi-diurnal oscillations in stability, we have utilised the 1-d turbulence closure model of Simpson and Sharples[1992] to achieve a synthesis of the various processes operating in the ROFI regime. The model uses an explicit scheme to integrate the equations of motion:

$$\frac{\partial u}{\partial t} = -\frac{1}{\rho}\frac{\partial P}{\partial x} + fv + \frac{\partial}{\partial z}\left(N_z\frac{\partial u}{\partial z}\right) \tag{4}$$

$$\frac{\partial v}{\partial t} = -\frac{1}{\rho}\frac{\partial P}{\partial y} - fu + \frac{\partial}{\partial z}\left(N_z\frac{\partial v}{\partial z}\right) \tag{5}$$

with x and y positive in the east and north directions respectively, and z increasing positively from the seabed. The second term on the right is the usual Coriolis forcing, and the third term is the effect of friction between the layers in transporting momentum vertically through the water column, with N_z the coefficient of vertical eddy viscosity. Boundary conditions for the momentum equations are a quadratic stress law at the bottom and wind stress at the sea surface, using the hourly wind speed and direction observations.

The horizontal pressure gradient terms in equations (4) and (5) are approximated in the form:

$$\frac{1}{\rho}\frac{\partial P}{\partial x} = g\left(\frac{\partial \eta}{\partial x}\right)_{tidal} + g\frac{\partial \overline{\eta}}{\partial x} + g(h-z)\left(\frac{\partial \rho}{\partial x}\right)_h \tag{6}$$

and similarly for the y component. The first term on the right of equation (6) is a tidally oscillating sea surface slope, the second two terms represent the effect of a depth-invariant horizontal density gradient in setting up a mean surface slope and driving a depth-dependent density circulation. The components of the horizontal density gradient are determined from observations of temperature and salinity. Calculation of the mean surface slope is achieved by specifying a zero net flow condition in the cross-shore direction.

Horizontal advection and vertical diffusion of salinity and temperature at each level are controlled by

$$\frac{\partial(s,T)}{\partial t} = -u\frac{\partial(s,T)}{\partial x} + \frac{\partial}{\partial z}\left(K_z\frac{\partial(s,T)}{\partial z}\right) \tag{7}$$

where $\partial(s,T)/\partial x$ are the horizontal salinity and temperature gradients, again assumed depth independent and taken from observations. K_z is the coefficient of vertical eddy diffusivity. There is no flux of salt or heat through the seabed, and no net flux of salt at the surface. Surface heating is specified in terms of the observed values of solar radiation, air temperature, relative humidity and wind speed following Gill [1982, p34]. Density ρ is derived from temperature and salinity using the standard equation of state [UNESCO, 1981].

A level 2 turbulence closure scheme is used to calculate vertical profiles of N_z and K_z as functions of local stability [Mellor and Yamada, 1974] via

$$N_z = S_M\, lq; K_z = S_H\, lq \tag{8}$$

where q^2 is the turbulent kinetic energy, l is a mixing length and S_M and S_H are stability functions which depend on the local gradient Richardson number [for details, see Sharples and Simpson, 1994].

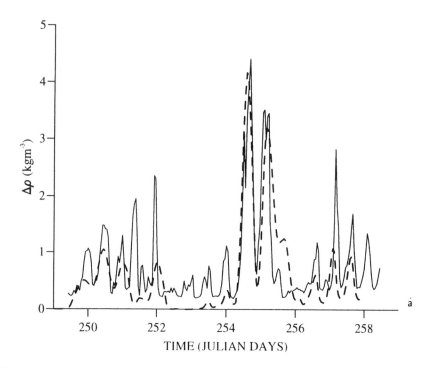

Figure 7. Stratification and displacement from the 1-dimensional turbulence closure model: (a) Density stratification; observations (continuos line) and the model (dashed line).

We have run this model with simplified forcing by surface slopes obtained from tide gauge observations at the mooring positions, the (predominant) cross-shore gradient derived from temperature and salinity measurements at moorings A and D and meteorological data provided by KMNI, from the Noordwijk tower (wind speed, air temperature, humidity) and Valkenberg (solar radiation).

Figure 7a,b shows the time series of stratification for the observations and the model. As in the 1990 case there is a good degree of correspondence between the two with the model exhibiting the main features of the observations, notably the two strong maxima in $\Delta\rho$ on Jdays 254-255 which are simulated with satisfactory timing and magnitude. The preceding and following periods of near zero stability are also reproduced by the model.

The model confirms the operation of the cross-shore straining mechanism in driving the semi-diurnal oscillations of stability. In Figures 7c,d we see that the differential displacement between surface and bottom predicted by the model are of comparable magnitude and similar phasing to those observed. The suppression of tidal straining by enhanced vertical mixing is clearly apparent, if somewhat, exaggerated in the model results.

The increase in cross-shore displacement, *i.e.* tidal straining, during periods of stratification is due to the modification in the ellipse properties. Hence, if the relative tidal displacement in Figures 7c,d are in approximate agreement, it is expected that the tidal ellipses will also be comparable. In Figure 8 we see fair agreement between model and observations, in amplitude distribution and timing of ellipses characteristics and stratification. The major axis is large at the surface from day 254, which will coincide with the springs period, while the ellipticity is negative during stratified periods with $\varepsilon = -0.2$. The difference in stratification

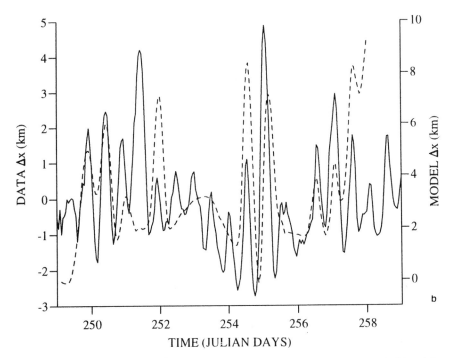

Figure 7. Stratification and displacement from the 1-dimensional turbulence closure model: (b) Relative displacement; observations (continuos line) and the model (dashed line).

and straining during periods of strong stirring might be due to the fact that the model overestimates the anticlockwise rotation during these periods in comparison with the observations. The values of phase and orientation are similar to those from the observed cross-shore and along-shore currents.

Discussion

On the basis of the above results we have confirmed the hypothesis, that the mechanism responsible for the strong semi-diurnal variability in stratification involves the sequence of interactions illustrated in Figure 9. Following a period of complete vertical mixing, when the isopycnels (Figure 9a) are vertical, the density gradients start to relax under gravity generating a coast parallel flow, due to the effect of the earth's rotation (Figure 9b). The stratification will then induce changes in the tidal ellipses with surface ellipses rotating clockwise and bottom ellipses rotating anti-clockwise, generating a strong cross-shore component of tidal shear (Figure 9c). The resulting off-shore tidal straining initiates oscillations in stability which combine with the mean stratification give the observed pattern of stability variation with the system coming close to complete mixing in each tidal cycle. When the added effect of wind and tidal stirring increases again (Figure 9d), vertical exchange is enhanced and the off-shore tidal shear is suppressed as the ellipses revert to near degenerate form and the stratification is broken down.

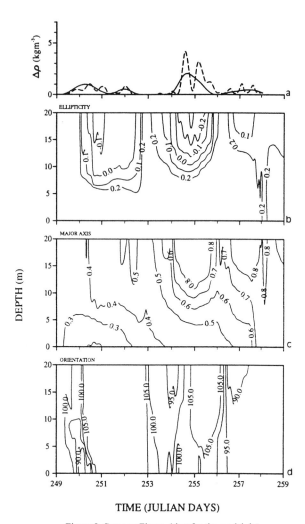

TIME (JULIAN DAYS)

Figure 8. Same as Figure 4 but for the model data.

The fact that the tidal straining, the semi-diurnal stratification and the changes in the tidal ellipses are reproduced by the model, in both magnitude and phase, lends further support to the proposal that the semi-diurnal stratification arises from the straining process and suggests that no other mechanisms are involved.

Acknowledgments. This work was conducted within PROFILE, an European funded project under the MAST programme, contract number MAS2-CT93-0054. A.J. Souza wishes to thank CONACYT, Mexico for support through a studentship.

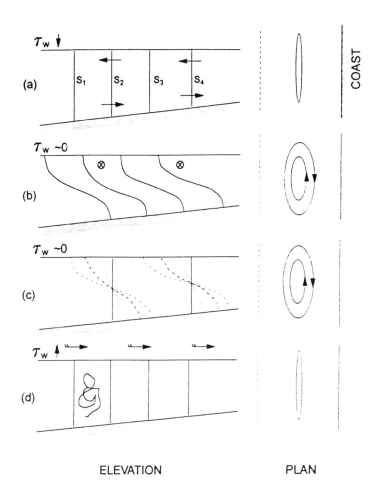

ELEVATION   PLAN

Figure 9. Process summary schematic showing the changes in ellipse format at surface and bottom due to the onset of stratification and the resulting semi-diurnal oscillations in stability driven by tidal straining. The symbolic windstress is shown in an arbitrary direction.

References

Bowden K.F. and S.H. Sharif El Din, Circulation and mixing processes in the Liverpool Bay area of the Irish Sea, *Geophys. J. of the R. Astro. Soc., 11*, 279-292, 1966.

Gill A.E., Atmosphere-Ocean Dynamics, Academic press, San Diego, 662, 1982.

Linden P.F. and J.E. Simpson, Modulated mixing and frontogenesis in shallow seas and estuaries, *Continent. Shelf. Res., 8*, 1107-1127, 1988.

Mellor G.L. and T. Yamada, A hierarchy of turbulence closure models for planetary boundary layers, *J. Atmos. Sci. 31*, 1791, 1806, 1974.

Ou H.W., Some two-layer models of shelf-slope front: Geostrophic adjustment and its maintenance, *J. Phys. Oceanogr., 13*, 1798-1808, 1983.

Prandle, D., The vertical structure of tidal currents. Geophys, *Astrophys. Fluid Dynamics, 22*, 29-49, 1982.

Sharples J. and J.H. Simpson, Semi-diurnal and longer period stability cycles in the Liverpool Bay R.O.F.I., *Continental Shelf Research*, 1994, (in press).

Simpson J.H., W.G. Bos, F. Schirmer, A.J. Souza, T.P. Rippeth, S.E. Jones, and D. Hydes, Periodic stratification in the Rhine ROFI in the North Sea, *Oceanologica Acta, 16*, 1, 23-32, 1993.

Simpson J.H., J. Brown, J. Matthews, and G. Allen, Tidal straining, density currents and stirring control of estuarine stratification, *Estuaries, 13*, 2, 125-132, 1990.

Simpson J.H., J. Sharples, and T.P. Rippeth, A Prescriptive Model of stratification induced by freshwater run-off, *Estuarine, Coastal and Shelf Science, 33*, 23-35, 1991.

Simpson J.H. and J. Sharples, Dynamically-active models in prediction of estuarine stratification, in *Dynamics and Exchange in Estuaries and Coastal Zone*, Coastal and Estuarine Studies, vol. 40, edited by D. Prandle, AGU, Washington, D.C., 1992.

Souza, A.J., and J.H. Simpson, The modification of the tidal ellipses by stratification in the Rhine ROFI, *Continental Shelf Research, 16*, 997-1007, 1996.

Uncles R.J., W.K. Gong, and J.E. Ong, Intratidal fluctuations in stratification within a mangrove estuary, *Hydrobiologia, 247*, 163-171, 1992.

UNESCO, Tenth Report of the Joint Panel on Oceanographic Tables and Standards, UNESCO Technical Papers in Marine Science No 36, UNESCO, Paris, 1981.

Visser A.W., A.J. Souza, K. Hesser, and J.H. Simpson, The influence of water column stratification on tidal current profile in a ROFI system, *Oceanologica Acta, 17*, 4, 369-381, 1994.

8

Detailed Flow Structure around a Thermohaline Front at the Mouth of Ise Bay, Japan

Tetsuo Yanagi, Xinyu Guo, Takashi Ishimaru and Toshiro Saino

Abstract

Detailed flow structure around a thermohaline front at the mouth of Ise Bay, Japan was observed with use of ADCP (Acoustic Doppler Current Profiler). The surface water converges at the transition zone between cold coastal water and warm off-shore water, sinks there and diverges like a skirt in the lower layer below the transition zone. The maximum surface convergence and sinking velocity are 5.0×10^{-4} s^{-1} and 0.25 cm s^{-1}, respectively.

1. Introduction

Quasi-steady thermohaline fronts develop on the shelf areas of the northwestern Pacific Ocean in winter [e.g. Nagashima and Okazaki, 1979, Yanagi, 1980, Lie, 1985]. The generation and maintenance mechanisms of such fronts have been investigated mainly with use of numerical models by Endoh [1977], Harashima and Oonishi [1981], and Akitomo et al. [1990]. From their results, the front is generated in a transition zone between cold and fresh coastal water and warm and saline oceanic water due to surface cooling and fresh water inflow from land but there is little density difference between both sides of the front because the salinity difference compensates the temperature difference. The front exists at nearly the same position from late autumn to early spring and plays an important role on the material transport [Yoshioka, 1988, Yanagi et al., 1989].

A remarkable thermohaline front is observed at the mouth of Ise Bay in winter and the distributions of water temperature and salinity around this front and their seasonal variations have been investigated [Sekine et al., 1992, Yanagi et al., 1995]. The detailed flow structure around this front, however, is not yet known. We report here the three-dimensional detailed flow structure around a thermohaline front at the mouth of Ise Bay, Japan.

2. Field observation

The field observations were carried out at the mouth of Ise Bay (Figure 1) by R.V. *Tansei-Maru* of the Ocean Research Institute, University of Tokyo, on 11 and 12 February 1995. The vertical profiles of water temperature and salinity were observed with use of CTD (Neil Brown Mark III B) at 9 stations from Sta. IF-2 to Sta. IF-10 shown in Figure 1 during 14:58 through 17:33 on 11 February 1995. The distance between successive observation stations is about 900 m. The horizontal distributions of water temperature at surface (1.0 m depth) and velocity at the depth of 3.5 m, 5.5 m, 7.5 m, 9.5 m, 11.5 m, 13.5 m, 15.5 m, 17.5 m and 19.5 m were

Buoyancy Effects on Coastal and Estuarine Dynamics
Coastal and Estuarine Studies Volume 53, Pages 97-106
Copyright 1996 by the American Geophysical Union

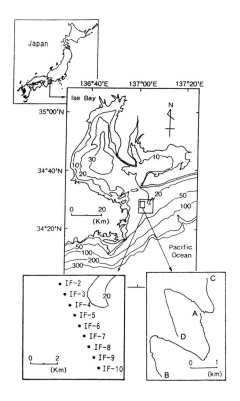

Figure 1. Ise Bay. Numbers show the depth in meter.

observed continuously along the observation lines from A to B (see Figure 1) during 10:56 to
11:44 on 12 February and from C to D during 12:28 to 13:08 on 12 February with use of a
thermistor and an ADCP (RD Instruments, 300 kHz). Averaged water depth along A-B and C-
D lines is about 23 m. Water temperature and velocity were measured at 15 seconds intervals,
which corresponds to a 15 m interval for R.V. *Tansei-Maru* steaming at about 2 kt. Raw data
were filtered with use of a box filter (60 m in horizontal length and 4 m in vertical length).

3. Results

Figure 2 is a NOAA 12 infra-red image, collected at 8:00 JMT on 11 February 1995,
which shows the position of the thermohaline front at the mouth of Ise Bay. The thermohaline
front, represented by a discontinuity line between cooler coastal water and warmer offshore
water, runs in a northeast to southwest direction at the mouth of Ise Bay. Due to cooling
through the sea surface in winter, water at the mouth of Ise Bay with moderate salinity becomes
the heaviest and sinks there. The colder coastal water cannot become heavy enough to sink
due to the fresh water inflow from land. Oceanic saline water also cannot become heavy
enough to sink because of the mixing with warm Kuroshio water. The thermohaline front
exists here from late autumn to early spring with short-term variation of the period of 10 to 15
days [Sekine et al., 1992].

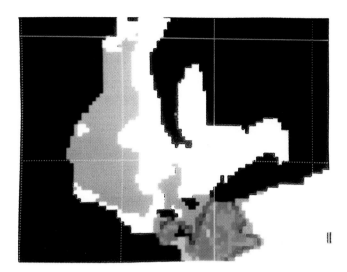

Figure 2. NOAA 12 infra red image collected at 8:00 JMT on 11 February 1995.

The observed vertical distributions of water temperature and salinity across this front are shown in Figure 3. The horizontal gradients of water temperature and salinity up to 3.0°C km-1 and 1.0 psu km-1, respectively, are observed in the surface layer.

On the other hand, the density gradient does not exist across the transition zone as seen in water temperature and salinity distributions. It appears that the surface water converges at the transition zone (near Sta. IF-5 in this case), sinks there and diverges like a skirt in the lower layer as shown in Figure 3. Such distribution patterns have been observed at a thermohaline front in the Kii Channel [Yoshioka, 1984] and at the mouth of Tokyo Bay [Yanagi et al., 1989].

The northward flood tidal current was dominant at the observation period along A-B and C-D lines on 12 February 1995 and the frontal structure was considered to be advected in a frozen manner by this tidal current. Therefore we first estimated the barotropic tidal current field at the observation period on 12 February. The horizontal distributions of vertically averaged velocity along A-B and C-D lines are shown in Figure 4. The averaged northward current is predominant at all observation points. The speed along the C-D line is larger than that along the A-B line. We assume that little alteration in vertical and horizontal frontal structure took place in a general northward barotropic tidal current field during our observation period of 48 minutes (A-B line) and 40 minutes (C-D line), respectively. Hence we shift the position of the observation points along A-B and C-D lines to the positions at the beginning of our field observation, that is, at 10:56 on 12 February along the A-B line and at 12:28 on 12 February along the C-D line, respectively; each observation point was moved to upstream direction by the distance calculated by multiplying the vertically averaged velocity at each point (shown in Figure 4) with the time difference between the beginning time and the observation time. The reconstructed horizontal distributions of surface (1 m depth) water temperature at the beginning of the observation are shown in Figure 5. Temperature values are objectively interpolated at each grid (100 m mesh size) with use of a hyperbolic function [Yanagi and Igawa, 1992] from the observed data along A-B and C-D lines. The maximum horizontal gradient of water temperature is 2.5°C / 200 m.

The horizontal distributions of anomaly velocity at each depth from the vertically averaged velocity are shown in Figure 6. They are also objectively interpolated in the same manner as in Figure 5. The southward flow is dominant in the upper layer and the northward flow in the lower layer. The colder coastal water and warmer oceanic water in the surface layer

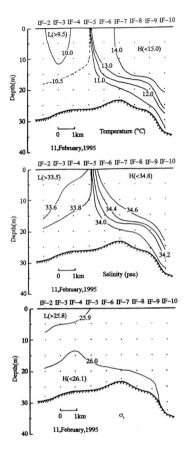

Figure 3. Vertical distributions of water temperature, salinity and sigma t across the thermohaline front at the mouth of Ise Bay.

seem to converge to the frontal zone. The water at the middle layer (11.5 m depth) stagnates. On the other hand, the colder coastal water and warmer oceanic water at the depth of 17.5 m seem to diverge from the transition zone.

The horizontal distributions of convergence and divergence Q, defined by Eq. (1), at each depth are calculated as.

$$Q = \frac{\partial u}{\partial x} + \frac{\partial v}{\partial y} \tag{1}$$

Here u denotes the eastward velocity and v the northward velocity. The relative error of observed velocity by ADCP is +/- 1 cm s^{-1} and that of positioning every 15 m by GPS (Global Positioning System) +/- 1 m because the sea surface was very calm during the observation. Therefore the accuracy of calculated Q is +/- 4 x 10^{-5} s^{-1} which is obtained by the relative accuracy of current / that of positioning.

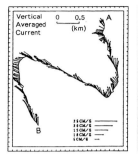 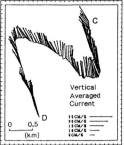

Figure 4. Vertically averaged currents along A-B and C-D lines.

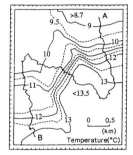 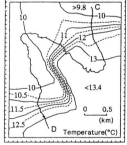

Figure 5. Horizontal distribution of water temperature (1 m depth) along A-B and C-D lines.

The evaluated horizontal distributions of Q at each depth are shown in Figure 7. The surface convergence with the value of -5.0×10^{-4} s^{-1} is seen just at the transition zone of water temperature shown in Figure 5 and the large divergence of 5.0×10^{-4} s^{-1} is seen at the depth of 17.5 m just below the surface convergence zone. The surface divergence zone exists next to the convergence zone at each depth of this observation area. The observed divergence or convergence is one-order of magnitude larger than those observed in the coastal sea [e.g. Akamatsu, 1975] or open ocean [e.g. Molinari and Kirwan, 1975] except at the frontal region [e.g. Tomosada et al., 1986]. The vertical velocity is estimated on the basis of the continuity equation and the Q shown in Figure 7. It is assumed that the vertical velocity W (positive upward) and horizontal divergence Q at the sea surface are 0.

It is shown as,

$$W_{z=3.5m} = \frac{1}{2} \ Q_{z=3.5m} \ x \ 350cm \tag{2}$$

Vertical velocities at the each depth are estimated successively in the same manner and shown as,

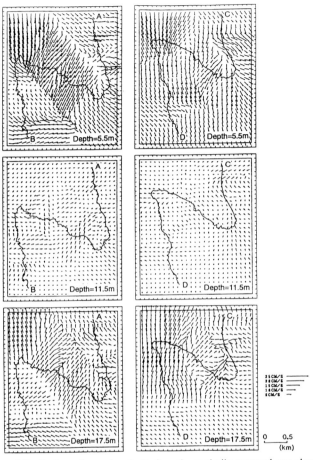

Figure 6. Horizontal distributions of anomaly velocity from the vertically averaged one along A-B and C-D lines.

$$W_{z=5.5m} = \frac{1}{2}\left(Q_{z=3.5m} + Q_{z=5.5m}\right) x \ 200 \ cm + W_{z=3.5m} \tag{3}$$

$$W_{z=11.5m} = \frac{1}{2}\left(Q_{z=9.5m} + Q_{z=11.5m}\right) x \ 200 cm + W_{z=9.5m} \tag{4}$$

$$W_{z=17.5m} = \frac{1}{2}\left(Q_{z=15.5m} + Q_{z=17.5m}\right) x \ 200 cm + W_{z=15.5m} \tag{5}$$

Q(×10⁻⁴sec⁻¹)

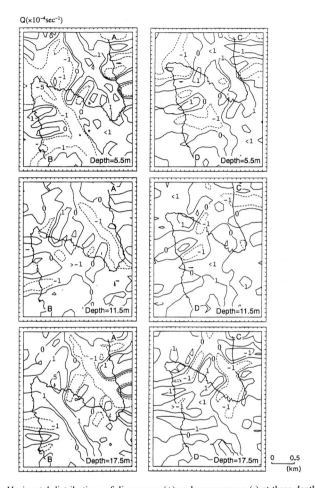

Figure 7. Horizontal distributions of divergence (+) and convergence (-) at three depths.

The accuracy of estimated vertical velocity is +/- 0.008 cm s⁻¹ on the basis of the accuracy of Q. The estimated horizontal distributions of vertical velocity at each depth are shown in Figure 8. The large downward velocity is seen at the depth of 11.5 m and it attains 0.25 cm s⁻¹.

Based on Figure 3, Figure 6 and Figure 8, the flow structure around the thermohaline front can be delineated. The surface coastal water and the surface oceanic water are advected to the thermohaline front at the speed of about 10 cm s-1. The downward velocity of 0.1 cm s⁻¹ is seen at the surface convergence zone. The maximum downward current appears at the middle layer just below the surface convergence zone and it is about 0.25 cm s⁻¹. Interestingly, an upward current coexists with the downward current at the depth of 11.5 m. This may indicate the existence of strong gravitational convection due to sea surface cooling at the transition zone which was suggested by Akitomo et al. [1990]. Below the mid-depth mixing layer, the mixed water diverges to both sides of the front with the speed of about 10 cm s⁻¹. This estimation of vertical velocity, although just a snapshot at the frontal region, is the first case where the directly observed current data were utilized to reconstruct a three-dimensional flow field around a coastal front.

W(cm/s)

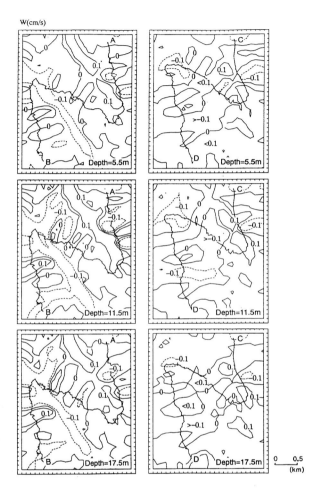

Figure 8. Horizontal distribution of upward velocity (+) an downward one (-) at three depths

The magnitude of vertical velocity obtained here is about twice as that estimated theoretically for the front driven by the cabling effect [Horne et al., 1978], and nearly the same as that estimated by diagnostic model calculation using the observed density distribution data at a thermohaline front at the mouth of Tokyo Bay [Yanagi et al., 1989]. The calculated vertical velocity in the mid-depth is not shown directly in Akitomo et al. [1990] but we can guess that their calculated vertical velocity is nearly the same as ours because they obtained nearly the same surface convergence values in their numerical model as ours. However, the calculated values mentioned above are all based on the vertical two dimensional analysis though our present observed current field shows the dominant three-dimensional structure. We have to carry out a three-dimensional numerical experiment in the near future in order to investigate the dynamics of three-dimensional current field around a thermohaline front.

4. Discussions

The material transport from the coastal sea to the open sea plays an important role in global material cycling [IGBP Report No. 25, 1993]. One of the main goals of IGBP/LOICZ (International Geosphere/Biosphere Programme, Land-Ocean Interaction in Coastal Zone) Project is to quantitatively estimate the material transport from the coastal sea to the open sea. To attain that goal, it is essential to evaluate the role of coastal fronts on the material transport, because there are various types of oceanic fronts in the coastal sea [Yanagi, 1987]. The material flux itself can be estimated simply by multiplying the material concentration and the current speed. It must be stressed that the variables required for this calculation are the material concentrations and the current speed averaged on relevant time and spatial scales because the temporal and spatial variations of coastal front is very large and such variability plays important role in the material transport [e.g. Akitomo, 1988]. The instantaneous three-dimensional flow structure around a thermohaline front shown here is the first step to clarify the time and spatial scales of material transport dynamics in the coastal sea.

The coastal fronts also play important roles in the coastal ecosystem [e.g. Kiorboe et al., 1988]. It is well known that coastal fronts are good spawning grounds of many fish species [Boucher et al., 1987], because phytoplankton and zooplankton are accumulated by the surface converging currents and serve as food necessary for the survival of fish larvae [Uye et al., 1992]. Given the maximum downward flow speed of 0.25 cm s^{-1}, larvae entrained into the downwelling stream will be transported to the bottom (- 23 m) in about 2.5 hours. In this case, only the larvae whose swimming speed exceeds the water movement speed may survive. Hence the instantaneous flow field should be taken into account when we consider the phenomena of relevant time and spatial scales such as biological processes in ecosystem dynamics around coastal fronts.

Acknowledgments. We thank to officers and crews of R.V. *Tansei-Maru* for their help in the field observation. This study was partially supported by the fund defrayed from the Ministry of Education, Science and Culture, Japan.

References

Akamatsu, H., On the oceanographic structure at the frontal zone in Japan Sea (I), *Umi to Sora, 50,* 123-136, 1975. (in Japanese).

Akitomo, K., A numerical study of a shallow sea front generated by the buoyancy flux: water exchange caused by fluctuation of the front, *J. Oceanogr. Soc. Japan, 44,* 171-188, 1988.

Akitomo, K., N. Imasato, and T. Awaji, A numerical study of a shallow sea front generated by buoyancy flux: generation mechanism, *J. Physical Oceanogr., 20,* 172-189, 1990.

Boucher, J., F. Ibanez, and L. Prieur, Daily and seasonal variations in the spatial distribution of zooplankton populations in relation to the physical structure in the Ligurial Sea Front, *J. Mar. Res., 45,* 133-173, 1987.

Endoh, M., Formation of thermohaline front by cooling of the sea surface and inflow of the fresh water, *J. Oceanogr. Soc. Japan, 33,* 6-15, 1977.

Harashima, A., and Y. Oonishi, The Coriolis effect against frontogenesis in steady buoyancy-driven circulation, *J. Oceanogr. Soc. Japan, 37,* 49-59, 1981.

Horne, E.P.W., M.J. Bowman, and A. Okubo, Crossfrontal mixing and cabling, in *Oceanic Fronts in Coastal Processes,* edited by M.J. Bowman and W.E. Esaias, pp. 105-113, Springer-Verlag, New York, 1978.

IGBP Report No. 25, 50 p, 1993.

Kiorboe, T., P. Munk, K. Richardson, V. Christensen, and H. Paulsen, Plankton dynamics and larval herring growth, drift and survival in a frontal area, *Mar. Ecol. Prog. Ser., 44,* 205-219, 1988.

Lie, H.J., Wintertime temperature-salinity characteristics in the south-eastern Hwanghae (Yellow Sea), *J. Oceanogr. Soc. Japan, 41,* 291-298, 1985.

Molinari, A., and A.D. Kirwan, Calculations of differential kinetic properties from Lagranjian observations in the western Caribbean Sea, *J. Physical Oceanogr., 5,* 483-491, 1975.

Nagashima, H., and M. Okazaki, Currents and oceanic condition at Tokyo Bay in winter, *Bulletin on Coastal Oceanography, 16,* 76-86, 1979 (in Japanese).

Sekine, Y., S. Kawamata, and Y. Satoh, Observation of coastal fronts in Ise Bay in early winter, *Bulletin on Coastal Oceanography, 29,* 190 - 196, 1992 (in Japanese with English abstract and captions).

Tomosada, A., K. Segawa, and K. Kuroda, Closely-spaced observations across the Kuroshio front south of the Shionomisaki, *Bull. Tokai Reg. Fish. Res. Lab., 120,* 11-25,1986 (in Japanese).

Uye, S., T. Yamaoka, and T. Fujisawa, Are tidal fronts good recruitment areas for herbivorous copepods?. *Fisheries Oceanography, 1,* 216-226, 1992.

Yanagi, T., A coastal front in the Sea of Iyo, *J. Oceanogr. Soc. Japan, 35,* 253-260, 1980.

Yanagi, T., Classification of "Siome", streaks and fronts, *J. Oceanogr. Soc. Japan, 43,* 149-158, 1987.

Yanagi, T., and T. Koike, Seasonal variation in thermohaline and tidal fronts, Seto Inland Sea, Japan, *Cont. Shelf Res., 7,* 149-160, 1987.

Yanagi, T., A. Isobe, T. Saino, and T. Ishimaru, Thermohaline front at the mouth of Tokyo Bay, *Cont. Shelf Res., 9,* 77-91, 1989.

Yanagi, T., O. Matsuda, S. Tanabe and S. Uye (1992) Interdisciplinary study on the tidal front in the Bungo Channel, Japan, in *Dynamics and Exchanges in Estuaries and the Coastal Zone,* edited by D. Prandle, pp. 617-630, AGU, Washington, DC, 1992.

Yanagi, T., and S. Igawa (1992) Diagnostic numerical model of residual flow in the coastal sea, *Bulletin on Coastal Oceanogr., 30,* 108-115,1992, (in Japanese).

Yanagi, T., X. Guo, T. Saino, and T. Ishimaru, Thermohaline front at the mouth of Ise Bay in winter, 1995, (to be submitted).

Yoshioka, H., Oceanic fronts in the coastal sea, *Bulletin on Coastal Oceanography, 21,* 110-117, 1984, (in Japanese).

Yoshioka, H., The coastal front in the Kii Channel in winter, *Umi to Sora., 64,* 79-111, 1988.

9

The Effect of Hyperpycnal Water on Tidal Exchange

Ned P. Smith

ABSTRACT

Current meter and hydrographic data from a tidal channel connecting the western side of Exuma Sound with the eastern margin of Great Bahama Bank are used to investigate the efficiency of tidal exchanges. The density of water leaving the bank is usually higher than the density of water arriving on the flood tide. Data from near the shelf break confirm that density currents cascade down the narrow shelf. Tidal period bursts of seaward-directed flow are recorded just above the bottom, and the net flow is seaward. Spikes appear in the plot of near-bottom density. Salinity is used as a tracer during two consecutive tidal cycles to show that 95 and 97.5% of the water entering the channel is from Exuma Sound, and not bank water associated with the previous ebb. Perturbation analysis is used to show that the time-varying salt transport, has a strong seasonal component that is directly related to the presence or absence of density currents.

Introduction

Strong gradients of temperature and especially salinity are common in estuarine and inner shelf waters. Salinity gradients form as a result of the juxtaposition of sea water and fresh water arriving as rainfall, freshwater runoff and groundwater seepage. With vertical mixing, horizontal salinity gradients arise from the different response of deep and shallow waters to spatially uniform freshwater gains and losses. In a similar way, horizontal temperature gradients can arise along a depth gradient because of the inverse response of the water column to spatially uniform sensible and latent heat fluxes. A relatively shallow water column will magnify the effect of a given heat gain or loss. Hydrographic time series from tidal channels connecting shelf and estuarine waters clearly demonstrate these gradients, as water of different salinity and temperature ebbs and floods past the study site.

The hydrography of waters surrounding the Exuma Cays of the Bahamas is unusual in the sense that continental runoff is negligible. Horizontal gradients of hydrographic variables arise solely as a result of differences in the local response to freshwater gains and losses and sensible and latent heat fluxes. While rainfall is spatially variable, especially in summer months, it is unlikely that spatial gradients in precipitation and evaporation are significant when averaged over time periods greater than a few weeks. The region experiences a pronounced net freshwater loss over the course of a year. Schmitt et al. [1989] reported a 175 cm annual excess of evaporation over precipitation (E-P). Seasonal variations in E-P include a winter maximum that is roughly four times the net loss characteristic of midsummer months. On average, Great Bahama Bank generates hyperpycnal water throughout the year, but especially in winter months, when E-P is greatest and when effects of lower water temperature reinforce effects of higher salinity.

Buoyancy Effects on Coastal and Estuarine Dynamics
Coastal and Estuarine Studies Volume 53, Pages 107-116
Copyright 1996 by the American Geophysical Union

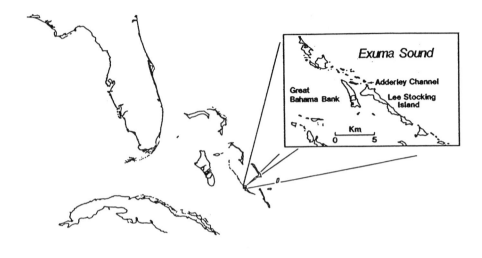

Figure 1. Map of the study area, showing Adderley Channel in the Exuma Cays along the western fringe of Exuma Sound. The two study sites are at the mouth of Adderley Channel and 0.2 km seaward, at the shelf break.

A series of field studies conducted from mid 1990 to mid 1992 [Pitts and Smith, 1993, 1994] investigated tidal exchanges in a region where deep-water and shallow-water environments are connected directly by tidal exchanges. The study area was between Exuma Sound and Great Bahama Bank (Figure 1) at the southern end of the Exuma Cays. Within one tidal excursion seaward of the Exuma Cays, water depths in Exuma Sound are on the order of 500 m; Great Bahama Bank extends for many tens of kilometers to the west with depths on the order of 3.5 m. The primary purpose of the field studies was to quantify long-term net transport through tidal channels serving as the connecting link between the bank and the sound.

Hydrographic data from Adderley Channel [Smith, 1995] exhibited great variability over complete tidal cycles. Relatively warm water left the bank on the ebb during summer months; cooler water was exported during winter months. High salinity water ebbed off the bank through much of the year, especially in late winter and early spring. Salinities as high as 40‰ were recorded at the end of the ebb cycle. High salinity water was virtually absent in the channel in late summer and early fall, when shoreward directed wind stress increased.

The purpose of this paper is to extend the earlier work by focusing on density and exploring the effect hyperpycnal water has on tidal exchanges at the mouth of the channel. It was hypothesized that hyperpycnal water recorded in the channel would cascade down the narrow shelf and over the wall, and thus not re-enter the tidal channel during the following flood cycle. To investigate density currents near the shelf edge, current and hydrographic data were recorded 0.5 m above the bottom, approximately 50 m from the top of a vertical wall that constitutes the shelf break. High-density water passing this study site is almost certainly lost to intermediate depths in Exuma Sound. Results confirm that hyperpycnal water descends the narrow shelf as a density current, increasing the effectiveness of tidal exchanges

Data

Data were obtained from two study sites. The first was near the mouth of Adderley Channel. The current meter was moored 2 m above the bottom in a water column that averaged 7 m deep. The study site was maintained from July 1, 1990 to June 30, 1991. Tidal conditions in the channel are summarized in Table 1 for the six principal constituents. M_2 tidal amplitudes at this location are 0.32 m and 0.67 m s^{-1}. The substantial phase lead of highest water level over strongest ebb speed indicates a predominantly standing wave pattern in the channel. The second study site was seaward of the mouth of the channel and about 50 m from the shelf break. The width of the shelf is on the order of 0.5 km. At the shelf study site, water depth was 25 m. The current meter was moored 0.5 m above the bottom, and the study site was maintained for an 84-day period of time from May 31 to August 22, 1992.

Table 1. Harmonic constants (amplitudes, η, in decibars or m s^{-1}, and local phase angles, κ, in degrees) of the principal tidal constituents calculated from current meter data and bottom pressures recorded in Adderley Channel. Positive current speeds represent ebb tide conditions.

	Tidal Constituent					
	M_2	S_2	N_2	K_1	O_1	P_1
A. Pressure						
1. η	0.32	0.05	0.08	0.09	0.07	0.03
2. κ	224	257	198	150	131	150
B. Current Speed						
1. η	0.67	0.11	0.15	0.06	0.05	0.02
2. κ	335	021	316	230	236	230

The current meter data and hydrographic data used in this study were provided by a General Oceanics Mark II recording inclinometer equipped with temperature and conductivity sensors. Temperature and conductivity were measured to accuracies of ±0.25°C and ±0.025 MS, respectively, according to instrument specifications. Conductivity, corrected for temperature, was converted to salinity [Perkin and Lewis, 1978], and density was calculated using the approach described by Millero and Poisson [1981]. Salinity was checked using measurements made at the study sites and in shelf waters unaffected by tidal exchanges. Salinity profiles were obtained using a Sea Bird Electronics Sea Cat Profiler.

Current speeds and directions from both study sites were decomposed into Cartesian coordinates; the along-channel component was used from the tidal channel station, and the across-shelf component was used from the shelf-break site.

Methods

Vertical current profiles made across the channel with a flow meter were combined with mid-channel current meter measurements to estimate volume transport for Adderley Channel [Smith, 1994]. Hydrographic data from the channel at the study site indicated that intense tidal mixing in the channel and wave mixing in Exuma Sound results in vertically and laterally homogeneous conditions at any given time. Using volume transport, in m^{3}s^{-1}, and salinity, in kg m^{-3}, salt transport can be expressed in kg s^{-1}. Perturbation analysis can then be used to quantify the relative importance of the mean and tide-dominated time-varying components of the total transport through Adderley Channel. Decomposing the hourly volume transport, v_i, and salinity, s_i, into the annual mean and the deviation from the mean, the instantaneous salt transport is expanded into four terms:

$$s_i v_i = S V + s_i' V + S v_i' + s_i' v_i' \qquad (1)$$

where S and V are the annual mean salinity and along-channel volume transport, and s' and v' are the deviations from the annual mean in the ith hour.

For determining the relative importance of mean and the time-varying transport, the annual mean of the s'v' term is usually calculated to make it comparable with the SV term. The right hand side of Equation (1) is immediately reduced to two terms, because the annual means of s' and v' are zero. Thus,

$$<s_i v_i> = S V + <s' v'>,$$ (2)

where the angle brackets represent the time average. Taking the average in this way, however, ignores useful information related to the temporal variability of the s'v' perturbation product. An alternate approach is to accumulate the s'v' terms and plot the cumulative net value as a function of time. For the mth hour of the study, the cumulative salt transport is given by

$$T_m = \sum_{i=1}^{m} s'_i v'_i$$ (3)

The slope of the plot equals the $< s' v' >$ term in equation (2), but the deviations from the straight-line slope reveal seasonal variability, as well as fluctuations over shorter time scales.

Salinity recorded in a tidal channel can also be used as a tracer to estimate percentages of bank and shelf waters in a water sample--if the two end points of the mixture are known. If the salinities of unmixed bank and shelf water are S_b and S_s, respectively, and the salinity in the tidal channel is S_c, then the fraction of the tidal channel sample that originated in Exuma Sound, F_s, is given by

$$F_s = \frac{S_b - S_c}{S_b - S_s}.$$ (4)

Values of F_s will range from 0.0 to 1.0. Calculations assume that salinity recorded in the tidal channel is representative of the entire channel. In situ hydrographic data support this assumption, although temporary stratification has been reported [Wilson, 1991] immediately following slack water at the end of the ebb tide cycle. Relatively low density sound water initially floods in as a surface layer before the water column is homogenized by energetic tidal mixing.

Homogeneous Exuma Sound water occurs within one tidal excursion of the mouth of Adderley Channel as a result of a long-term net inflow through Adderley Channel [Smith, 1995]. Hydrographic gradients are common over Great Bahama Bank, however, and this makes it difficult to establish the value for S_b needed to evaluate equation (4). Nevertheless, at times relatively constant salinity water leaves the bank during much of the latter part of the ebb tide cycle. With values for both S_b and S_s, one can use hourly measurements of S_c to evaluate F_s. Using the current meter data to estimate the total volume of water entering the channel during the mth hour of the flood tide, one can quantify the volume of Exuma Sound water as a fraction of the total. Summing over the entire flood tide cycle quantifies the efficiency of the exchange over the flood half of the tidal cycle, E_f:

$$E_f = \sum_{m=1}^{n} \frac{F_s V_m}{V_m},$$ (5)

where V_m is the total volume of water entering the channel during the mth hour of the flood tide.

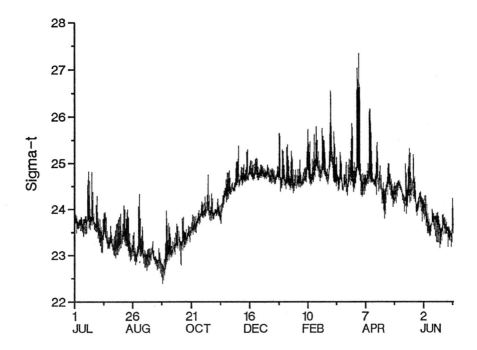

Figure 2. Sigma-t values calculated from conductivity and temperature measurements made at the mouth of Adderley Channel, July 1, 1990 through June 30, 1991.

Results

Figure 2 shows a time series of density in the form of sigma-t from the Adderly Channel study site obtained during the first study period. Superimposed onto the annual cycle and onto low-frequency variations over time scales on the order of 1-2 weeks are fluctuations associated with tidal exchanges. Tidal period deviations from the seasonal and low-frequency non-tidal variations occur in the form of transient "spikes" that increase sigma-t values by as much as two and a half units (2.5 kg m^{-3}). Density spikes are most prominent in late winter months and, to a lesser extent, in mid summer.

The relationship between density and along-channel flow in Adderly Channel is demonstrated in Figure 3, which contains time series of density and along-channel volume transport for a one-week period in early February, 1991. Ebb currents are defined to be positive. Careful comparison of the two curves shows that highest densities at the mouth of the channel coincide with slack water after the ebb. Density decreases quickly with the turn of the tide, and for much of the flood half of the tidal cycle density is low and relatively uniform. The asymmetry of the density plot over tidal cycles suggests that water entering on the flood tide is not the same water that left on the preceding ebb.

Current and hydrographic data from the shelf-break station provide information on the seaward movement of hyperpycnal water leaving Adderley Channel (Figure 4). A plot of sigma-t from the second field study shows decreasing values during the first month of the study. This represents the last of the late spring and early summer warming. Density spikes are less

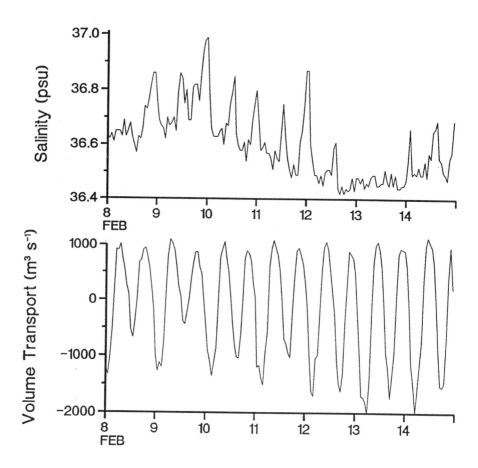

Figure 3. Composite of along-channel volume transport ($m^3 s^{-1}$) and salinity (psu) recorded at the mouth of Adderley Channel, February 8-14, 1991.

prominent, indicating that some vertical mixing occurs as hyperpycnal water cascades down the narrow shelf on the ebb tide. Nevertheless, transient spikes are evident, especially early and late in the study.

Figure 5 shows the across-shelf current component recorded at the shelf-break site during the same time period. Seaward flow is favored over landward flow as a result of density currents coming out of Adderley Channel. Weak flow into the channel occurs during the flood tide cycle, but for this time period the mean seaward and landward current speeds are +5.1 and -2.4 cm s^{-1}, respectively. Current meter data from Adderley Channel are not available for this time period, but the mean current speed during June, July and August of the 1990-1991 study was an inflow of about -6 cm s^{-1}. Thus it appears that the export of hyperpycnal water by density currents changes a flood-dominant condition in the channel into a locally ebb-dominant condition in near-bottom layers across the narrow shelf. Onshore winds characteristic of the study area may have contributed by encouraging a downwelling condition, but this would not occur at tidal periodicities.

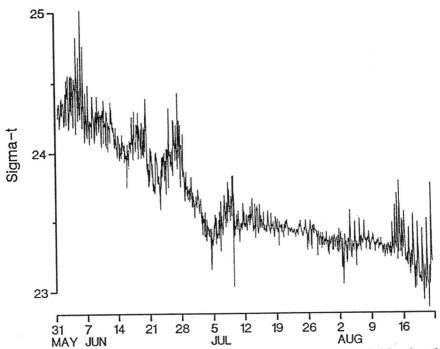

Figure 4. Sigma-t values calculated from conductivity and temperature measurements made 0.5 m above the bottom at the shelf-break study site, May 31 through August 22, 1992.

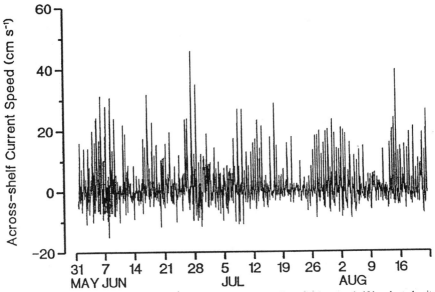

Figure 5. Across-shelf current speeds (cm s⁻¹) recorded 0.5 m above the at the bottom shelf-break study site, May 31 through August 22, 1992. Positive current speeds indicate seaward flow.

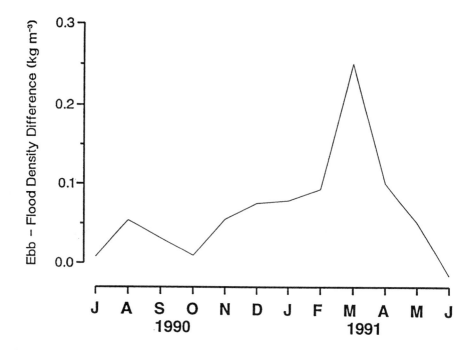

Figure 6. Monthly mean values of the difference between ebb tide densities and flood tide densities (in kg m^{-3}). Study months 1 through 12 refer to July 1990 through June 1990.

The annual variation in the export of hyperpycnal water is shown in Figure 6, which contains the monthly mean values of the differences between ebb and flood densities. On average, ebb densities are between 0.05 and 0.1 kg m^3 higher than flood densities. Greatest differences occur from February through April, when latent heat losses are greatest [Schmitt et al., 1989], when shallow bank waters are cooler, and when rainfall is at an annual minimum.

Salinity can be used as a tracer to quantify the percentage of shelf water in the channel. Combining this fraction with the hourly volume transport gives the transport of shelf water onto the bank. Equations (4) and (5) were used to quantify F_s for the two flood tide cycles on March 31, and to distinguish between Exuma Sound water entering the channel and bank water returning on the flood. Results suggest that 95% and 97.5% of the flood tide volume was Exuma Sound water not related to the previous ebb.

The role that density currents play in exporting salt from Great Bahama Bank on the ebb tide can be visualized by summing the perturbation products defined by equation (2). Results are shown in Figure 7. From the start of the record through late August, the s'v' values are predominantly positive, indicating a net export of salt, and the plot of cumulative values slopes upward. From early September through late December, the plot is essentially flat, indicating little net salt transport through the channel by tidal exchanges. During this same time period, sigma-t spikes are reduced relative to those recorded earlier and especially later in the study (Figure 2). In late February, sigma-t spikes increase in magnitude, the accumulation of exported salt increases correspondingly, and the slope of the curve in Figure 7 steepens. The similarity of the gross features of Figures 2 and 7 supports the logical assumption that hyperpycnal water leaving the bank on the ebb tide is lost to Exuma Sound via density currents, thereby enhancing the tidal exchange of water between the bank and the sound.

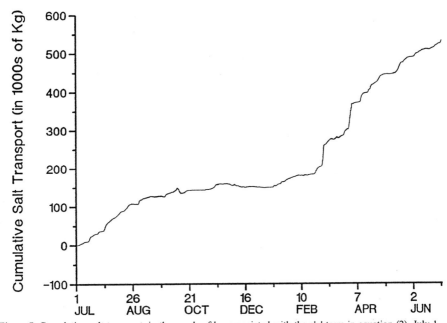

Figure 7. Cumulative salt transport, in thousands of kg, associated with the s'v' term in equation (2), July 1, 1990 through June 30, 1991. Positive accumulations indicate an export of salt from Adderley Channel.

Discussion

It is likely that there is a distinct seasonality in the existence and magnitude of density currents, and thus in the efficiency of tidal exchanges. The earlier Adderley Channel transport study [Smith, 1995] found that the tide-induced seaward mass transport was influenced partly by the differential response to seasonal heating in bank and sound waters, and partly by seasonally varying winds in the study area. During fall and early winter months, strong shoreward winds force Exuma Sound water onto the bank well beyond one tidal excursion. At those times, tidal exchanges move sound water back and forth through the channel, and hyperpycnal bank water is not exported to Exuma Sound. At times of relatively weak shoreward winds, nontidal inflow decreases. Longitudinal mixing by tidal currents brings hypersaline conditions to within one tidal excursion of the mouth of the channel, and density currents occur at the end of the ebb tide. Wind effects are alternately reduced and enhanced over seasonal time scales by sensible heat gains and losses that warm and cool bank water relative to sound water.

Data from near the shelf break show that density spikes are reduced somewhat due to mixing on the ebb tide. Nevertheless, density currents are sufficient to effect a long-term net transport into Exuma Sound. This occurs directly seaward of a channel that has an annual net inflow [Smith, 1995]. It follows that the net inflow must be fed from near-surface and intermediate layers over the inner shelf. Applications of these shelf transport patterns will not be pursued here, but the erosive effects of a seaward near-bottom volume transport should be significant as a mechanism for sediment transport. Similarly, the landward transport in near-surface layers, as well as the near-bottom seaward transport may be significant within the context of larval transport.

Data presented in this study focus on the effect that density currents have on the efficiency of tidal exchanges over the shelf. As noted above, however, this is the end product of a more complex series of cause-and-effect relationships. The presence or absence of density currents in shelf waters is a direct consequence of the availability of hyperpycnal water over the bank. That, in turn, is directly related to wind forcing in Exuma Sound. When strong, shoreward wind stress floods sound water onto the bank, hyperpycnal water retreats, density currents are temporarily absent in shelf waters, and tidal exchanges are temporarily less efficient.

A second effect of a quasi-steady export of hyperpycnal water should be noted in passing, although it lies outside the intended scope of this paper. The nearshore circulation along the western boundary of Exuma Sound is an unusually persistent south-to-north flow. Pitts and Smith [1994] have presented results that included mid-depth current measurements over a 402-day period of time from late October 1991 through late November 1992. During that time, the resultant current speed was just under 6 cm s^{-1}, and reversals in direction were infrequent and brief. The current meter record has not been decomposed into thermohaline and wind-driven components, but it is unlikely that wind forcing by itself would result in such a persistent south-to-north transport. Rather, it is postulated that the quasi-steady export of hyperpycnal water through Adderley Channel and other tidal channels along the Exuma Cays results in a band of high-density water along the western side of Exuma Sound. This would maintain a landward-directed baroclinic pressure gradient and force northward flow. The spatial and temporal characteristics of such a feature of the hydrography of Exuma Sound, as well as the impact it might have on the general circulation of Exuma Sound, are subjects for follow-up studies.

Acknowledgments. Support for this study was provided by the NOAA National Undersea Research Program through the Caribbean Marine Research Center on Lee Stocking Island, Bahamas. Special thanks to Patrick Pitts, who played a central role in the field study.

References

Millero, F.J. and A. Poisson, International one-atmosphere equation of state of seawater, *Deep-Sea Research, 28A*, 625-629, 1981.

Perkin, R.G. and E.L. Lewis, The practical salinity scale 1978: fitting the data, *IEEE Journ. of Oceanic Engineering, OE-5*, 9-16, 1980.

Pitts, P.A. and N.P. Smith, Annotated summary of temperature and salinity data from the vicinity of Lee Stocking Island, Exuma Cays, Bahamas, Caribbean Marine Research Center, Tech. Rept. Series No. 93-3, Caribbean Marine Research Center, Vero Beach, FL, 43 pp., 1993.

Pitts, P.A. and N.P. Smith, Annotated summary of current meter data from the vicinity of Lee Stocking Island, Exuma Cays, Bahamas, Caribbean Marine Research Center, Techical Report Series No. 94-4, Caribbean Marine Research Center, Vero Beach, FL, 29 pp., 1994.

Schmitt, R.W., P.S. Bogden, and C.E. Dorman, Evaporation minus precipitation and density fluxes for the North Atlantic, *Journ. of Phys. Oceanogr., 19*, 1208-1221, 1989.

Smith, N.P., Long-term Gulf-to-Atlantic transport through tidal channels in the Florida Keys, *Bull. Mar. Sci. 54*, 602-609, 1994.

Smith, N.P., Observations of steady and seasonal salt, heat and mass transport through a tidal channel, *J. Geophys. Res. ,100*, 13713-13718, 1995.

Wilson, P.A., Density cascading: implications to carbonate bank and periplatform sedimentation (Northern Bahamas and Southern Florida, Masters Thesis, Dept. of Geology and Geophysics, Louisiana State University, Baton Rouge, LA, 213 pp., 1987.

10

The Effect of Tidal Motion on the Salinity Distribution in Lower Delaware Bay

Kuo-Chuin Wong

Abstract

Delaware Bay is a partially to weakly stratified estuary on the east coast of the United States. A set of intensive hydrographic observations are used to examine the lateral salinity structure across the bay mouth. Results indicate the presence of significant lateral variability for the tidally averaged distribution. The evidence suggests the existence of two branches of buoyant estuarine outflow along the shores separated by more saline water in the deep channel doming upward toward the surface in the middle of the mouth. The tidal motion can produce significant intratidal variability in the lateral structure. It appears that the salinity distribution on the adjacent continental shelf and the presence of the Delaware coastal current offshore from Cape Henlopen are important factors in the determination of the intratidal variability at the bay mouth. During the ebb tide both the lateral and vertical salinity gradients are strengthened through the combined effects of differential tidal advection, tidal straining, and tidal stirring. In contrast, the salinity structure becomes very diffused and the lateral and vertical gradients are minimized at the end of the flood tide. During the time when the lateral gradients are intensified, the low salinity water along the shores and the more saline water in the middle of the bay may be separated by sharp frontal boundaries with up to 5 psu variation in salinity over a distance of 150 m. Both the sharpness of the frontal boundaries and the positions of the fronts can change significantly with the phase of the tide.

Introduction

The determination of the factors controlling the salinity distribution in estuaries has been a problem of long standing. It is generally known that the salinity distribution can be influenced by a variety of processes over a broad spectrum of time scales. These include tidal advection and diffusion [Kjerfve and Knoppers, 1991; Valle-Levinson and Wilson, 1994 a,b], atmospherically induced subtidal variability [Kjerfve, 1986; Smith, 1988], and long-term variations in the river discharge [Schroeder, 1978; Schroeder and Wiseman, 1986].

The present study focuses on the effect of tidal motion on the intratidal salinity variability in a coastal plain estuary. The recent studies of Simpson et al. [1990] and Sharples et al. [1994] showed that estuarine stratification within a tidal period may be strongly influenced by the flood-ebb cycle of the tidal current. Furthermore, the work of Huzzey [1988] and Huzzey and Brubaker [1988] in the York River estuary indicate that the lateral salinity structure across the estuary may also undergo substantial variation over a tidal cycle. In light of these developments, the present study uses a set of intensive CTD observations to examine the characteristics of the intratidal variabilities in the degree of stratification and the lateral salinity gradients across the mouth of Delaware Bay. Furthermore, high resolution thermosalinograph data are used to examine the scale of lateral variability and detect the presence of tidal fronts.

Buoyancy Effects on Coastal and Estuarine Dynamics
Coastal and Estuarine Studies Volume 53, Pages 117-137
Copyright 1996 by the American Geophysical Union

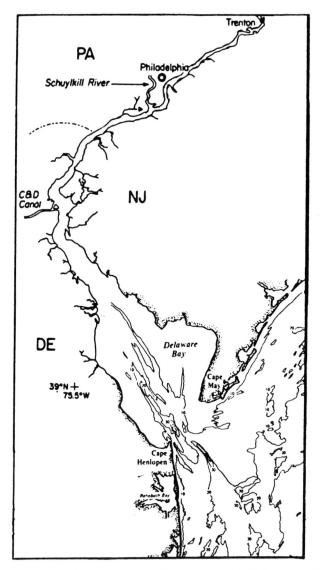

Figure 1. The location map of Delaware Bay and the adjacent continental shelf. Bathymetry is in meters.

Background and Data Sources

Delaware Bay (Figure 1) is a major coastal plain estuary located on the east coast of the United States. It communicates with the Atlantic Ocean through a mouth 18 km in width between Cape May, New Jersey and Cape Henlopen, Delaware. The Delaware River, gauged at the head of the estuary at Trenton, New Jersey, contributes approximately 60% of the total fresh water discharge.   The Schuylkill River, entering through Philadelphia, Pennsylvania, contributes another 15%. No other single source is responsible for more than 1% of the total discharge [Sharp, 1983]. More than 95% of the fresh water enters the bay in its narrow reaches north of the Chesapeake and Delaware (C&D) Canal. The mean depth of the bay is about 8 m, with very complex bathymetry in the lower bay.   At the bay mouth the most prominent bathymetric feature is the deep channel off Cape Henlopen. The isobaths around the deep channel are mostly in the along-bay direction. There are extensive shoals around Cape May (Figure 1). The orientation of the isobaths around the shoals has a significant across-bay component.

Tides in Delaware Bay are dominated by the semi-diurnal M_2 motion. Figure 2 shows the near-surface and near-bottom M_2 tidal ellipses at the mouth. The M_2 currents are essentially rectilinear. The orientation of the tidal ellipses indicates that the M_2 currents are aligned with the direction normal to the bay mouth transect between Cape Henlopen and Cape May. This suggests the presence of across-isobath flow over the shoals around Cape May while the tidal currents near Cape Henlopen essential flow along the isobath. The amplitude of the near-surface M_2 current undergoes substantial variation across the mouth, decreasing from about 70 cm/s near Cape Henlopen to 50 cm/s near Cape May. The amplitude of the near-bottom M_2 current shows a similar reduction across the bay mouth. In addition to the variation in amplitude, the phase of the M_2 current also exhibits spatial variation across the bay, with the current off Cape May leading that off Cape Henlopen by about 20°.

Between June 12-13, 1990, the near-surface (0.5 m) salinity distribution in the lower part of Delaware Bay and the adjacent continental shelf was mapped by a thermosalinograph as the R/V *Cape Henlopen* followed a saw-toothed shiptrack in the study area. Furthermore, a RD Instrument ADCP (307 KHz) was used to measure the currents around the bay mouth. On June 14, 1990, intensive hydrographic surveys were conducted across the mouth. CTD profiles, at roughly 1.8 km intervals, were repeatedly taken as the vessel traversed across the bay mouth back and forth at about 75-minute intervals for a total of 11 transects over a 12.5 hour period. The CTD transects were maintained at such sampling intervals over a semidiurnal tidal cycle so that both the tidally averaged distribution and the intratidal variability could be resolved. In addition, the near-surface salinity and temperature distributions across the bay mouth over the tidal cycle were measured by a thermosalinograph with a spatial resolution of 150 m. The weather was very calm throughout the survey so atmospheric forcing did not play an important role in modifying the observed hydrographic distributions.

Results

Based on CTD profiles taken repeatedly at 10 stations across the bay mouth over a tidal cycle, Figure 3 shows the tidally averaged salinity distribution across the mouth as well as the standard deviations around the tidal averages. The open triangles in this figure mark the positions of the CTD stations. The view is upbay, with the Delaware shore to the left and the New Jersey shore to the right. At depths above 15 m the isohalines dome up toward the surface in the central part of the bay mouth. This suggests that the salinity in the central region is higher than that along either the Delaware shore or the New Jersey shore. The bay mouth can thus be roughly divided into three regions at depths shallower than 15 m and a fourth region in the deep channel based on hydrographic properties shown in Figure 3. Figure 4 shows a schematic diagram of the location of the four regions across the mouth. In region I, a branch of low salinity water exists along the Delaware shore from Cape Henlopen to a point about 8 km offshore. Relatively high salinity water occupies region II in the central portion of the bay mouth between 8 to 12 km offshore from Cape Henlopen. Another branch of low salinity water exists in region III near Cape May. The deep channel at depth greater than 15 m represents region IV, and the salinity in this region shows near uniform distribution. As a whole, the tidally averaged distribution indicates that high salinity water resides in the deep channel and domes up toward the surface in the central portion of the bay mouth to the right of

M2 Tidal Current Ellipses   Sfc

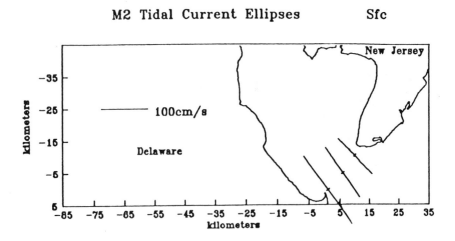

M2 Tidal Current Ellipses   Bot

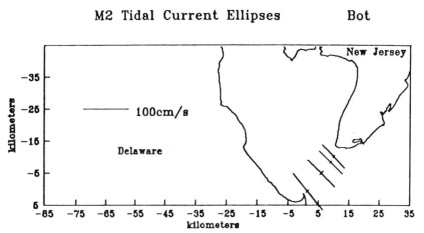

Figure 2. The M_2 tidal current ellipses at the mouth of Delaware Bay. The upper panel shows the near surface currents and the lower panel shows the near bottom currents. (from Wong and Moses-Hall, 1995).

the channel as one looks upbay. This high salinity water separates two branches of water with lower salinity, one along the Delaware shore and the other along the New Jersey shore.

The observed tidally averaged salinity distribution does not fit the conventional idea of a two-layer estuarine gravitational circulation with a low salinity outflow in the surface layer and a high salinity inflow in the lower layer. With the influence of the Coriolis effect, the conventional idea allows for the tilting of the interface between the inflow and the outflow and the deflection of the brackish outflow against the left shore as one looks upbay [Pritchard, 1989]. This would produce a lateral structure, as the low salinity outflow is more concentrated on the left shore and the high salinity inflow is more concentrated on the right shore as one looks upbay. The doming of the isohalines to the right of the deep channel

Salinity Mean

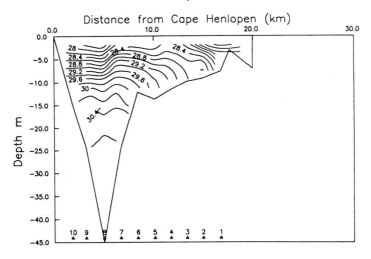

Salinity Standard Deviation

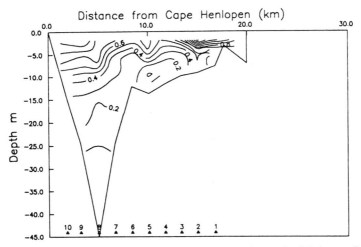

Figure 3. The tidally averaged (mean) salinity distribution across the mouth of Delaware Bay and the standard deviations around the mean values. Salinity is in psu.

(Figure 3) is perhaps an indication of the deflection of the high salinity inflow by the Coriolis effect. However, the Coriolis effect alone cannot explain the presence of two separate branches of low salinity water, particularly the branch along the right shore as one looks upbay. The fact that almost all of the river discharge enters the system in the upper bay also precludes the possibility that the separate branches of low salinity water might be produced by significant amount of fresh water input along the shores of the lower bay.

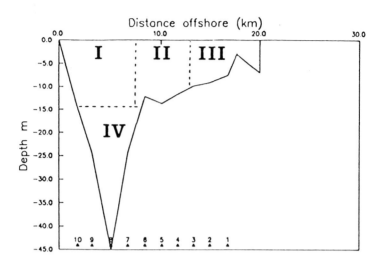

Figure 4. A schematic diagram indicating the four regions across the bay mouth. Buoyant estuarine discharges are found in regions I and III along the shores.

Fischer [1972, 1976] was the first to suggest that the considerable lateral variation in bathymetry typically found in drowned river valley estuaries may produce significant lateral structure in the gravitational circulation. Inflow is expected to be concentrated in the deeper portions of the cross-section, as the upstream baroclinic pressure gradient increases with depth below the water surface. In contrast, outflow is expected to be concentrated in the shallower regions along the shores. Recently, Wong [1994] developed an analytical model of estuarine gravitational circulation and found that the solutions are very sensitive to the lateral bathymetry across the system. For an estuary with a rectangular cross-section, the gravitational circulation exhibits a two-layer structure with surface outflow and bottom inflow. When the estuary carries a triangular cross-section with greater depth in the middle and shallower areas along the shores, the gravitational circulation exhibits considerable lateral structure. Two separate branches of outflow are now found in the shallow areas along the shores and inflow is mainly focused in the deeper area over the channel. Furthermore, a flow reversal with depth occurs only over a limited part of the cross-section. This suggests that the residual circulation may be dominated by a lateral structure rather than a vertical structure. Since inflow is normally associated with high salinity water and outflow with low salinity water, the lateral variation in bathymetry across the bay may be a plausible cause for the presence of two separate branches of low salinity water along the shores.

The lower panel in Figure 3 gives the standard deviations around the tidally averaged salinity distribution across the mouth. The standard deviations provide a measure of the strength of the intratidal variability over a tidal cycle. It can be seen that high intratidal variabilities are found in region I off Cape Henlopen and region III off Cape May. It appears that the regions with relatively low tidally averaged salinity exhibit the strongest intratidal variation in salinity over a tidal cycle. In contrast, intratidal variability is weakest in region IV below 15 m in the deep channel. This is the region where the residual inflow of the adjacent continental shelf water is expected to occur.

At this point, it is instructive to examine the intratidal variability in each of the four regions more closely. For the sake of brevity, the intratidal salinity variation at one representative station in each region will be examined. The contour diagram in the upper panel of Figure 5 shows the temporal variations in salinity at station 2 located in region III off Cape May. The ordinate of Figure 5 gives depth in meters while the abscissa gives time in hours.

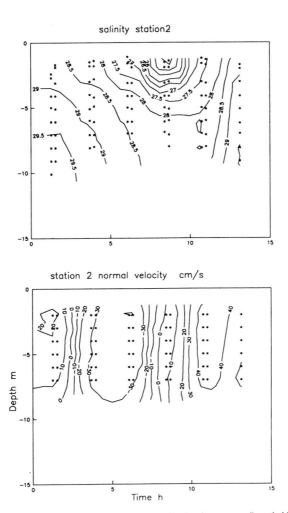

Figure 5. The temporal variations of the vertical salinity distribution (upper panel) and tidal currents normal to the bay mouth transect at station 2 (lower panel). Station 2 is located in region III along the New Jersey shore. Salinity is in psu.

The asterisks in this figure indicate the points where data are available. This station is located in the region where the New Jersey branch of the low salinity water resides. The near-surface salinity varies considerably over the tidal cycle. The highest surface salinity (28.5 psu) is reached 2 hours into the survey while the lowest surface salinity (24.5 psu) is reached 6 hours later. The vertical structure also exhibits large variation over time. The water column is well mixed over much of the tidal cycle, with surface to bottom salinity difference typically below 1 psu. However, during a period of about 3 hours centered around hour 9, the water becomes more stratified. At hour 9 the surface (24.5 psu) to bottom (28.5 psu) salinity difference is 4 times larger than that at hour 3.

It is reasonable to assume that the intratidal variation in salinity should be linked to the temporal variation in the tidal current over the tidal cycle. Based on the harmonic constants

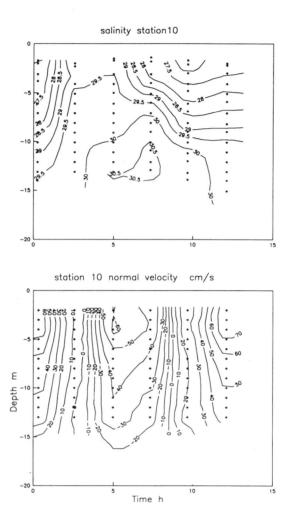

Figure 6. The temporal variations of the vertical salinity distribution (upper panel) and tidal currents normal to the bay mouth transect at station 10 (lower panel). Station 10 is located in region I along the Delaware shore. Salinity is in psu.

derived by Münchow et al. [1992] and Moses-Hall [1992], the tidal currents at the CTD stations are simulated for the time period when the hydrographic surveys were conducted. The simulated tidal currents consist of the dominant M_2 constituent as well as four other weaker constituents (N_2, S_2, K_1, and O1). The lower panel of Figure 5 shows the temporal variation of the predicted tidal current component normal to the bay mouth transect at station 2. Positive value indicates flooding current into the bay and negative (-) value indicates current ebbing out of the bay. A comparison of the salinity and current variations indicates that the salinity reaches its maximum value at about 2 hours into the survey. This time coincides with the time of slack water (zero velocity) after flood. From that point in time the tidal current begins to ebb and the salinity decreases steadily until a minimum value is reached at the end of the ebb cycle

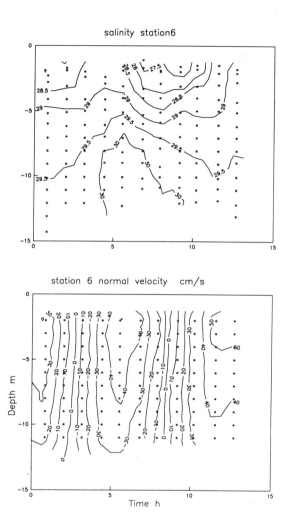

Figure 7. Same as in Figure 6, except for station 6 located in the central portion of the bay mouth (region II).

(slack water after ebb). The tidal current then begins to flood again, resulting in an increase in salinity. The pattern of the overall variation in salinity is thus consistent with the advection of an ambient salinity gradient by the oscillating tidal current. In addition to the overall variation in salinity over time, the vertical salinity structure also changes significantly with the phase of the tide. The water column is virtually well mixed except for a 3 hour period centered around the time of slack water after ebb. This ebb-tide stratification is probably caused by tidal straining of the horizontal density gradient [Simpson et al., 1990; Sharples et al., 1994]. The amplitude of tidal current is larger at the surface than at depth. During the ebb tide this vertical current shear would produce a larger downstream tidal excursion at the surface of the water column than at the bottom. Given the typical longitudinal salinity gradient in an estuary with decreasing salinity as one moves upstream, this straining process

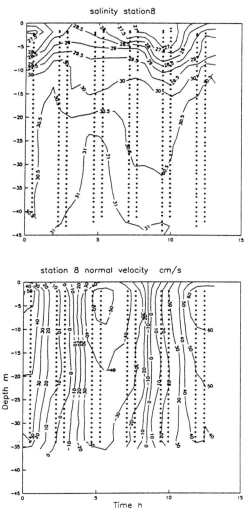

Figure 8. Same as in Figure 6, except for station 8 located over the deep channel.

results in lower salinity water at the surface than at the bottom. The stratification is further enhanced at the time of slack water after ebb, as tidal mixing is minimized at that time. With the beginning of the flood cycle, the ebb-induced stratification breaks down due to the reversal in the straining process and the increase in tidal mixing.

The upper panel of Figure 6 shows the vertical structure of salinity at station 10 over a tidal cycle. Station 10 is located in the Delaware branch of the low salinity water (region I). The lower panel shows the corresponding tidal current at this station. The beginning of the survey corresponds to 1 hour after maximum flood. The salinity increases from that point in time up to the time of slack water after flood. The tidal current then begins to ebb and the salinity decreases until a minimum value is reached at the time of slack water after ebb. Surprisingly, salinity remains relatively unchanged throughout the first half of the next flood cycle. Near the surface the salinity varies from about 27.5 psu at the beginning of the survey to 29.5 psu at slack water after flood. The surface salinity then decreases again to 27.0 psu at slack water after ebb. Near the bottom the intratidal variation is about 1 psu. In terms of the vertical structure, the water column is well mixed only during a 3 hour period centered around the time of slack water after flood. A surface to bottom salinity difference of 2.5 to 3 psu exists throughout the rest of the tidal cycle. It is interesting to note that the degree of stratification during part of the first flood cycle is about the same as that at the time of slack water after ebb.

Station 6 is located in region II. This region is situated in the central portion of the bay mouth where the tidally averaged salinity is higher than that along the shores (regions I and III). Figure 7 shows the temporal variations in salinity and current at this station. Generally speaking, the intratidal variability in this region carries qualitative features which are about halfway between those found in regions I and III. For example, similar to region III, maximum stratification occurs at the time of slack water after ebb. Furthermore, the salinity increases and the degree of stratification decreases during the first half of the second flood cycle. However, there is also evidence that similar to region I, the degree of stratification is higher during part of the first flood cycle than the early phase of the following ebb cycle. The overall intratidal variation in salinity in region II is smaller than that in either region I or III. Near the surface, for example, the salinity varies by only about 1.5 psu over the tidal cycle.

Figure 8 shows the temporal variations in salinity and current at station 8. This station is situated right over the deep channel. The top 15 m of this station belongs to region I, and the intratidal salinity variation there is very similar to that at station 10 described earlier. At depth greater than 15 m, the salinity in region IV undergoes very weak intratidal variation (of the order of 0.5 psu) despite the fact that their still exists vigorous tidal currents in the deeper part of the channel. The deep channel is apparently the main conduit for the high salinity shelf water to enter the bay. The weak intratidal variability there suggests that the intrusion of the relatively homogeneous shelf water can be found in the deeper part of the channel beyond one tidal excursion length from the bay mouth.

A comparison of Figures 5 and 6 indicates that the intratidal variation in salinity is greater in region III off Cape May (station 2) than region I off Cape Henlopen (station 10). In contrast, the tidal current in region I is substantially stronger than that in region III. This suggests that the horizontal salinity gradient near Cape Henlopen should be significantly smaller than that near Cape May. Figure 9 shows the near surface salinity distribution in the lower bay and the adjacent continental shelf. The contours are constructed based on thermosalinograph data taken along a saw-toothed shiptrack in this area. Since it took more than a day to complete the surface mapping, a certain amount of tidal distortion is inevitable. This is a common problem facing shipboard observations over an extensive area in the estuarine and coastal environment with tidal influence. However, the contours still provide a useful quasi-synoptic description of the surface salinity distribution in the study area. Figure 9 indicates that the two branches of low-salinity water along the shores extend well beyond the mouth into the interior of the lower bay. Near the surface the high salinity water is not situated directly over the deep channel. This is consistent with the CTD observations which show the isohalines dome up from the deeper part of the bay toward the surface to the right of the channel. This rightward leaning tendency is perhaps associated with the deflection of the high salinity inflow due to the Coriolis effect. Upon exiting the bay mouth, the Delaware branch of the low salinity water turns and forms a southward flowing coastal current along the Delaware-Maryland-Virginia coasts. The presence of the brackish Delaware Bay outflow can be detected some 100 km south of the bay mouth. The surface salinity distribution suggests that the width of the Delaware coastal current is of the order of 20 km. Based on ADCP measurements and geostrophic calculations, Wong and Münchow [1995] found the typical speed of the coastal current to be about 10-20 cm/s.

The hydrographic observations do not provide conclusive evidence for the fate of the New Jersey branch of the low salinity water after it leaves the bay. Based on the surface salinity distributions, this branch of the low salinity water appears to flow across the mouth, as the isohalines around Cape May are essentially parallel to the bay mouth transect.

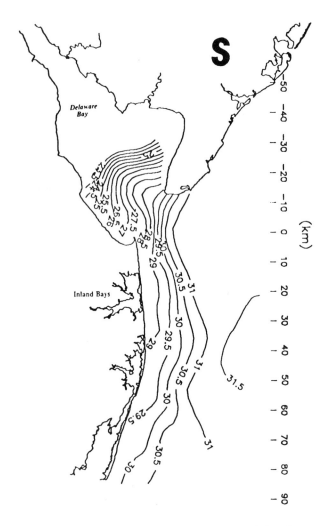

Figure 9. The surface salinity distribution in the lower Delaware Bay and the adjacent continental shelf. Salinity is in psu.

However, it is not clear at all what happens to this across-bay flow of brackish water. Part of it could be mixed with the incoming shelf water and part of it might make it across the bay mouth and joins the Delaware coastal current. There is some direct evidence which indicates a significant across-bay flow near Cape May. Figure 10 shows the residual circulation at the bay mouth as derived from the ADCP data taken during the mapping of surface salinity. The tidal variations in the currents are removed based on the method described in Wong and Münchow [1995]. Even though the 307 KHz ADCP nominally carries a vertical resolution of 2 m, the top 2 bins near the surface and the near bottom bins did not provide usable data. In the shallower part of the bay mouth the ADCP provides only 1 or 2 usable readings through the water column. The ADCP is therefore used to map out horizontal current distribution only, and the vectors in Figure 10 represent the residual currents at the bin centered at 5 m from

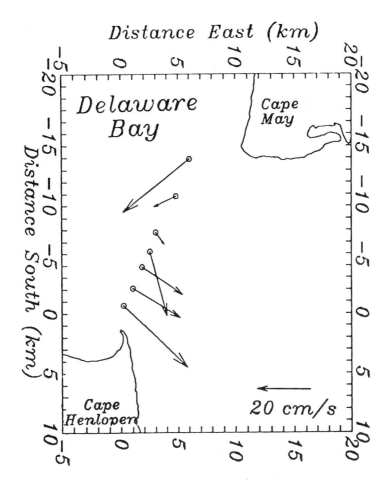

Figure 10. The distribution of ADCP-derived residual currents (at 5 m depth) across the bay mouth. Strong across-bay (but along isobath) flow is observed off Cape May. Strong along-isobath flow out of the bay is observed off Cape Henlopen.

the surface. The results show that there is a strong outflow off Cape Henlopen, consistent with the salinity distributions shown in Figure 9. Near Cape May, however, the residual current has a strong across-bay component. It appears that the residual currents essentially flow in the along-isobath direction. Near Cape Henlopen this lends to a flow out of the bay. Near Cape May, the presence of extensive shoals (Figure 1) steers the residual flow across the bay. This is in sharp contrast to the semidiurnal tidal currents which flow across the isobath in that area. It thus appears that whether the flow would be strongly steered by the local bathymetry depends on the time scales involved. For residual currents, the long time scale permits the flow to adjust to the local bathymetry.

The differences between the horizontal salinity gradients off Cape Henlopen and Cape May may thus be caused by the manner in which the Delaware and the New Jersey branch of the low salinity water exit the bay. Within a tidal excursion, the horizontal salinity gradient is relatively small at the bay mouth near Cape Henlopen since the brackish water of the

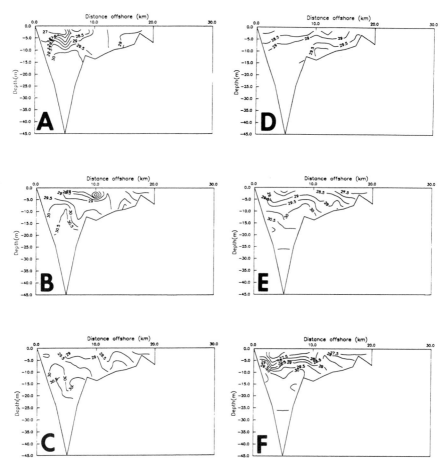

Figure 11a. The lateral salinity distributions across the bay mouth at 6 different stages of the tidal cycle (marked as transects A through F). Salinity is in psu.

Delaware coastal current can be found offshore. The horizontal salinity gradient is larger near Cape May since the saline ambient shelf water is directly impressed against the bay mouth transect there. The temporal variation in the vertical salinity structure in region III near Cape May can be largely explained by the combined effect of tidal straining and tidal stirring. However, the intratidal salinity variation in region I near Cape Henlopen cannot be easily explained by these effects. This is especially evident for the situation between the final stage of the ebb cycle and the early stage of the following flood cycle (Figure 6). Perhaps the interaction between the bay and the Delaware coastal current contributed to the formation of some of the observed features. Additional observations are required to resolve these issues.

The data suggest that there are large differences in the characteristics of the intratidal salinity variations among the four regions across the mouth of Delaware Bay. The lateral structure of salinity across the bay mouth may thus change significantly with the phase of the tide. At any instant in time the lateral structure may be quite different from the tidally averaged distribution shown in Figure 3. It is therefore instructive to examine the salinity distributions

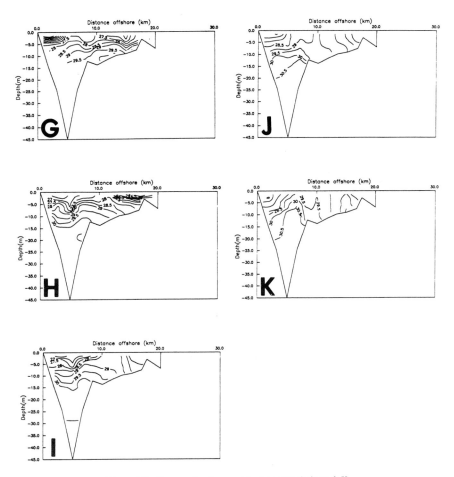

Figure 11b. Same as in a), except for transects G through K.

associated with the 11 separate CTD transects, hereafter marked as transects A through K (Figures 11a and 11b).

At the beginning of the 12.5 hour survey, the tidal current is entering the final phase of a flood cycle. Transect A indicates that there is a branch of low-salinity water which extends to a depth of about 10 m along the Delaware shore. This low salinity water is confined between Cape Henlopen and a point roughly 8 km offshore. There is a well defined gradient separating the low salinity water in this region and the more saline water in the deeper part of the channel as well as that in the middle of the bay. Near the New Jersey shore there is indication that the salinity there (29.0 psu) is slightly lower than that in the middle of the bay (29.5 psu). The most distinctive feature of this transect is the sharp differences between the situations near the Delaware and New Jersey shores. During this stage of the flood cycle, the presence of a pool of brackish water in the Delaware coastal current off Cape Henlopen probably accounted for the fact that the branch of low salinity water is still prominent at the mouth near the Delaware shore. On the other hand, the inflow of the saline shelf water with the flooding current near

Cape May tends to diminish the presence of the branch of low salinity water along the New Jersey shore.

As flooding continues the intensity of the branch of low salinity water along the Delaware shore decreases significantly, resulting in a reduction in the lateral variability across the mouth. Furthermore, there is indication in transect B that the core of the Delaware branch of the low salinity water appears to drift away from Cape Henlopen. This repositioning may be caused by the complex interaction between the flooding tidal current and the Delaware coastal current. At the end of the flood cycle (transect C) the salinity gradients across the bay mouth become more diffused and the lowest salinity water (28.0 psu) can be found along the New Jersey shore.

The onset of the following ebb cycle brings low salinity water from the interior of the bay toward the mouth. This is especially evident for the shallow areas along the New Jersey shore (transect E). The effect of the outflowing ebb current acting upon the ambient salinity structure in the lower bay tends to sharpen the difference in salinity between the water along the shores and the water in the middle of the bay mouth. As a result of this differential advection of salt, the lateral salinity gradient increases steadily until the end of the ebb cycle (transects G and H). Furthermore, the combined effect of tidal straining and reduced tidal mixing also maximizes the degree of stratification within the two branches of low salinity water at the time of slack water after ebb. Given the intensification of the lateral and vertical salinity gradients, the buoyant water along the shores can be most easily identified at the end of the ebb cycle. It is interesting to note that at this time (transect H) the salinity of the water near the New Jersey shore is lower than that near the Delaware shore.

As the tide turns again in the next flood cycle (transect I), the salinity along the New Jersey shore increases with the intrusion of the high salinity shelf water off the mouth of the bay. With the inflow of more saline water, tidal straining now works in reverse to rapidly break down the vertical structure. The situation near the Delaware shore is different. Despite the slight increase in salinity and a small decrease in the degree of stratification, the qualitative structure of the branch of low salinity water near Delaware remains more or less intact. This may in part be due to the presence of the Delaware coastal current off Cape Henlopen. The inflow associated with the flooding tide would bring water from the Delaware coastal current into the bay, thus maintaining the identity of the low salinity water at the bay mouth near the Delaware shore, at least during the early stage of the flood cycle. Yet another plausible explanation is that the Coriolis effect would favor the branch of buoyant outflow along the Delaware shore of the bay relative to the branch along the New Jersey shore. The stronger buoyant outflow along the Delaware shore would thus be more effective in mitigating the effect of the incoming tidal current during the early stage of the flood cycle. The observations at the beginning of the survey suggest that the buoyant outflow at the mouth along the Delaware shore would eventually be overwhelmed during the final stage of the flood cycle. At that time the lateral salinity gradient across the mouth is minimized.

The contour diagrams shown in Figure 11 are derived from CTD observations with a spatial resolution of 1.8 km. Any salinity feature with a horizontal scale smaller than that cannot be resolved. This approach is inadequate in providing information on small scale processes such as fronts. To remedy the situation, Figure 12 provides the near surface salinity and temperature distributions across the mouth. These data were obtained with a thermosalinograph as the vessel traversed back and forth across the mouth at a spatial resolution of 150 m.

Figure 12 shows that there are a lot of small scale features across the mouth of the bay. At any given phase of the tide, these features may involve salinity variations in excess of 1 psu over a distance of 150 m to 300 m. However, the intensity of the small scale features appears to be greatest at either the mid stage of a flood cycle or the end of an ebb cycle. Near the peak of the flood cycle (transect A) there is a sharp front separating the low salinity water along the Delaware shore from the high salinity water in the middle of the bay. Across this frontal boundary the salinity changes by more than 3 psu over a distance of only 150 m. The formation of this front can perhaps be attributed to the convergence of the high salinity inflow

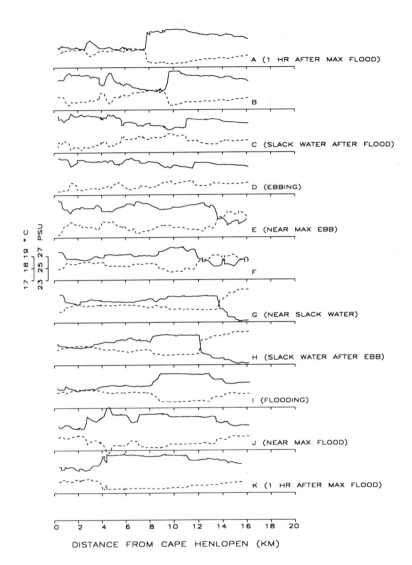

Figure 12. The surface salinity (solid line) and temperature (dashed line) distributions across the bay mouth taken at 11 different stages of the tidal cycle (marked as transects A through K).

and the buoyant estuarine discharge along the Delaware shore. The position of the front may change rapidly over time, depending on the details in the balance between the intruding saline water and the buoyant estuarine discharge. This is probably a tidal intrusion front similar to those found in the estuary of the river of Seiont at Caernarfon [Simpson and Nunes, 1981] and the Conway estuary [Nunes and Simpson, 1985; Simpson and Turrell, 1986].

At the end of the ebb cycle there is a very sharp front separating the low salinity water along the New Jersey shore and the saline water in the middle of the bay. At slack water after ebb the salinity varied by almost 5 psu over a distance of only 150 m. Recently, Huzzey

[1988] and Huzzey and Brubaker [1988] showed that differential tidal advection on top of an ambient salinity field with a constant longitudinal gradient could produce significant lateral variability in the York River estuary. Tidal currents in the York River estuary are faster over the deep channel than in the shallower areas along the shores. As a result, the stronger ebb current produces lower salinity in the area over the channel relative to the shallow areas along the shores. The lateral salinity gradient reverses during the flood tide, with higher salinity values in the middle of the estuary where the flood current is stronger. The situation at the mouth of Delaware Bay is more complicated, as the oscillating tidal currents can act on an ambient salinity field with a significant lateral structure both inside and outside of the bay. However, the differential advection of low salinity water clearly played a major role in the formation of the front off the New Jersey shore. From the interior of the lower bay, the ebb current draws on the pool of low salinity water along the New Jersey shore towards the bay mouth. This tends to sharpen the lateral salinity gradient between the brackish water near shore and the more saline water in the central portion of the mouth. Furthermore, the combined effect of tidal straining and reduced tidal stirring also maximizes the degree of stratification within the branch of low salinity water along the shore. In contrast, the water in the central portion of the bay experienced relatively weak intratidal variations in both the overall salinity value and the degree of stratification. These factors apparently contributed to the sharpness of the frontal boundary at the surface.

The temperature and salinity distributions in Figure 12 indicate that there is an inverse correlation between the two, with low salinity corresponding to high temperature, and vice versa. The low salinity estuarine outflow can trace its origin to the inland waterways further up the estuary. During the spring the inland water warms up faster than the water on the continental shelf. The inverse T-S correlation thus indicates the interaction of the warm, brackish estuarine water and the cool, saline shelf water at the bay mouth. To further examine this relationship, Figure 13 shows the T-S diagrams constructed from the thermosalinograph data collected from the individual transects (A through K) as well as a diagram showing a composite of all the data. The results show that the T-S relationship remains linear throughout the tidal cycle, with a 1°C decrease in temperature corresponding to a 2.2 psu increase in salinity. The slope of temperature versus salinity is apparently insensitive to the time of the day in which the data were collected. The data show that temperature can vary by 1.5°C to 2°C across the salinity fronts. This T-S correlation opens the possibility of using remote sensing techniques to detect the presence of these fronts in estuaries.

Conclusion

The present study shows that the tidally averaged salinity distribution across the mouth of Delaware Bay exhibits a lateral structure. Two branches of low salinity water are found in the shallow water along the shores separated by high salinity water doming up from the deep channel toward the surface in the central portion of the bay. This lateral structure does not fit the conventional idea of a two layer gravitational circulation with a low salinity surface outflow and a high salinity inflow in the lower layer. The presence of two separate branches of low salinity water also suggests that the deflection of the buoyant outflow via the Coriolis effect cannot provide an adequate explanation for the observed lateral structure. A plausible explanation lies in the interaction between the longitudinal density gradient and the lateral bathymetric profile which typically carries shallow areas along the shores and much deeper areas in the middle of the bay. The depth-dependent nature of the baroclinic pressure gradient would tend to focus the inflow over the deep channel and the outflow along the shores. The outflow along the Delaware shore of the bay has been confirmed by ADCP measurements (Figure 10). However, the observed currents near the New Jersey shore do not indicate the presence of significant current into or out of the bay mouth. Instead, the residual current off Cape May is largely in the across-bay direction. One may postulate that the residual outflow along the New Jersey shore may be steered by the local topography around the extensive

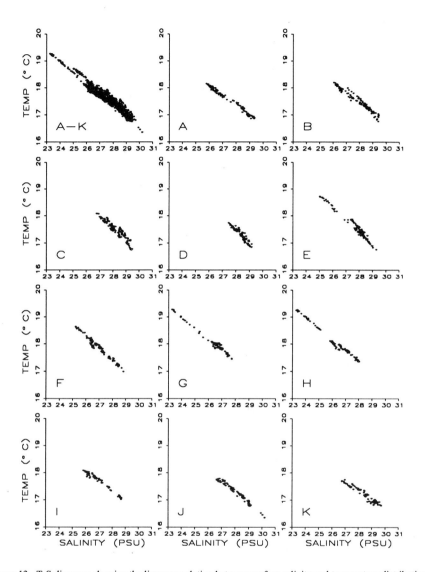

Figure 13. T-S diagrams showing the linear correlation between surface salinity and temperature distributions along the 11 separate transects (A through K). The T-S diagram in the upper left corner shows a composite of all the thermosalinograph data taken during the semidiurnal tidal cycle.

shoals as it exits the bay. The across-bay current off Cape May may simply reflect along-isobath flow. An earlier study conducted by Münchow and Garvine [1993] offshore of the bay mouth also indicates the presence of strong depth-averaged outflow in a narrow region off the Delaware shore (the Delaware coastal current) and significant across-bay current off the New Jersey shore. However, their study also shows a strong depth-averaged current into the bay right next to the Delaware coastal current. That may simply reflect the influence of the inflow from the shelf in the deep channel. Figure 10 does not show any significant inflow, as the currents only reflect near-surface (5 m) flow field. It is apparent that additional studies are required to resolve these issues.

The results further show that there are large intratidal variabilities in the salinity distributions. Generally speaking, the differential advection of salt tends to sharpen the lateral salinity gradients during the ebb tide. The combined effect of tidal straining and reduced tidal stirring further strengthens the vertical gradients toward the end of the ebb cycle. The situation during the flood tide is more complicated. During the early stage of the flood cycle the inflow of saline shelf water rapidly breaks down the vertical and lateral gradients between the low salinity water along the New Jersey shore and the surrounding water. However, the complex interaction between the flooding tidal current, the water in the Delaware coastal current just off shore of the bay mouth, and the buoyant estuarine discharge maintains sharp lateral and vertical gradients between the low salinity water along the Delaware shore and the surrounding water. Toward the end of the flood cycle the incoming tide finally draws sufficient saline water into the bay and the salinity structure across the bay mouth becomes diffused, with very weak lateral and vertical gradients.

In conclusion, the present study reveals the presence of highly time-dependent lateral and vertical salinity structure across the mouth of Delaware Bay. The salinity distribution at the mouth of the estuary is important because it provides information as to how the estuary is coupled to the adjacent continental shelf. It is also important to studies aiming at the numerical simulation of physical processes within the estuaries, as the salinity distribution at the mouth is often required as the open boundary condition. The high spatial and temporal variabilities in the salinity distributions would tend to make the task of a realistic simulation more difficult.

Acknowledgments. This study was funded by the National Science Foundation under grant OCE-9000158. Partial support was also provided by NOAA Sea Grant under grants NA116RG0162 (project R/ME-11) and NA56RG0147 (project R/ME-16). The dedication of Captain Donald McCann and the crew of the R/V *Cape Henlopen* is appreciated. Dr. Andreas Münchow provided the tidal current predictions. Drs. Leslie Bender and Joy Moses-Hall provided valuable assistance during the cruise. Timothy Pfeiffer provided excellent technical support throughout the course of the survey. Claudia Ferdelman developed the software for some of the graphics. Sheila Rollings typed the manuscript.

References

Fischer, H. B., Mass transport mechanisms in partially stratified estuaries, *J. Fluid Mech.*, *53*(4), 671-687, 1992.

Fischer, H. B., Mixing and dispersion in estuaries, *Annu. Rev. of Fluid Mech.*, *8*, 107-133, 1976.

Huzzey, L. M., The lateral density distribution in a partially mixed estuary, *Estuarine Coastal Shelf Sci*, *9*, 351-358, 1988.

Huzzey, L. M., and J. M. Brubaker, The formation of longitudinal fronts in a coastal plain estuary, *J. Geophys. Res.*, *93*(C2), 1329-1334, 1988.

Kjerfve, B., Circulation and salt flux in a well mixed estuary, in *Physics of Shallow Estuaries and Bays, Coastal Estuarine Stud.*, vol. 16, edited by J. van de Kreeke, pp. 22-29, Springer-Verlag, New York, 1986.

Kjerfve, B., and B. A. Knoppers, Tidal choking in a coastal lagoon, in *Tidal Hydrodynamics*, edited by B. Parker, pp. 169-181, John Wiley & Sons, New York, 1991.

Moses-Hall, J. E., Observed tidal, subtidal and mean properties in a laterally variable coastal plain estuary, Ph.D. dissertation, University of Delaware, Newark, Delaware, 338 pp, 1992.

Münchow, A., and R. W. Garvine, Dynamic properties of a buoyancy-driven coastal current, *J. Geophys. Res.*, *98*(C11), 20,063-20,077, 1993.

Münchow, A., A. K. Moses, and R. W. Garvine, Astronomical and nonlinear tidal currents in a coupled estuary shelf system, *Cont. Shelf Res.*, *12*(4), 471-498, 1992.

Nunes, R. A., and J. H. Simpson, Axial convergence in a well-mixed estuary, *Estuarine Coastal Shelf Sci.*, *20*, 637-649, 1985.

Pritchard, D. W., Estuarine classification--a help or hindrance, in *Estuarine Circulation*, edited by B. J. Neilson, A. Kuo and J. Brubaker, pp. 1-38, Humana Press, Clifton, New Jersey, 1989.

Schroeder, W. S., Riverine influence on estuaries: a case study, in *Estuarine Interactions*, edited by M. L. Wiley, pp. 347-364, Academic Press, New York, 1978.

Schroeder, W. S., and W. J. Wiseman, Low-frequency shelf-estuarine exchange processes in Mobile Bay and other estuarine systems on the northern Gulf of Mexico, in *Estuarine Variability*, edited by D. A. Wolfe, pp. 355-367, Academic Press, New York, 1986.

Sharp, J. H., Introduction to science chapters, in *The Delaware Estuary: Research as Background for Estuarine Management and Development*, edited by J. H. Sharp, pp. 1-7, University of Delaware and New Jersey Marine Sciences Consortium, Newark, Delaware, 1983.

Sharples, J., J. H. Simpson, and J. M. Brubaker, Observations and modelling of periodic stratification in the upper York River estuary, Virginia, *Estuarine Coastal Shelf Sci.*, *38*, 301-312, 1994.

Simpson, J. H., and R. A. Nunes, The tidal intrusion front: an estuarine convergence zone, *Estuarine Coastal Shelf Sci. 13*, 257-266, 1981.

Simpson, J. H., and W. R. Turrell, Convergent fronts in the circulation of tidal estuaries, in *Estuarine Variability*, edited by D. A. Wolfe, pp. 139-152, Academic Press, New York, 1986.

Simpson, J. H., J. Brown, J. Matthews, and G. Allen, Tidal straining, density currents, and stirring in the control of estuarine stratification, *Estuaries*, *13*(2), 125-132, 1990.

Smith, N. P., The Laguna Madre of Texas: hydrography of a hypersaline lagoon, in *Hydrodynamics of Estuaries*, Vol. II, edited by B. Kjerfve, pp. 31-40, CRC Press, Boca Raton, Florida, 1988.

Valle-Levinson, A., and R. E. Wilson, Effect of sill bathymetry, oscillating barotropic forcing and vertical mixing on estuary/ocean exchange, *J. Geophys. Res.*, *99*(C3), 5149-5169, 1994a.

Valle-Levinson, A., and R. E. Wilson, Effects of sill processes and tidal forcing on exchange in eastern Long Island Sound, *J. Geophys. Res.*, *99*(C6), 12,667-12,681, 1994b.

Wong, K.-C., On the nature of transverse variability in a coastal plain estuary, *J. Geophys. Res.*, *99*(C7), 14,209-14,222, 1994.

Wong, K.-C., and A. Münchow, Buoyancy forced interaction between estuary and inner shelf: observation, *Cont. Shelf Res.*, *15*(1), 59-88, 1995.

11

Formation of the Columbia River Plume - Hydraulic Control in Action?

Cynthia N. Cudaback and David A. Jay

Abstract

The Columbia River Plume is a large dynamic feature with a significant effect on circulation along the Northeast Pacific shelf and slope and on local and cross-shelf transport of nutrients, pollutants and sediments. Essential characteristics of this plume are determined at its origin in the constricted entrance to the river, where the meeting of salt and fresh water creates a highly stratified flow and large vertical shear. The channel is constricted laterally by stone jetties in two locations and vertically by a shallow sill. The effect of these constrictions has been examined by analysis of velocity and density data from the Columbia River entrance area. Along-channel sections and time-series data have been compared with results of a hydraulic control model incorporating a sinusoidally varying barotropic current [Helfrich, 1995], and show qualitative similarities with the theoretical results. Significant differences between theoretical and experimental results arise from a complicated time dependence of barotropic forcing, frictional effects and lateral variations in along-channel currents. Barotropic current strength varies with river flow and unequal semidiurnal tides, causing alternate two-layer and unidirectional flows. Turbulent mass transfer between the layers forms a thick interfacial layer, and bottom friction forces peak flood currents into the interfacial layer. Finally lateral variations in current speed cause lateral differences in time-series measurements of internal Froude number. Results of hydraulic control theory can account for some but not all velocity and density observations in the Columbia River entrance channel.

1. Rationale and Objectives

The Columbia River Plume is a large dynamic feature with a significant effect on circulation along the Northeast Pacific shelf and slope and on local and cross-shelf transport of nutrients, pollutants and sediments. The plume is formed on the Oregon-Washington border by outflow from the Columbia River, which carries 3/4 of the fresh water entering the Pacific Ocean between San Francisco and the Canadian border. This outflow combines with the plumes of other large rivers in British Columbia and Alaska to form a vast surface freshening delineated by the 32.5 psu isopleth [Barnes et al 1972]. Within this greater surface freshening, the plume is formed as a series of distinct fresh water pulses (<22 psu) on successive ebbs rather than as a steady outflow (Figure 1). Essential characteristics of the plume are determined at its origin in the constricted entrance to the river (Figure 2). This area is subject to strong river currents and semidiurnal tides with a substantial diurnal inequality. The meeting of salt and fresh water creates a highly stratified flow and large vertical shear. The channel is constricted laterally by stone jetties in two locations and vertically by a shallow sill. The questions at hand are: what are the effects of these constrictions and do they act as hydraulic

Buoyancy Effects on Coastal and Estuarine Dynamics
Coastal and Estuarine Studies Volume 53, Pages 139-154

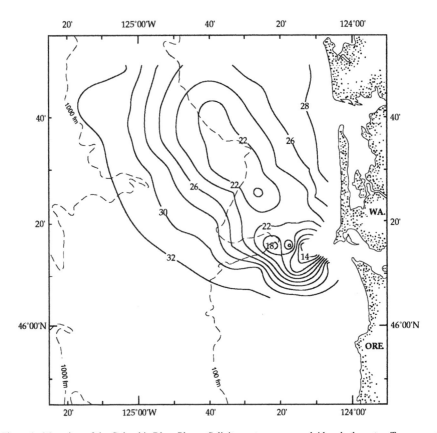

Figure 1. Map view of the Columbia River Plume. Salinity contours are overlaid on bathymetry. Two separate pulses of less than 22 psu are visible, representing the outflow from two different ebbs.

controls? We will attempt to answer these questions by analysis of velocity and density data from the Columbia River entrance area.

The organization of this paper is as follows. In Section 2, we discuss classical steady-state hydraulic control theory and its limitations when applied to the Columbia River entrance channel and discuss work on time-varying barotropic forcing by Helfrich, [1995] will be described. Section 3 describes the physical characteristics of the entrance channel, data collection and data reduction techniques. In Section 4, we present along-channel sections of velocity and salinity and examine their qualitative comparison with the results of hydraulic control theory. A qualitative discussion of time series measurements is followed by a comparison of internal Froude numbers resulting from a model by Helfrich [1995] with those calculated from our data.

2. Hydraulic Control Theory

The sill and lateral constrictions in the Columbia River entrance area affect both the relatively fresh outgoing surface layer and the incoming salt water mass, suggesting a physical interpretation in terms of two-layer hydraulic control theory. This theory was originally

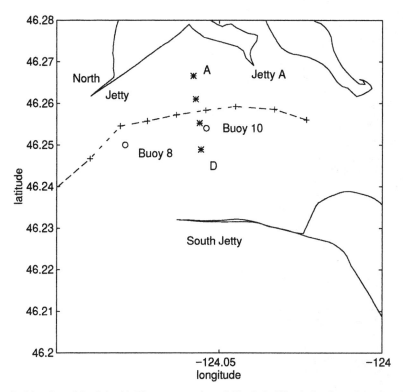

Figure 2. Map view of the Columbia River entrance channel. The dashed line is the along-channel trackline, and the '+' signs are the stations used in the along-channel sections in 1993. Stars represent the section occupied as a time-series in 1992.

developed from studies of exchange through locks. When a lock gate separating infinite reservoirs of fresh and saline water is opened, baroclinic forces drive a two-layer flow, which eventually settles to a steady two way exchange [Wood, 1970]. The steady two-layer exchange flow can be described with a relatively simple two-dimensional (x,z), rigid lid theory. For hydrostatic flow, along-channel variations in width and depth must be gradual [Wood, 1970]. The fluid must be non-rotating and inviscid, with negligible mixing between the layers. Furthermore, each layer must be homogeneous and unsheared. For this purely baroclinic case, also called internal forcing, Wood [1970] predicted a smooth s-shaped layer interface.

Armi and Farmer [1986] and Farmer and Armi [1986] extended the two-layer exchange concept to include an imposed barotropic flow, also called external forcing, such as might be found in narrow estuarine channels. To describe exchange flows in the presence of sill and constrictions, they introduced the internal Froude number, G.

$$G^2 = F_1^2 + F_2^2 = u_1^2 / g'h_1 + u_2^2 / g'h_2 \tag{1}$$

where g' is reduced gravity

$$g' = g(\rho_2 - \rho_1)/\rho_1 \tag{2}$$

Subscripts 1 and 2 refer to the upper and lower layers, ρ is the average density of a given layer, h is layer thickness and u is the average along-channel current speed in the layer. In the absence of constrictions or barotropic flow, G=1 everywhere in the channel. With an imposed steady barotropic flow, regions of supercritical (G>1) and subcritical (G<1) flow are created, separated by points of hydraulic control where the layer interface is displaced vertically. The location of these control points depends on the strength of barotropic inflow, as measured by q_{bo}.

$$q_{b0} = \frac{u_{b0}}{\sqrt{g'H}}$$

(3)

where u_{bo} is the barotropic current speed, H is the total water depth and g' is reduced gravity. If $q_{b0} <=1$, the hydraulic control point (G=1) occurs at the narrowest point of a lateral constriction or at the crest of a sill. If $q_{b0} >1$, the control point occurs downstream of the constriction [Largier, 1992]. One attractive aspect of classical steady-state hydraulic control theory is that the entire flow is controlled by conditions at a few control points.

Classical steady-state theory, however, is unable to explain crucial aspects of the circulation in the Columbia River entrance. First, the barotropic forcing (riverine and tidal) is strong and varies on several time scales. The river flow changes both seasonally and on a storm time scale, and the tidal currents are variable both on daily and tidal monthly basis. The formation of the plume as a series of distinct pulses on ebb is further evidence of the tidal time dependence. Second, bottom friction and turbulent mixing at the interface have a significant effect on both water masses and play a major role in tidal and sub-tidal circulation. Turbulent mixing due to interfacial shear causes both mass and momentum exchange between the layers. The mass exchange thickens the interface between the upper and lower layer, forming a new interfacial layer of thickness between ~1/3 and ~3/4 of the water depth. Bottom friction reduces near-bed flows, so the maximum flood current is usually seen in the interfacial layer. The turbulent momentum transfer between the layers is also a forcing mechanism for internal tidal asymmetry, a form of tidal rectification that may contribute substantially to two-layer exchange [Jay and Musiak, 1996]. Finally, cross-channel depth variations and channel curvature induce both lateral circulation and lateral variation in internal Froude number. Lateral variations in along channel flow will be discussed below, and lateral circulation and forcing will be discussed in a future paper.

The first problem, time-dependent forcing, has been partially addressed by Helfrich [1995], whose recent modeling work extends hydraulic control theory to include a barotropic current which varies sinusoidally in time, e.g., a pure semidiurnal tide. The published model does not include a steady flow like that of the Columbia River. Helfrich [1995] suggests parameters to estimate the strength of barotropic forcing and the importance of time dependence. Strength is measured by the same parameter q_{b0}, but u_{b0} is now understood to be the maximum speed of a sinusoidally varying current. Time dependence is estimated by comparison of the tidal period with the time for an internal wave to propagate across the sill. If the propagation time were very much less than a tidal period, steady state theory would still be valid [Largier, 1992]. Helfrich [1995] expresses this concept with his parameter

$$\gamma = \frac{T\sqrt{g'H}}{L}$$

(4)

the tidal period over the internal adjustment time, where T is the tidal period and L is the length of the strait. The quasi-steady-state approximation is valid for $\gamma > 30$, or very slow tides over a short sill [Helfrich, 1995]. The Columbia River entrance area (about 10 km long and 20 m deep, see Figure 2) is subject, by these criteria, to very strong forcing with significant time dependence; $\gamma \sim 4 - 6$ and q_{b0} is between 1 and 1.5.

Helfrich [1995] found a number of important effects associated with strong, tidally driven barotropic currents. In a two-layer exchange flow, a fast barotropic current (q_{bo}>=1) can block the opposing layer, creating a one-layer flow. Over the tidal cycle, the current speed changes, so that single-layer currents at peak flood and peak ebb alternate with two-layer circulation near slack water. Another model result is that, for a constant γ, the average exchange transport increases with increasing q_{bo}. For a constant q_{bo}, the maximum exchange is obtained in the quasi-steady limit of very large γ. Finally, the time dependence makes it impossible to determine the flow based only on fluid properties at the control points. Complete information on the geometry of the strait is needed.

Helfrich [1995] predicted the shape of the density interface and the evolution of the internal Froude number, G, for two geometries and several combinations of forcing strength and time dependence. The first geometry, a simple lateral constriction, produces only minor perturbations on the classical s-shaped salinity interface. The time series of G shows some interesting differences between weak and strong barotropic forcing, which will be discussed below. The more complicated model geometry, a combination of a sill and constriction is, unfortunately, not comparable with our measurements. Helfrich's sill is on the fresh side of the constriction, whereas in the Columbia River entrance, the sill is seaward of the constriction.

The lack of mixing effects in classical hydraulic control theory has been addressed by Chao and Paluskiewicz [1991], who used a three-dimensional primitive equation model to study the hydraulics of a density current near an isolated sill. A surface density current was released at one end of the model channel and allowed to run to steady state with no barotropic current. If mixing was allowed, there was downward entrainment upstream of the sill and the salt-fresh interface lost definition. Downstream of the sill, the interface remained sharp. Lawrence [1985, 1990] observed the reverse pattern in flume experiments with steady two-layer unidirectional flow over an obstacle. A hydraulic jump was observed downstream of the obstacle, consistent with strong barotropic inflow. Upstream of the obstacle, the lower layer currents were faster than the upper layer, and the interface was clearly defined. Between the sill and the jump, changing shear generated Kelvin-Helmholtz billows filling about 1/3 of the water column. Both of these studies involved relatively weak unidirectional currents and mixing was confined to discrete sections of the flow. By contrast, in the Columbia River entrance, strong reversing currents and significant turbulent mixing throughout the flow may wipe out any sharp layer interface.

Other problems in the application of hydraulic control theory are related to basic assumptions, some of which are not amenable to simple extensions of theory. The rigid-lid theory cannot accommodate wavelike barotropic transport, which is an important effect in tidal channels. The effects of along-channel variations in cross-section on transport are expressed in terms of Green's law, modified to account for the effects of frictional damping [Jay, 1991]. In a slowly-converging channel, frictional effects dominate over topographic effects, and tidal transport varies with the square root of the channel width and depth. In the Columbia, whose cross-section varies by about 15%, barotropic transport thus varies by about 8%. In a silled fjord, along-channel variations are much more significant. These variations cannot be reproduced with a rigid-lid theory.

In the Columbia, lateral variations in transport are more significant than along-channel variations. Hydraulic control theory, being inherently two-dimensional, cannot account for these lateral variations. The entrance channel is tightly curved, bending almost 90 degrees from northeast to southeast in 10 km on the landward approach, with a 6 km radius of curvature. The channel cross section is a complex curve resulting from a combination of natural scouring by higher velocity currents along the outside of the bend and dredging to maintain a navigable ship channel. The various channel constrictions cause some lateral divergence and convergence of currents. This complicated topography induces both across-channel circulation and lateral variations in along-channel current velocity.

Although we expect *a priori* that lateral currents and lateral variations in along-channel currents will be quite significant, we will first ignore these effects in order to compare the observed density and along-channel velocity fields with the predictions of hydraulic control theory. Thus, we will first treat our data as if they were collected in a two-dimensional system, then examine the lateral variations.

3. Methods

Data collection

The various topographic features in the Columbia River entrance (Figure 2) may be expected to have distinct and important effects on the velocity and salinity fields. Lateral and vertical constrictions include the seaward ends of the North and South Entrance Jetties (km 0) and the crest of the sill (km 1), where there is also a mild lateral constriction. Finally, an intruding jetty (Jetty A, km 4) forms the strongest lateral constriction, and the resulting strong currents have scoured a deep hole. These features are superimposed on the channel curvature described in the previous section. Along-channel sections chosen to show the effects of these features will be labeled with the above distances. Two data sets of velocity and salinity were collected in the Columbia River entrance area, to study the effects of time dependent forcing. Time series measurements were made in May of 1992 and along-channel sections in September and October of 1993.

Both research cruises were made aboard *R/V Snowgoose*, carrying a Conductivity Temperature Depth (CTD) profiler and an Acoustic Doppler Current Profiler (ADCP). Position was determined by GPS and the ships orientation by a gyrocompass with a synchronized interface. The Ocean Sensors CTD was deployed at stations located 500-800 m apart. Averaging of the 100-Hz sensor output was set to yield data at 8 Hz, providing better than 0.2 m vertical resolution. For calculation of the gradient Richardson number, as described below, the CTD data were averaged to 1 m vertical resolution. The R. D. Instruments 1.2 MHz ADCP was in constant operation. One sample, or acoustic "ping" is an acoustic signal sent by the ADCP and reflected by particles moving with the flow. The water velocity relative to the instrument is calculated from the Doppler shift of the returning signal. The signal reflecting off the bottom measures the speed of the vessel over ground, so the absolute water speed may be calculated. 60 to 70 acoustic pings are averaged together over a 20 second period to minimize random errors. In a vessel moving at 8 knots, this gives a horizontal resolution of <100 meters. The data presented here were averaged again over the 5-10 minute duration of each CTD cast. In this area , subject to large vertical shears, the maximum vertical resolution of 1 meter is used. Errors of a few cm s^{-1} can be induced by vertical shear, but the water velocities of order 1 m s^{-1} were significantly larger than the error.

Data Reduction

Classical hydraulic control theory assumes two inviscid layers with no mixing between the layers. By contrast, in the Columbia River entrance channel, there is a great deal of turbulent mixing between the layers, and the layer interface is not sharply defined. The stability of the water column is estimated using the total gradient Richardson number (Ri_{gt}).

$$Ri_{gt} = \frac{N^2}{S^2 + N^2}$$

(5)

where

$$N^2 = \frac{-g}{\rho_0}\left(\frac{\partial \rho}{\partial z}\right)$$

(6)

is the square of the buoyancy frequency and

$$S^2 = \left(\frac{\partial u}{\partial z}\right)^2 + \left(\frac{\partial v}{\partial z}\right)^2$$

(7)

is the square of the vertical shear. This formulation of the gradient Richardson number, developed by Geyer [1988], is based on the need to include both mean flow shear and internal wave shear. Mean flow shear is resolvable by the ADCP, but internal wave shear is not. N^2 can be measured at the higher vertical resolution of the CTD measurements, and if equipartition of kinetic and potential energy is assumed, N^2 serves as an estimate of the kinetic energy in a saturated internal wave field. In this formulation, Ri_{g_t} varies between zero (unstable) and one (stable), and the critical value is about 0.3. Above this critical level, turbulence cannot be maintained by the flow.

Calculation of the internal Froude number, (G, equation 1) is somewhat problematic. During flood and ebb, the strong barotropic flow overwhelms the baroclinic circulation, and currents in the entrance channel are unidirectional. However, the salinity distribution is essentially two-layer during most of the tidal cycle, so it seems reasonable to apply the concept of two-layer circulation [see also Helfrich, 1995]. In our data reduction, we calculated G only when each layer was at least 2 meters thick, thereby avoiding large spikes produced by the 'pinching off' of a layer. Near peak flood, at some places in the channel, both salinity and velocity were one-layer, and calculation of G would have been meaningless.

There are several possible ways to define the layer interface; for example the depth of maximum stability (N^2) or maximum Ri_{g_t} might be used. In practice, as both of these quantities are calculated from vertical gradients, there is enough noise in the vertical profiles that concise selection of a maximum is impractical. Inspection of the measured salinity sections (Figures 3b and 5b) and time series (Figure 7b) reveals a strong salinity gradient around 22-26 psu. The 22 psu contour also marks the lateral boundaries of the plume pulses in Figure 1, so it may be used as a delineator for the forming plume. To calculate G, we chose the 24 psu isopleth as the layer interface.

4. Observations

Along-channel Sections

We compare the results of hydraulic control theory with measured along-channel sections of velocity and salinity. Observations were made in repeated along-channel crossings of the entrance bar in September and October of 1993. We present here two of the sections, representing conditions at flood onset (Figures 3 and 4) and peak flood (Figures 5 and 6). Each section was run in a seaward direction and took about two hours, so the western (left) end of each represents a later stage in the tidal cycle than does the eastern (right) end. At peak ebb, when plume-pulse formation is well underway, no observations could be made, due to extremely rough sailing conditions on the bar. Therefore, none of our sections show seaward currents at all depths.

The two salinity sections have some common features (Figures 3b and 5b). In general we find an orderly progression from salt (>30 psu) to fresh (~8 psu) water in layers separated by an s-shaped interface (~22-26 psu). The interface is actually a 5-10 meter thick layer due to turbulent mixing, but the basic shape is familiar from hydraulic control theory. Comparison among all the measured sections shows this interface moving landward on flood and seaward on ebb (not plotted here). An interesting detail revealed in these sections is that the leading edge of the salt wedge actually crosses the sill as a series of discrete pulses, which produce a multi-layer intruding water mass (as opposed to a single salt wedge) farther upstream.

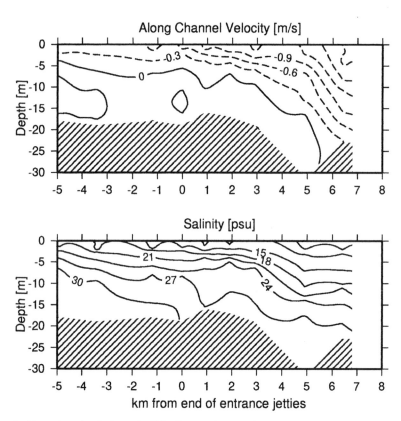

Figure 3. Along-channel velocity and salinity for a section crossing the Columbia River bar at the start of flood, Oct 19, 1993. In (a) contours of along-channel velocity show two-layer circulation, flooding at the bottom and ebbing at the surface.. Salinity contours (b) also show a two-layer structure.

The section most resembling the results of two-layer hydraulic control theory was taken an hour after low water (Figure 3). This circulation is possible only when flood currents oppose the river flow and are of roughly comparable strength. The system is then briefly subject to weak barotropic forcing. The bottom layer is flooding and the surface layer is still ebbing slightly, indicating the end of the plume pulse. The plume lifts off the river bottom just landward of the sill, due to a combination of topographic uplift at the sill and hydraulic control at Jetty A. Both stratification and mixing are apparent directly above the sill (km 1, Figure 3b). Note the sharp halocline between the 15 and 24 psu contours (about 5 m apart) as well as the strong mixing between the 24 and 27 psu contours.

Stability, as measured by Ri_{gt} , was estimated from the above data and plotted as a set of vertical profiles (Figure 4a). The dashed lines overlaid on the profiles indicate Ri_{gt} = 0.3, or critical stability. At flood onset, the entire section is essentially critical, except near the bed at km 2, just landward of the area of mixing. The 24 psu isopleth was used as a layer interface to calculate the internal Froude number. Consistent with the strong ebb currents landward of the sill, G is strongly supercritical in this area, and critical seaward of the sill (Figure 4b).

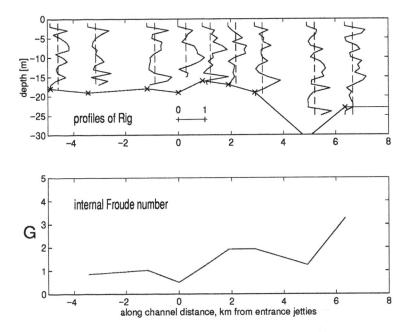

Figure 4. Diagnostic variables calculated from data in Figure 3, at flood onset. Profiles of Gradient Richardson number (a), indicate that most of the water is near critical stability. The internal Froude number (b) is critical seaward, and supercritical landward of the sill.

During peak flood, the strong barotropic forcing ($q_{bo} \sim 1$) is expected to drive a one-layer flow [Helfrich, 1995], or a two-layer unidirectional flow with faster currents in the bottom layer [Lawrence, 1985, 1990]. The observed current (Figure 5a, taken four hours after low water) is unidirectional, but strongest at middepth. This three-layer flow is due to a combination of baroclinic forcing (which increases with depth), turbulent mixing at the original layer interface and friction at the river bottom. The mixing creates a thick interfacial layer, and bottom friction retards the flood current, forcing the maximum current up into the interfacial layer [see also Geyer, 1985, 1988]. Note that the region of currents >1.2 ms⁻¹ (Figure 5a) roughly coincides with a salinity band of 18-30 psu (Figure 5b). The middepth current accelerates upon passing the slight constriction of the entrance jetties, and accelerates further over the sill. The current then dives over the sill and reaches its maximum velocity in the constriction at Jetty A. These effects, although not directly predicted by hydraulic control theory, demonstrate the importance of the constrictions.

The salinity section (Figure 5b) suggests the presence of a blocked plunge line related to the strong inflow [Largier, 1992]. Such fronts were observed during this section, being advected with the flood currents. The presence of salmon sport fishermen in the area indicates the convergence of nutrients, plankton and salmon on the fronts. Above the sill, the 18-24 psu salinity contours all come to the surface within a two kilometer long region. Two-layer flow is thus seen only landward of the sill, where G is strongly supercritical (Figure 6b). Profiles of Ri_{gt} measured at peak flood (Figure 6a) indicate slightly higher stability than at flood onset. There appears to be a stable mid-depth layer overlying well-mixed water for all profiles landward of the sill. The stable layer is somewhat thinner over the sill, which may indicate topographically enhanced mixing.

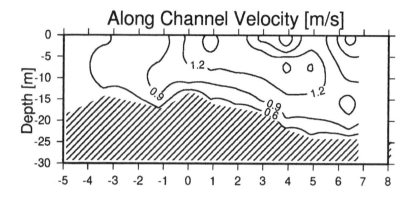

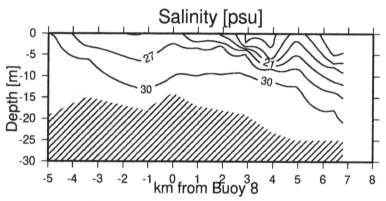

Figure 5. Along-channel velocity and salinity for a section crossing the Columbia River bar at peak flood, Oct 18, 1993. In (a), contours of along-channel velocity show that the current is unidirectional, and strongest at middepth. The 18-24 psu salinity contours in (b) touch the surface over the sill, indicating a plunging inflow.

Time series Measurements

Two-layer circulation in an estuarine channel can be predicted under sinusoidal tidal forcing and in the absence of friction [Helfrich, 1995]. At slack water, estuarine baroclinic circulation should dominate, giving a two layer flow (surface seaward, bottom landward) at the onset of either flood or ebb. Strong tidal currents can overwhelm this circulation, giving unidirectional flows at peak flood and peak ebb. For sinusoidal forcing, peak ebb and peak flood currents should be the same strength. In the absence of bottom friction, ebb currents should be strongest near the surface and flood currents should be strongest at the bottom. The fresh water advected seaward on ebb is expected to thicken the surface layer, while the dense flood current should thicken the lower layer. In the absence of interfacial mixing, both layers should maintain a uniform density (Ri_{gl} nearly zero) and the interface should be quite stable (Ri_{gl} nearly one).

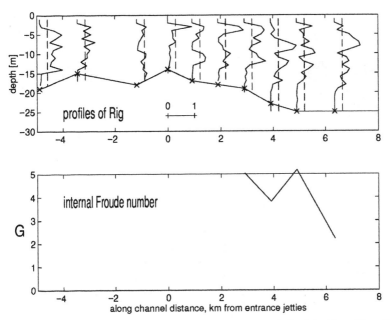

Figure 6. Diagnostic variables calculated from data in Figure 5, at peak flood. In (a) profiles of Gradient Richardson number suggest the presence of a mid-depth stable layer. The layer is thinner over the sill, indicating topographically enhanced mixing. Internal Froude number (b) is only measured landward of the sill, where it is strongly supercritical.

Time series of salinity and velocity were measured on May 25, 1992 at four stations near Buoy 10, just landward of the crest of the sill (Figure 2). This is the site of both intensified flood currents and ebb plume liftoff, and is therefore a good place to study intra-tidal variability. We were able to measure time series varying in length from 12 to 18 hours or 1/2 to 3/4 of a diurnal tidal cycle. The longest time series was collected at station D, near the south side of the channel.

The measured along-channel currents at station D (Figure 7a) show the effect of both strong river forcing and bottom friction. Peak ebb currents (~1200 PST and 2400 PST) are about 50% stronger than peak flood currents (~1800 PST). Ebb currents are stronger in the upper layer, but are significant at all depths. Ebb onset is sudden and almost simultaneous throughout the water column, creating unidirectional currents instead of the predicted two-layer flow (~0900 PST and 2100 PST). Both the strength and sudden onset of the ebb currents are due to the strong river flow, which enhances the ebb current but opposes the flood. In contrast, flood onset takes over an hour to propagate from the bed to the surface. The flood current, being driven by denser oceanic waters, starts in the lower layer (~1400 PST). As described above, bottom friction impedes lower layer flow, pushing the maximum flood current upward with time. The time series shows the full vertical progression of the flood current.

Both the salinity time series (Figure 7b) and the Ri_{gt} time series (Figure 7c) show evidence of interfacial mixing. The salinity interface, which moves up and down tidally, is about 5-10 m thick, depending on the stage of the tide. This interfacial layer coincides with a persistent stable layer in the Ri_{gt} time series. Due to a reduction in vertical shear, the stable layer thickens to fill the water column at slack water. Thus the system alternates between stable and well mixed on a quarter-diurnal period. This combination of quarter and semidiurnal

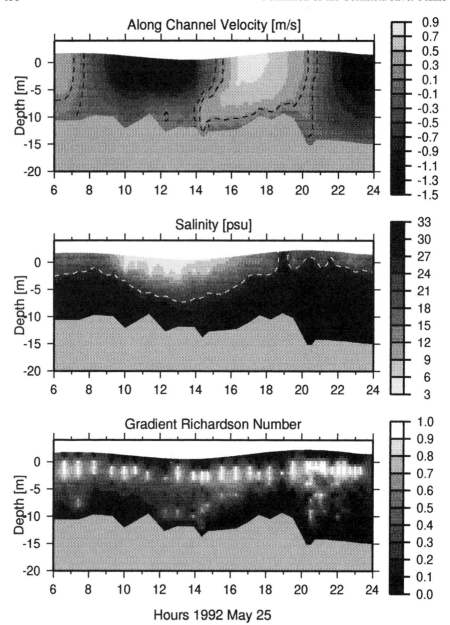

Figure 7. Along-channel velocity, salinity and gradient Richardson number for a time series measured at the crest of the sill (Buoy 10), near the south side of the channel, May 25, 1992. The dashed line in (a) is the zero-velocity contour. Note that the flood currents start in the lower layer and move upwards, but the ebb currents are strong at all depths from onset. The dashed line in (b) is the 24 psu isohaline. Note that the fresh upper layer thickens on late ebb and thins on late flood. Gradient Richardson number is plotted in (c) A relatively stable mid-depth layer is thick at slack water and thinnest at peak ebb, indicating internal tidal asymmetry.

signals in Ri_{gt} indicates internal tidal asymmetry and lies outside the assumptions of hydraulic control theory [Jay and Musiak, 1996].

Both the along-channel section at peak flood (Figure 5a) and the time series (Figure 7a) suggest the importance of bottom friction and interfacial mixing to along-channel transport. In the classical two-layer model, most flood transport is in the saltier lower layer, so significant amounts of salt are advected landward. The observed transport is concentrated higher in the water column and carries less salt. Thus the estuary should be significantly fresher than would be predicted from a two-layer model with the same exchange volume.

Lateral Variations

In our low-order comparison of observations and the predictions of hydraulic control theory, we ignored lateral variations and examined only the effects of turbulent mixing and time varying barotropic currents. However, currents in the Columbia River entrance also vary laterally due to channel curvature and other topographic effects. The time series discussed up to now was measured south of mid-channel, near the inside edge of the curve (station D). Salinity and velocity time series measured on the north side of the same channel section (station A) were qualitatively similar to those from station D, but represent different physical circumstances. Station D is near mid-channel, where the water is deep and the currents fast. Station A is in much shallower water, and the currents are slower. These differences are clearly seen in a comparison between internal Froude number time series calculated at the two stations (Figure 8). For both measured time series, there is a clear pattern of long supercritical periods during ebb and flood, interspersed with brief subcritical periods at slack water. At station A, G varies between 0.5 near slack water and 1.5 at peak flood and ebb (Figure 8a). At station D, the amplitude of the supercritical Froude numbers varies with tidal period, being about 3 on lesser ebb, 2 on flood and up to 4 on greater ebb (Figure 8b). The dramatic difference between the two time series is due to both spatial and temporal variations in barotropic forcing strength (which includes the effect of depth). These temporal variations, due to the strong river flow and unequal semidiurnal tides, have a significant effect on local dynamics.

We compare the two measured time series of G with results from Helfrich's [1995] model. Our observations were made about 1 km seaward of the constriction, and should be qualitatively comparable with model predictions of G at the constriction. Model input parameters are: time dependence γ, tidal forcing strength q_{bo} and mean current q_m (e.g. a river current). The magnitude of γ depends on the definition of the length scale L; it has a precise mathematical definition in a model, whereas in a real channel, L must be estimated from topography. In this section of the channel, $\gamma \sim 4 - 6$, which is comparable with Helfrich's [1995] choice of $\gamma = 4$. Total forcing strength can be expressed [Helfrich, personal communication] as

$$q_b(t) = q_{bo} \sin(\omega t) + q_m \tag{8}$$

In a real channel, q_{bo} and q_m are estimated from the tidal current speed and the total water depth (3). At station A, $q_{bo} \sim 0.85$ and $q_m \sim 0.15$, whereas at station D, $q_{bo} \sim 1.2$ and $q_m \sim 0.2$. Measured G at these stations are compared with two model runs. Helfrich's original model used $\gamma = 4$ and $q_{bo} = 1.0$, parameters which approximate the average tidal barotropic forcing in the Columbia River entrance, but do not include the river currents. At our request, Dr. Helfrich made a special model run using the same tidal forcing as above, with the addition of $q_m = 0.3$ to represent strong river flow.

Although $q_{bo} \sim 1$ at both stations, indicating strong barotropic forcing, a subtle difference in forcing can lead to significant differences in G. If $q_b > 1$ at any time, the barotropic currents

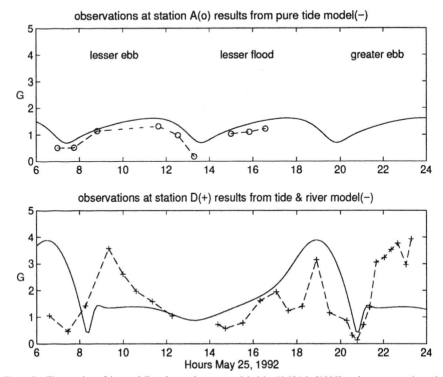

Figure 8. Time series of internal Froude number, as modeled by Helfrich [1995] and as measured on the north (station A) and south (station D) sides of the channel. In (a) observations from station A compare well with model results using $g = 4.0$, $q_{b0} = 1.0$ and $q_m = 0.0$. In (b) observations from station D are compared with model results using $g = 4.0$, $q_{b0} = 1.0$ and $q_m = 0.3$. Observed peaks in G are due to strong ebb currents, whereas peaks in the model results indicate the lower layer pinching off.

can overcome the baroclinic forcing, so that one layer is pinched off and the net flow becomes unidirectional. As h_i becomes very small, the layer Froude number

$$F_i = \frac{u_i}{\sqrt{g' h_i}}$$

(9)

becomes quite large. This effect should be more prominent at station D, where the maximum q_b is ~1.4, than at station A, where the maximum q_b is ~ 1.0. In our observations, the layer interface is carefully chosen so that neither layer ever gets less than 2 meters thick, and the pinching-off effect on G is tempered.

Comparisons between model and observation are shown in Figure 8. The measured G time series from station A closely resembles the model results without river forcing (Figure 8a), whereas the observations at station D are more comparable with the model results including river currents (Figure 8b). This difference shows the importance of lateral variations in along-channel circulation. There are also significant differences between the observations at station D and the river-forced model results, which indicate the complementary effects of current speed and layer depth in calculating the layer Froude numbers. The large observed peaks in G at station D (+ signs) are attributed to the strength of the river-enhanced peak ebb current (large u_1 -- large F_1.). The modeled G time series (solid line) has peaks with amplitudes close to those measured at station D, but the maximum theoretical G leads peak ebb by about 3 hours. The model maxima occur at the end of flood, as the bottom layer is pinched off by the oncoming ebb (small h_2 -- large F_2). Both observations and model results are valid, but slight differences in the method of calculating G produce significantly different results.

In summary, comparison between theory and observation reveals three significant limitations of the hydraulic control model. First, the lack of mixing effects gives an incorrect prediction of the flood velocity profile, which would lead to an over-estimate of landward salt transport. Second, significant lateral variations cannot be recreated by a two-dimensional model. However, it might be possible to model the internal hydraulic properties of each streamline individually, and present them in parallel to approximate the three-dimensional flow. Third, even in very similar flows, subtle differences in the method of calculating G may produce different views of the internal hydraulics. This effect does not imply a limitation in the usefulness of a model, but only a caution about interpretation.

5. Conclusions

The physics of plume formation in the Columbia River entrance can be partially explained in terms of hydraulic control theory. The basic importance of sills and constrictions is clear, as indicated by the ebb plume liftoff and flood current intensification between the sill and constriction. The time dependent hydraulic control model of Helfrich [1995] predicts qualitative features of the flow, but there are some significant quantitative differences between theory and observation.

The presence of a river current causes the following effects:
- Ebb is stronger and longer in duration than flood.
- Ebb onset is sudden and simultaneous at all depths (one-layer flow).
- Two-layer flow is observed only at the start of flood.

Bottom friction and turbulent mixing cause the following effects:
- The layer interface is of finite thickness (>5m) at all stages of the tide.
- The interface thickness increases to the water depth (~20m) at slack water and decreases at peak flood and peak ebb; mixing varies at twice the tidal frequency.
- Bottom friction impedes flood currents in the lower layer, so flood mass transport occurs at mid-depth or above.
- Therefore, flood currents should advect less salt landward than would be predicted from a two-layer model.

These are the observed lateral variations in along-channel circulation:
- Time series of salinity and velocity measured at opposite sides of the channel are qualitatively similar.
- Observed time series of internal Froude number vary laterally by a factor of 3 at peak ebb, because of lateral variations in depth and subtle lateral variations in velocity and density.
- Measured G at the north side of the channel appears to be affected only by tides.
- Observed G on the south side of the channel has large peaks attributed to the strength of the river-enhanced ebb current.
- A model prediction using river and tidal currents together does not match the observed time series. Instead, large peaks in G are due to the lower layer pinching off at late flood.

Acknowledgments. Thanks go to Karl Helfrich, who not only provided model results which were then in press, but also ran his model again using parameters consistent with Columbia River entrance conditions. Andy Jessup provided the ship time used to measure the along-channel sections, an excellent exchange for the loan of an instrument. Edward Flinchem and Andrew Newell made helpful comments on the manuscript, as did Greg Lawrence and an anonymous reviewer. Financial support for the first author was provided by an ONR graduate student fellowship. This work has been also supported by NSF grants OCE-8918193, Columbia River Plume studies and OCE-8907118, the Columbia River Estuary Land Margin Ecosystem Research Program. Secondary data reduction was supported by ONR grant N00014-94-1009.

References

Armi, L., and D. M. Farmer, Maximal two-layer exchange through a contraction with barotropic net flow, *J. Fluid Mech., 164,* 27-51, 1986.

Barnes, C. A., A. C. Duxbury, and B. Morse, Circulation and selected properties of the Columbia River effluent at sea, in *The Columbia River Estuary and Adjacent Ocean Waters,* edited by A. T. Pruter and D. L. Alverson, University of Washington Press, 1972.

Chao, S-Y, and T. Paluskiewicz, The hydraulics of density currents over estuarine sills, *J. Geophys. Res., 96,* 7065-7076, 1991.

Farmer, D. M., and L. Armi, Maximal two-layer exchange over a sill and through the combination of a sill and contraction with barotropic flow, *J. Fluid Mech., 164,* 53-76, 1986.

Geyer, W.R., The time-dependent dynamics of a salt wedge, Ph.D thesis, Univ. of Wash. Seattle, 1985.

Geyer, W. R., The advance of a salt wedge front: Observations and dynamical model, in *Physical Processes in Estuaries,* edited by J. Dronkers and W. van Leussen, pp. 295-310, Springer-Verlag, New York, 1988.

Helfrich, K. R., Time dependent two-layer hydraulic exchange flows, *J. Phys. Oceanogr., 25,* 359-373, 1995.

Jay, D. A., Green's law revisited: tidal long-wave propagation in channels with strong topography, *J. Geophys. Res., 96,* 20585-98, 1991.

Jay, D. A., and J. D. Musiak, Internal tidal asymmetry in channel flows: origins and consequences, in C. Pattiaratchi, ed. *Mixing Processes in Estuaries and Coastal Seas,* an American Geophysical Union *Coastal and Estuarine Sciences Monograph,* 219-258, 1996.

Largier, J. L., Tidal Intrusion Fronts, *Estuaries, 15,* 26-39, 1992.

Lawrence, G. A., The hydraulics and mixing of two-layer flow over an obstacle, *Hydraul. Eng. Lab. Rep. UCP/HEL-85/02,* 122pp., Univ. of Calif., Berkeley, 1985.

Lawrence, G. A., On the hydraulics of Boussinesq and non-Boussinesq two-layer flows, *J. Fluid. Mech., 215,* 457-480, 1990.

Wood, I. R., A lock exchange flow, *J. Fluid. Mech, 42,* 671-687, 1970.

12

The Role of Thermal Stratification in Tidal Exchange at the Mouth of San Diego Bay

D.B. Chadwick, J.L. Largier, and R.T. Cheng

Abstract

We have examined, from an observational viewpoint, the role of thermal stratification in the tidal exchange process at the mouth of San Diego Bay. In this region, we found that both horizontal and vertical exchange processes appear to be active. The vertical exchange in this case was apparently due to the temperature difference between the bay water and ocean water. We found that the structure of the outflow and the nature of the tidal exchange process both appear to be influenced by thermal stratification. The tidal outflow was found to lift-off from the bottom during the initial and later stages of the ebb flow when barotropic forcing was weak. During the peak ebb flow, the mouth section was flooded, and the outflow extended to the bottom. As the ebb flow weakened, a period of two-way exchange occurred, with the surface layer flowing seaward, and the deep layer flowing into the bay. The structure of the tidal-residual flow and the residual transport of a measured tracer were strongly influenced by this vertical exchange. Exchange appeared to occur laterally as well, in a manner consistent with the tidal-pumping mechanism described by Stommel and Farmer [1952]. Tidal cycle variations in shear and stratification were characterized by strong vertical shear and breakdown of stratification during the ebb, and weak vertical shear and build-up of stratification on the flood. Evaluation of multiple tidal-cycles from time-series records of flow and temperature indicated that the vertical variations of the flow and stratification observed during the cross-sectional measurements are a general phenomenon during the summer. Together, these observations suggest that thermal stratification can play an important role in regulating the tidal exchange of low-inflow estuaries.

Introduction

We have recently examined some aspects of the tidal exchange process at the mouth of San Diego Bay, CA. The horizontal and vertical structure of the exchange were studied to gain insight into the predominant mechanisms controlling the export of water-borne contaminants out of the bay. Petroleum hydrocarbons from chronic sources within the bay present a significant environmental concern, and also provide a useful tracer for distinguishing bay water from oceanic water. Previous studies of the tidal exchange process have focused on two general mechanisms. In a coastal embayment, with negligible freshwater inflow, horizontal asymmetries between the ebb and flood tides lead to net tidal exchange through a process termed "tidal pumping" [Stommel and Farmer, 1952; Signell and Butman, 1992]. For instance, the ebb flow from an inlet is often focused by the topography and takes the form of a tidal jet. However, the inflow on the flood tide may be drawn from any direction and thus often takes the form of a sink-flow. The material exchange that takes place between the bay and the ocean thus is limited to the region where the sink flow overlaps the jet. In estuaries with significant

Buoyancy Effects on Coastal and Estuarine Dynamics
Coastal and Estuarine Studies Volume 53, Pages 155-174

freshwater inflow, density driven exchange is significant and may dominate the net tidal exchange [Dyer, 1974].

At the mouth of San Diego Bay, we found that both horizontal and vertical exchange processes appear to be active. This was somewhat surprising in view of the absence of freshwater inflow to the bay during most of the year. The vertical exchange in this case was apparently due to the temperature difference between the bay water and ocean water. The resulting temperature-dominated density gradient led to two related effects which influenced the bay-ocean exchange including, (1) a period of stratified exchange between the outer bay and the ocean, primarily during the ebb-flood transition, and (2) the lift-off of the ebb flow as it exited the mouth of the bay. The first effect can be expected to enhance the tidal exchange directly, while the lift-off may lead to structural variations in the outflow which indirectly effect the exchange due to tidal pumping.

In this paper, we set out to provide a description of this tidal exchange process at the mouth of San Diego Bay, and investigate the role of thermal stratification in modifying the exchange. The first section describes the limited existing knowledge of the circulation and hydrography of San Diego Bay, followed by a description of the field effort undertaken for this study. From the field observations, we develop a descriptive view of the summer conditions within the bay as a whole, and in the mouth region in particular. The lift-off of the ebb flow and the subsequent vertical-exchange during the reversal to flood are resolved and evaluated on the basis of the outflow Froude number. A qualitative comparison of vertical and horizontal exchange is made on the basis of the residual transport at the mouth section, and a temperature-tracer mixing diagram. The generality of the observed vertical structure is evaluated in terms of tidal cycles of vertical velocity-shear and stratification from moored time-series measurements of velocity and temperature near the mouth. A mechanistic description of the development of these "thermal estuary" characteristics and comparison to other low-inflow estuaries is provided in a companion paper [Largier, et al., 1995a].

Description of the Study Site

San Diego Bay is relatively long and narrow, 25 km in length and 1-3 km wide, forming a crescent shape between the City of San Diego on the north, and Coronado Island/Silver Strand on the south (Figure 1). Exchange with the ocean is limited to a single channel at the mouth. This north-south channel is about 1.2 km wide, bounded by Point Loma on the west and Zuniga jetty on the east, with depths between 5-15 m. The bay hydrograph is characterized by small annual freshwater inflow (mean rainfall 26 cm·yea^{r-1}), with zero inflow during most summers. The currents within the bay are dominated by a mixed diurnal-semidiurnal tidal forcing, with a dominant semidiurnal component [Peeling, 1974]. The tidal range from MLLW to MHHW is about 1.7 m with extreme tidal amplitudes close to 3 m. Recent bottom mounted ADCP measurements at the mouth indicate current velocities ranging from 0.2-0.8 m·s^{-1} [Cheng and Gartner, 1993]. Near the mouth, historical observations [Hammond, 1976, Navy, 1950, Smith, 1970], and the ADCP data suggest a net inflow at the bottom, with the bottom flood-tide velocities leading the surface velocities by about 30 to 90 minutes. Hammond [1976] attributed this inflow to a net northward bottom transport in the ambient flow outside the bay, based on seabed drifter trajectories. However, the ADCP records indicate net landward flow near the bottom is enhanced during periods of neap tide, suggesting that baroclinic circulation may be significant during these episodes.

Surface and bottom water temperatures in San Diego Bay exhibit a seasonal cycle with an amplitude of about 8-9°C [Smith, 1972]. During the summer period net heat flux to the bay and limited exchange with the ocean lead to the development of both vertical and longitudinal temperature/density gradients in the bay. Smith [1972] found maximum vertical temperature gradients of about 0.5 °C·m^{-1} during the summer. Typical longitudinal temperature gradients of about 7-10 °C over the length (~0.3-0.5 °C·km-1) of the bay have been reported [Federal

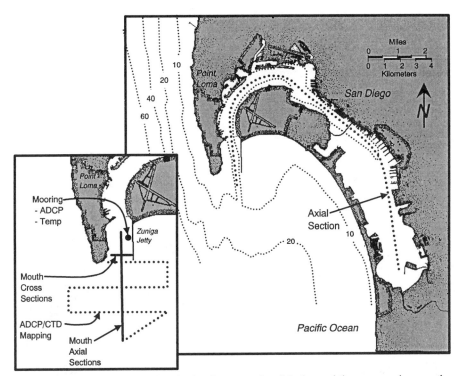

Figure 1. Plan view of San Diego Bay showing the topography of the bay and the survey region near the mouth. Also shown are survey transects for the bay axial sections (Jan 14 and May 17, 1994), mouth axial sections (May 19, 1994), mouth cross sections (May 18, 1994), offshore mapping (May 19, 1994), and the location of the ADCP (June 22 - July 23, 1993) and thermistor mooring (July 29 - August 22, 1993). Depths contours are in meters.

Water Pollution Control Administration, 1973, Largier et al., 1995b] during the summer months. Spectral analysis of a year-long time series of surface and bottom temperature showed significant peaks at the diurnal and semidiurnal tidal periods, as well as at periods of 8.2, 5.1, and 4.2 (surface only) hours [Smith, 1972]. The 8.2 hour response was explained as an interaction between the diurnal and semidiurnal tidal constituents, while the 5.1 and 4.2 hour periods were attributed to a possible sieche.

Field Measurements

A series of observations were undertaken during 1993-94 to examine the dynamics which regulate the exchange of bay water with ocean water in the region of the mouth. The observations included bottom-moored ADCP and thermistor records, ADCP/CTD surveys across the mouth of the bay and in the region of the outflowing jet, and axial CTD surveys of the entire bay.

In conjunction with the CTD measurements, a flow-through fluorometer was used to measure hydrocarbon concentrations using an ultra-violet fluorometric (UVF) technique as described by Katz and Chadwick [1991]. The fluorometer was calibrated against a standard diesel fuel (marine) and concentrations were reported as diesel fuel marine equivalent (dfme).

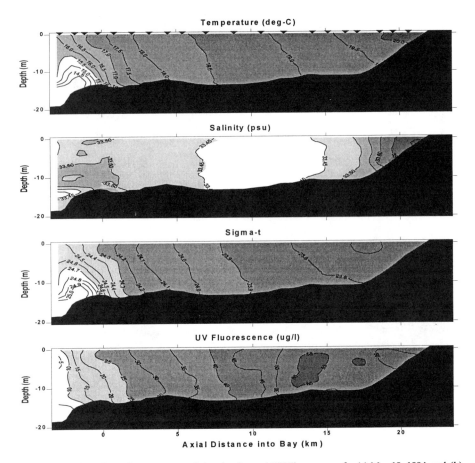

Figure 2. Axial sections of temperature, salinity, density, and UV Fluorescence for (a) May 19, 1994, and (b) January 14, 1994. Contours were developed from a series of vertical profiles performed over a 4-5 hour period. Profile locations are indicated by triangles. The May section shows the evidence of a "thermal estuary" condition with longitudinal and vertical density gradients dominated by the temperature field. The January section shows the breakdown of this condition after winter cooling and a weak "inverse" condition due to the salt field. Both sections show the dominate source of hydrocarbons from the southern region of the bay (> 15 km from the mouth).

The fluorometer showed good linear response over a range of 1-500 ug·l⁻¹. Because the fluorescence signal is a combined response to a number of compounds which fluoresce under UV excitation, it is subject to error if the relative composition of these compounds changes. However, previous work in San Diego bay indicated that the fluorescence is dominated by pyrogenic hydrocarbons, primarily associated with leachate from creosote impregnated pier pilings, and that the fluorescence characteristics were consistent throughout the bay [Katz and Chadwick, 1991; Katz et al., 1995].

During the period June 22 to July 23, 1993, a broadband 1.2 MHz ADCP was deployed upward-looking at a bottom depth of about 15 meters just inside the mouth. Data were recorded continuously at a 10-minute interval, providing a temporal record of the vertical flow structure at resolution of 1 meter. Self-recording thermistors were deployed at buoy stations along the axis of the bay at 9 locations, and temperatures were recorded every 9.6 minutes during the period 29 July to 22 August, 1993. A transect across the mouth of the bay was repeated 31 times through an ~12 hour, semidiurnal, symmetrical tidal cycle on 18 May, 1994. A 1.2 MHz narrow-band shipboard ADCP and towed CTD/UVF system were used to measure water properties and flow at the section. The CTD/UVF system was profiled while underway to obtain a continuous series of 12 profiles across the mouth. A series of nine surveys of the region outside the mouth were performed the following day (19 May, 1994), during tidal conditions similar those of the cross-mouth transects. The surveys consisted of a series of transverse and axial transects in the region extending from ~1 km inside the mouth to ~5 km offshore from the mouth. Current meters and thermistors were deployed on sub-surface moorings at three locations off the mouth during the period 17-19 May, 1994. A series of CTD/UVF vertical profiles were obtained along the axis of the entire bay on 17 May, 1994. Transect lines and mooring locations are shown in Figure 1.

Results & Discussion

San Diego Bay   Axial-Sections

Typical summer conditions for the axial temperature, salinity, density, and UVF tracer distributions in San Diego Bay were developed from the vertical profiles of 17 May, 1994 (Figure 2a). The plots show several interesting points. The density field during this period was dominated by the effects of temperature except at the head of the bay where hypersaline conditions had begun to develop [see also Largier et al., 1995a]. The overall longitudinal temperature difference in the bay was about 4-5°C, with a local gradient near the mouth of about $0.5°C \cdot km^{-1}$. The temperature/density field progressed from well mixed in the inner bay, to a partially mixed condition in the outer bay where the proximity of the cooler ocean water was evident in the enhanced vertical stratification. The vertical gradient in the mid to inner bay was generally less than $0.05°C \cdot m^{-1}$, while near the mouth, the gradient increases to about $0.3°C \cdot m^{-1}$. The UVF distribution clearly indicated the primary source of hydrocarbons in the inner bay, and also illustrated the influence of stratification in the outer bay on the tracer distribution. The hydrocarbon influx is considered to be relatively steady, resulting primarily from the leaching of polycyclic aromatic hydrocarbons from creosote impregnated pier pilings [Katz et al, 1995.]. Unstratified winter conditions from a similar axial transect on 14 January, 1994 are shown in Figure 2b for comparison. The thermal structure was absent in winter, while the hypersalinity still persisted resulting in a weak inverse density gradient. This period is typically followed by rain events which lead to sub-ocean salinity in the inner bay.

Outflow Mapping

Surface water conditions off the mouth were examined with a series of near-synoptic maps developed from the surveys of 19 May, 1994 (see Figure 1). The UVF tracer (Figure 3) and velocity fields (not shown) indicated that the outflow, which started as a radial structure from the mouth, developed into a low-aspect jet-like structure with a streamwise scale of about 5 km. When the flow reversed to flood, the outflow water was "pinched off" laterally at the mouth, and then drifted to the east under the influence of the longshore currents while continuing to disperse. This view of the horizontal exchange is conceptually consistent with the tidal pumping model suggested by Stommel and Farmer [1952], but it is significantly distorted by the longshore flow and the topographical complexity of the study area.

In addition to the ebb-flood asymmetry of the outflow and inflow regions, the observations indicate that the exchange between the bay and the ocean may also be enhanced

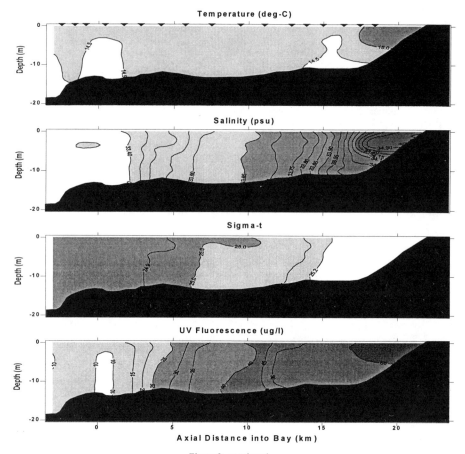

Figure 2. continued

by the inertia of the outflowing jet which continues to carry bay water offshore after the transition to flood flow. This implies that, for some period, the water drawn into the bay will be preferentially drawn in either from the edges of the jet where the inertia of the ebb flow is weak, or potentially from beneath the jet where bottom friction and density gradients also counteract the inertia of the jet. This appears to be the cause of the lateral "pinch off" of the outflow illustrated in Figure 3. The vertical exchange is explored in depth in the following sections.

Mouth Sections

An alternate view of the exchange process, and especially of the potential thermal-buoyancy effects, was obtained from the transverse and axial sections performed near the mouth where the stratification was strongest. A tidal sequence of the transverse sections for velocity, temperature, and UVF was developed from a subset of the cross-sectional transects from the 18 May, 1994 survey (Figure 4a-c). Velocity data from the ADCP was generally limited to within about 1.5 m from both the surface and bottom boundaries, and the overall transect was

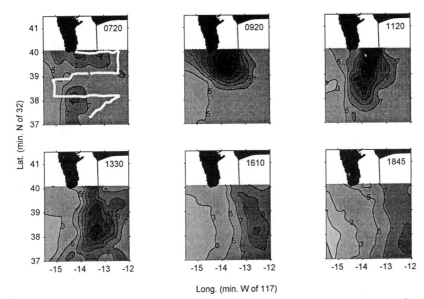

Figure 3. Surface water UVF distributions (@ 1 meter depth) off the mouth of San Diego Bay from the mapping surveys of May 18,1994. The pulse of bay water which issues from the mouth during the ebb is pinched off laterally during the transition to flood and subsequently advected eastward by the longshore current.

restricted to within about 100-200 m of the shore. Twelve temperature and UVF profiles were obtained during the tow-yo's performed on each cross section. The vertical limits of these profiles generally extended from about 1 m below the surface to within 1 m of the bottom. Velocity, temperature, and UVF data were grid interpolated using a universal kriging technique [Olea, 1974]. Temperature and UVF were extrapolated freely to the boundaries of the sections, while velocity was forced to zero at the solid boundaries and left free at the surface boundary.

The tidal sequence began at the slack flood and progressed through the ebb/flood cycle to the subsequent slack flood. We found that the ebb-flow velocities were significantly stronger at the surface with vertical shear $\Delta U_z = O(0.2\ \mathrm{m \cdot s^{-1}})$ during much of the ebb period. The vertical structure of the shear displayed a nearly linear gradient from surface to bottom. The transition from ebb to flood was characterized by a period of vertical exchange flow lasting about 2 hours during which the surface water continued to ebb while the bottom water had reversed to flood. The flood tide velocities were significantly more uniform in the vertical, and the flood-ebb transition of the surface and bottom velocities occurred approximately in phase. From the temperature profiles, we found the thermocline at a high-water depth of about 6 meters. The vertical temperature difference was about 3˚C, leading to a density difference of about 0.5 kg·m⁻³. The pronounced thermocline present at slack high-water broke down during the ebb flow and then rebuilt during the subsequent flood. During the flood tide, the thermocline also displayed a significant cross-stream slope, presumably due to the influence of a relatively strong (10-15 knot) westerly wind during the afternoon. The UVF sections showed the bay water concentrated in the surface layer during the ebb, the appearance of ocean water initially at the channel sides associated with the "pumping" process, and the subsequent return-flow of mixed bay/ocean water at mid-depth.

The axial sections through the mouth region showed quite clearly the thermal structure associated with the observed variations in velocity and stratification at the mouth (Figures 5-

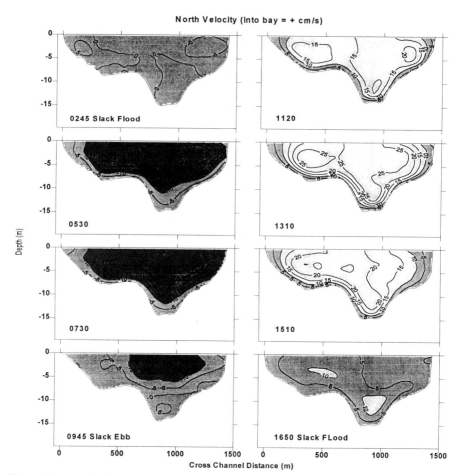

Figure 4. Cross-sectional contours of (a) axial velocity, (b) temperature, and (c) UVF from the tidal cycle transects of May 17, 1994 looking into the bay. Transect times are shown on each plot. Velocity sections were interpolated from shipboard ADCP measurements at 1 meter vertical spacing between about 1.5 meters below the surface and 1.5 meters above the bottom. Positive velocities flow into the bay. Temperature and UVF sections were interpolated from a series of 6 vertical "tow-yo" casts (12 profiles), an example is shown the first plate of the temperature sections. From the velocity sections, note the strong vertical velocity shear during the ebb flow, the period of exchange flow during low-water, and the absence of vertical shear during the flood.

6). At slack high water, a cold thermal-wedge could be seen extending into the bay, enhancing the vertical density gradient near the mouth. In the early ebb, the velocity was weak and the outflow lifted off as it encountered the thermal wedge. As the ebb flow developed in strength, the thermal wedge was broken down by the strong vertical shear, and advected seaward, the southward velocity extending to the bottom. As the ebb weakened, the flow again separated from the bottom, developing into a vertical exchange flow during the transition from ebb to flood. During the flood tide, the cold thermal wedge rebuilt and intruded into the bay with a steep frontal feature reminiscent of a density current.

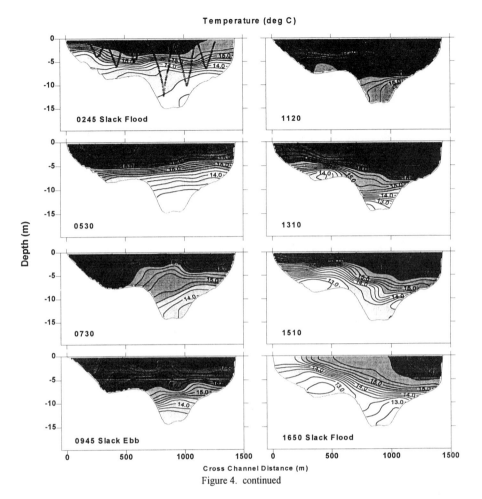

Temperature (deg C)

Figure 4. continued

Buoyancy Effects

The internal hydraulic balance of the thermal-driven buoyancy forces and the barotropic tidal forcing for the flow is characterized by the Froude number [Armi and Farmer, 1986] as

$$F_o = \frac{u_o}{\sqrt{g'h_o}}$$

where u_o is the flow velocity, g' is the reduced gravity, and h_o is the depth. Neglecting frictional effects and mixing, Armi and Farmer showed that strong tidal-forcing of a simple two-layer exchange flow through a topographic expansion leads to a progression of flow conditions. Beginning with a two-layer, baroclinic exchange-flow during the slack tide, the flow structure progresses to an arrested-wedge condition as the barotropic forcing increases, and finally to a fully-flooded condition in which one layer fills the entire section. In this progression, the transition from two-layer exchange-flow, to the arrested conditions occurs when $F_o \sim 1$.

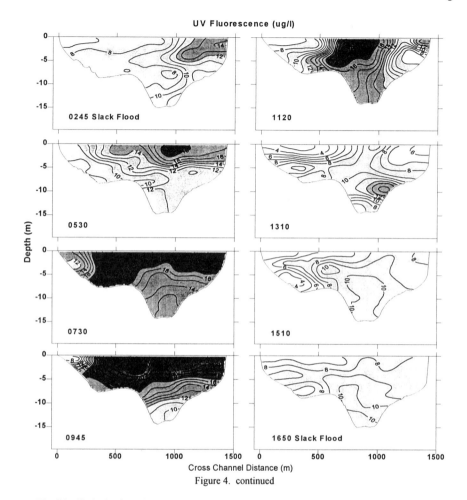

Figure 4. continued

The liftoff of a buoyant jet over a sloping bottom has been examined in the laboratory by Safaie [1979] and in the field by Hearn et al. [1985]. In both cases it was found that the liftoff depth varied in proportion to $F_o^{1/2}$. The empirical relation from Safaie expresses the liftoff depth as

$$h_1 = 0.914 \frac{u_o^{1/2} h_o^{3/4}}{(g_o')^{1/4}}$$

where h_1 is the liftoff depth. To determine when the liftoff initiates, the liftoff depth may be set to h_o and the above equation solved in terms of F_o to give

$$F_o = \frac{1}{0.914^2} \approx 1.2$$

Thus we might expect that liftoff near the entrance should occur when the inertial and buoyancy forces approach a balance with a Froude number in the range of 1-1.2.

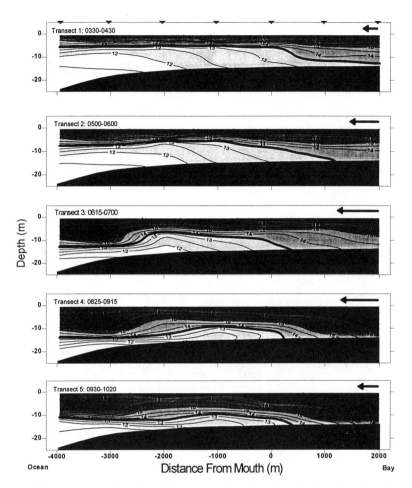

Figure 5. Axial temperature sections through the mouth of San Diego Bay. Sections are interpolated from vertical CTD profiles obtained at 1 kilometer spacing (see triangles) on May 17, 1994. Transect times are shown on each plot. The bold arrows indicate the relative magnitude and direction of the barotropic tidal current. Note the cold thermal wedge present at high-water, its breakdown at the mouth during the ebb, and its re-establishment during the subsequent flood.

For the San Diego Bay mouth region, the topographic expansion studied by Armi and Farmer [1986] is replaced by the hydrodynamic expansion of the outflowing tidal-jet in a situation similar to those studied by Safaie [1979] and Hearn et al. [1985]. For purposes of a simple analysis, the flow velocity was determined for both the sectional average and the sectional maximum, and the density difference was calculated based on the difference between the sectional average and the sectional maximum. The tidal evolution of the velocity structure, the Froude number calculated for both average and maximum sectional velocities, and density gradient are shown if Figures 7a-7c. Examining the progression of the Froude number and velocity based on the observations, we found that during the early ebb, the flow was weak and the Froude number was subcritical ($F_o < 1$) suggesting that during this period eliminate of the

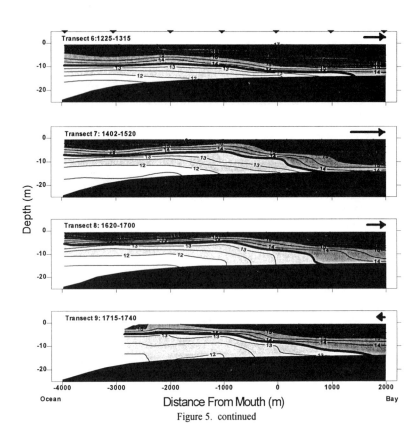

Figure 5. continued

outflow liftoff should occur. During the mid-stage ebb, the flow velocity increased, while the density difference remained approximately constant leading to a period of supercritical flow ($F_o>1$) during which we expect the mouth section to be flooded. As the ebb weakened, the flow again transitioned to subcritical. The flow remained subcritical during the transition from ebb to flood when two-way vertical exchange was observed at the mouth section. Comparing this sequence with the axial flow observations in Figure 6, we found that indeed liftoff did appear to occur near the mouth during the early and later stages of the ebb flow, and two-way exchange was observed during the subcritical period of the ebb-flood transition. The weak vertical exchange during the flood-ebb transition (see Figure 4a) might be explained by the effects of bottom friction in damping the velocity of the lower "layer" during the flood tide. This damping was apparent in the axial velocity sections through the mouth (Figure 6).

Tidal Exchange

To evaluate the effect of thermal stratification on tidal exchange, tidal residual sections were constructed for axial velocity, UVF concentration, and UVF transport, based on the 31 cross sectional transects on 18 May, 1994 (Figure 8). Residuals were estimated by averaging the sectionally interpolated velocity, concentration, and transport at each grid point through the duration of the symmetrical, semi-diurnal tidal period. By choosing a symmetrical tide during which the ebb and flood tidal range were approximately equal, tidal contributions to the estimated residual circulation were minimized. From this analysis, we found that the

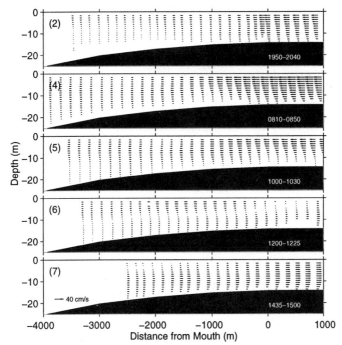

Figure 6. Axial velocity sections through the mouth of San Diego Bay. Sections are based on shipboard ADCP measurements obtained during the mapping surveys on May 18, 1994. Transect times are shown on each plot, and the numbers in parentheses indicate the corresponding temperature transect from Figure 5 based on similar tide stage. Strong vertical shear and liftoff of the outflow are observed during the early stage ebb. During the period of strong ebb, the flow remains bottom-attached, but appears to liftoff again as the ebb weakens prior to flood. Note the vertical exchange during the transition from ebb to flood, and the strong flow at depth during the early stages of the flood.

residual velocity displayed a significant vertical structure, with a tidally averaged surface outflow of $O(0.05 \text{ m·s}^{-1})$ and bottom inflow of $O(0.1 \text{ m·s}^{-1})$. Lateral variations in the residual flow field were also present with enhanced outflow on the eastern portion of the channel. This is inconsistent with Coriolis effects and could be attributed to upstream topography, wind-driven lateral circulation, or cross-channel variations in depth. The tidal residual UVF section also showed significant vertical variation with highest concentrations near the surface $O(15 \text{ μg·l}^{-1})$ and lower concentrations at depth $O(10 \text{ μg·l}^{-1})$. From the resulting net tidal transport of UVF we found that the residual field was characterized by a clear vertical-exchange structure.

A qualitative comparison of vertical and horizontal exchange can be made on the basis of a Temperature-UVF mixing diagram. For this purpose, we identified three primary water masses, the bay water which was warm and high UVF, the surface ocean water which was warm and low UVF, and the sub-thermocline (seasonal thermocline at ~20 m) ocean water which was cold and low UVF (Figure 9). The mixing line described by sectional averages from the mouth transects followed a distinctive cycle through the tide. Starting at slack high water, we found cold, low UVF water associated with the intrusion of ocean water into the mouth region. As the ebb flow developed, temperature and UVF both increased due to the influence of outflowing bay water. At the slack low water, the temperature and UVF reached a maximum. On the reversal to flood, we found that initially the UVF levels dropped dramatically while the

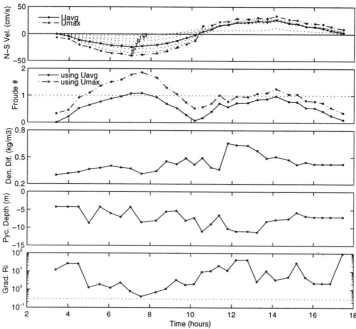

Figure 7. Tidal cycle evolution of (a) cross sectional average and cross sectional mean velocity and velocity at various depths, (b) Froude number based on sectional average and sectional maximum velocity, (c) density difference between the sectional mean density and the sectional maximum density, and (d) pycnocline depth, and (e) gradient Richardson number calculated at the pycnocline depth. All values were developed from laterally averaged profiles from the channel region of the mouth transects (cross channel distance 800-1000 m). Strong vertical shear is present during the ebb and a period of exchange flow occurs during the ebb-flood reversal, corresponding to transitions from critical to sub-critical Froude number. Note the weak vertical shear and development of stratification during the flood and the correspondingly high Richardson numbers.

temperature remained high. This is indicative of the lateral exchange of surface ocean water and bay water associated with the tidal pumping process described by the horizontal mapping surveys. At a point approximately midway through the flood tide, the temperature dropped rapidly, indicating the vertical intrusion of the sub-thermocline, oceanic water associated with the vertical exchange. This cold water intrusion carried with it slightly elevated UVF levels, possibly associated with vertical mixing during the outflow. It is important to recognize that the mixing diagram is not necessarily indicative of true mixing, since the points are derived from sectional averages, but describes mixing in the sense of bulk tidal exchange of waters between the bay and ocean.

Generalization to a Longer Time-Series

To examine the generality of this process, we evaluated the one-month time series from the bottom mounted ADCP, and the vertical thermal structure from a five day period of similar tides (Figures 10-11). We found that the vertical structure and response to thermal buoyancy and tidal barotropic forcing of the flow at this station near the mouth was similar to that suggested by the transverse sections at the mouth. Strong vertical shear observed consistently during ebb tides (0.1-0.3 $m \cdot s^{-1}$) was generally absent during the floods. Reversal of the bottom

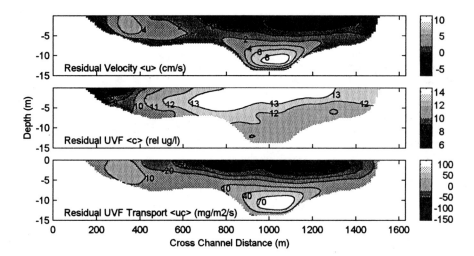

Figure 8. Tidally-averaged residuals from the cross sectional transects of May 17, 1994 for (a) velocity (positive into the bay), (b) UVF concentration, and (c) UVF transport (positive into the bay). The residual velocity is outward at the surface and inward at depth and on the western side of the channel. High UVF is concentrated in the surface, leading to a net vertical exchange with strong outward transport at the surface, and a weaker return of material primarily through the deep part of the channel where the intrusion of the thermal wedge is most pronounced.

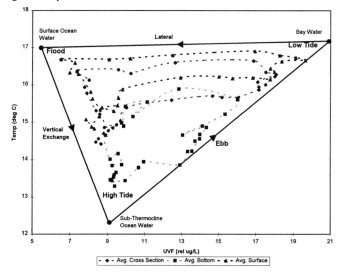

Figure 9. Temperature-UVF mixing diagram based on the cross sectional measurements at the mouth on May 18, 1994. Curves are shown for the complete cross-sectional average, surface section ($z < 6$ m) average, and bottom section average ($z > 6$ m). During the ebb, the section transitions from cold, low-UVF oceanic water to warm, high-UVF bay water. On the flood, lateral exchange (tidal pumping) is evident as warm, low-UVF surface ocean water first enters the section, followed by a period of vertical exchange as the cold, low UVF thermal wedge intrudes.

currents during the ebb-flood transition tended to occur before the surface reversal, especially during neap tide periods. In addition, vertical shear during the neaps was comparable to or greater than during the springs, suggesting enhanced baroclinic circulation during periods of weak tidal mixing.

The tidal variation of surface-bottom temperature difference during the 5 day period 29 July through 2 August, 1994 (Julian date 211-215) indicated that the stratification often builds and peaks during the transition from ebb to flood flow. Flood tide periods generally exhibited stratified conditions. On strong ebb-flows, the thermal stratification often broke down entirely, while on weaker ebbs, only a partial breakdown occurred. This is generally consistent with the cross sectional observations during a moderate ebb flow in which a partial breakdown of the stratification was observed (Figure 7c). This cycle of stratification is in contrast to the recent results of Nunes-Vaz and Simpson [1994] where the effect of straining of the longitudinal density gradient by the vertical shear (SIPS) leads to the build up of stratification during the ebb and a breakdown during the flood. Since the straining component is negative during the flood flow, the stratification cycle observed at the mouth of San Diego Bay could be explained by either advection of the non-zero longitudinal stratification gradient, or on the basis of variations in local vertical mixing rates.

Stratification and Shear

The unusual nature of the stratification cycle described above bears some further analysis. While the observed shear was consistent with a classical gravitational circulation pattern, the cycle of stratification appears to be inconsistent with previous observations in many estuaries [Burau et al., 1994; Simpson et al. 1991]. One possible explanation for this is that the stratification cycle is controlled by advection of the non-zero longitudinal gradient of the vertical stratification, a term that is commonly neglected in one-dimensional analyses of shear/stratification cycles. An estimate of the stratification resulting from SIPS can be obtained following Burau et al. [1994] as

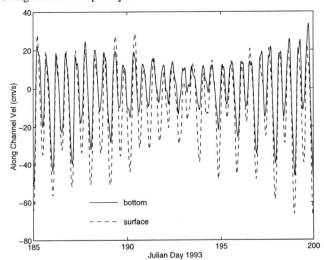

Figure 10. Time-series evolution of surface and bottom velocity from the bottom mounted ADCP during the period June 3 - June 18, 1993. Strong vertical shear is apparent during all ebb tides. The bottom velocity tends to lead the surface velocity during the ebb-flood reversal, especially during periods of neap tides (see Julian day 190-192). Vertical shear during the neaps is comparable to or greater than during the springs, suggesting enhanced baroclinic circulation during periods of weak tidal mixing.

$$\frac{\partial(\rho_z)}{\partial t} = -u_z\rho_x$$

where ρ_z is the vertical density gradient, ρ_x is the longitudinal density gradient, and u_z is the vertical velocity gradient. In a similar fashion, the stratification due to advection can be estimated from

$$\frac{\partial(\rho_z)}{\partial t} = u\frac{\partial(\rho_z)}{\partial x}$$

Thus the ratio of advectively induced stratification to strain induced stratification, signified here as Ω, follows as

$$\Omega = \frac{u\rho_{zx}}{u_z\rho_x}$$

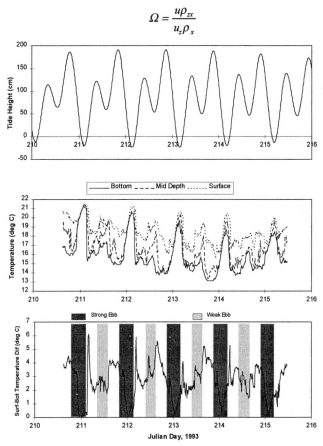

Figure 11. Time-series variations in thermal stratification from the thermistor string at the mouth during the period July 29 - 2 August, 1993. Tides are similar to those shown for days 185-190 in the ADCP record in Figure 10. On the strong ebb flows, the thermal stratification breaks down entirely, followed by a rapid rebuilding during the ebb-flood transition. On the weak ebb flows, the stratification may break down partially or remain unaffected.

Evaluating this ratio for the mouth region of San Diego Bay we have $u \sim 0.5$ m·s^{-1}, $\rho_{zx} \sim 10^{-5}$ kg·m^{-5}, $u_z \sim 0.02$ s^{-1}, and $\rho_x \sim 10^{-4}$ kg·m^{-4}. This gives a value for Ω of about 2.5, suggesting that advection of the non-zero longitudinal stratification gradient may contribute more to stratification than SIPS in this case.

An alternative possibility is that the stratification cycle is controlled by tidal variations in vertical mixing. In this case, the ebb-enhanced vertical shear in the flow would act to break down stratification during the ebb tide, and the absence of shear during the flood would allow the stratification to rebuild. The relationship between vertical mixing and stratification is often characterized by the gradient Richardson number defined as

$$Ri_g = \frac{g\rho_z}{\rho u_z^2}$$

Previous studies have shown that a reasonable local stability criteria is satisfied when $Ri_g > 0.3$ [Geyer and Smith, 1987; Rohr and Van Atta, 1987]. The gradient Richardson number was calculated for the cross sectional data shown in Figure 4. Values were obtained for each transect from the laterally averaged velocity and density profiles within the main channel region (800-1000 m). For each profile, the depth of the pycnocline was determined by locating the maximum vertical gradient in density. The Richardson number was then obtained from the vertical density and velocity gradients at the pycnocline depth. The results of this calculation are shown in Figure 7d-7e for the tidal evolution of the pycnocline depth and Richardson number respectively. The pycnocline depth is found near the surface at slack high water, progresses towards the bottom during the ebb flow, and rises back toward the surface during the flood. We found that during the early ebb flow, Ri_g exceeded the stability threshold but was decreasing. In the strongest portion of the ebb tide Ri_g approached the stability threshold with lowest values of about 0.3-1 occurring near peak flow when the vertical shear was at its maximum. During the transition from ebb to flood tide Ri_g increased to values well above the stability threshold and remained high during the remainder of the flood tide. Since the tidal variation in the vertical density gradient was fairly small, most of the variation in Ri_g can be attributed to tidal variations in vertical shear. These results suggest that tidal variations in vertical mixing due to flood-ebb asymmetries in vertical shear may also contribute to the observed stratification cycle.

Finally, while the observations suggest a significant coupling between thermal gradients and gravitational circulation, it is not clear that the measured values of the tidally averaged vertical shear can be attributed entirely to this phenomenon. Assuming a steady state balance between the longitudinal density gradient and internal friction, Officer [1976] derived a theoretical formulation for the circulation induced by a longitudinal density gradient as

$$u(z) = \frac{g\rho_x h^3}{48\rho N_z}\left[8\left(\frac{z}{h}\right)^3 - 9\left(\frac{z}{h}\right)^2 + 1\right]$$

where N_z is the vertical eddy viscosity. This has extrema at $z/h = 0$ and $z/h = 3/4$ so that the maximum shear due to the imposed longitudinal density gradient can be estimated as

$$D_z u = u\left(\frac{z}{h} = 0\right) - u\left(\frac{z}{h} = \frac{3}{4}\right) = \frac{9g r_x h^3}{256 r N_z}$$

From the San Diego Bay observations we can estimate $\rho_x \sim 10^{-4}$ kg·m^{-4}, $\rho \sim 1024.5$ kg·m^{-3}, and $h \sim 15$ m, while a typical value for N_z may be taken as 0.005 m^2·s^{-1} [Officer, 1976]. The resulting estimate for the vertical shear for San Diego Bay gives $\Delta_z u \sim 0.02$ m·s^{-1}, approximately an order of magnitude less than the observed tidally averaged shear.

This finding suggests that either the vertical circulation is due to some other mechanism, or that the gravitational circulation is enhanced by vertical stratification which limits the vertical mixing (N_z). An estimate of the relationship between vertical mixing and stratification may be obtained from the analysis of Munk and Anderson [1948] as

$$N_z = N_{zo}\left(1 + 3.33 Ri_g\right)^{-3/2}$$

where N_{zo} is the eddy viscosity for neutral stability. Taking $N_{zo} = 0.005$ m^2·s^{-1} and the observed value of $Ri_g \sim 1$, the revised estimate for the vertical shear is $\Delta u_z \sim 0.23$ m·s^{-1}. This analysis suggests that the observed shear is within a range consistent with the observed longitudinal and vertical thermal gradients.

Concluding Remarks

We have examined, from an observational viewpoint, the role of thermal stratification in the tidal exchange process at the mouth of San Diego Bay. We found that the structure of the outflow, and the nature of the tidal exchange process both appear to be influenced by thermal stratification. The tidal outflow is found to lift-off from the bottom during the initial and later stages of the ebb flow when barotropic forcing is weak. During the peak ebb flow, the exit Froude number becomes supercritical, the mouth section is flooded, and the outflow extends to the bottom. As the ebb flow weakens, the Froude number transitions to subcritical, and a period of two-way exchange occurs, with the surface layer flowing seaward, and the deep layer flowing into the bay. The structure of the tidal residual flow and the residual transport of a measured tracer (UVF) are strongly influenced by this vertical exchange. Exchange appears to occur laterally as well in a manner consistent with the tidal pumping mechanism described by Stommel and Farmer [1952]. Tidal cycle variations in shear and stratification are characterized by strong vertical shear and breakdown of stratification during the ebb, and weak vertical shear and build-up of stratification on the flood. Evaluation of multiple tidal cycles from time-series records of flow and temperature indicate that the vertical variations of the flow and stratification observed during the cross-sectional measurements are a general phenomenon during the summer. Analysis of the stratification cycle suggests that it may be regulated by a combination of factors including advection of the non-zero longitudinal gradient of the vertical temperature stratification, and by tidal variations in vertical mixing. Analysis of the observed shear indicates that the tidally averaged vertical circulation probably results from the observed longitudinal density gradient which is enhanced by vertical stratification which limits vertical mixing. Together, these observations suggest that thermal stratification can play an important role in regulating the tidal exchange of low-inflow estuaries.

Acknowledgments. This work was supported by grants from the Naval Command, Control, and Ocean Surveillance Center (#ZW86524A01), the California Regional Water Quality Control Board (IAA #1-188-190-0), and the California Department of Boating and Waterways (IAA #93-100-026-13). Field work and data processing was made possible by support from Chuck Katz, Andy Patterson, Brad Davidson, Kimball Millikan, Ron George, and Jeff Gartner.

References

Armi, L., and D. Farmer, Maximal two-layer exchange through a contraction with barotropic net flow, *J. Fluid Mech., 164,* 27-51, 1986.

Burau, J.R., S. Monismith, and M. Stacey, Hydrodynamic transport and mixing processes in Suisun Bay, in *Proceedings of the Annual Meeting of the Pacific Division AAAS*, 19-23 June 1994, San Francisco State Univ., (in press), 1994.

Dyer, K.R., The salt balance in stratified estuaries, *Est. Coast. Mar. Sci., 2*, 275-281, 1974.

Gartner, J.W., R.T. Cheng, and K.R. Richter, Hydrodynamic characteristics of San Diego Bay, California, Part II, recent hydrodynamic data collection, *Eos Trans. AGU, 75*(3), 60, 1994..

Geyer, R.W., and J.D. Smith, Shear Instability in a highly stratified estuary, *J. Phys. Oceanogr., 17*, 1668-1679, 1987.

Hammond, R.R., Seabed drifter movement in San Diego Bay and adjacent waters, Naval Undersea Research and Development Center, San Diego, Calif., *Tech. Rep. TP507*, 1976. .

Hearn, C.J., J.R. Hunter, J. Imberger, and D. van Senden, Tidally induced jet in Koombana Bay, Western Australia, *Aust. J. Mar. Freshw. Res., 36*, 453-479, 1985.

Katz, C.N., and D.B. Chadwick, 1991. Real-time fluorescence measurements intercalibrated with GC-MS, in *Proceedings, Oceans 1991, 1*, 351-358, 1991.

Katz, C.N., L. Skinner, and D.B. Chadwick, In-situ leach rate measurements from creosote impregnated pier pilings, in *Proceedings, Oceans 1995, 3*, 1722-1729, 1995.

Largier, J.L., C.J. Hearn, and D.B. Chadwick, 1995a. Density structures in low-inflow, seasonal estuaries, aubmitted to: *Proceedings of the 7th International Biennial Conference on Physics of Estuaries and Coastal Seas*, November, 1994, Woods Hole, Mass., 1995a.

Largier, J.L., J.T. Hollibaugh, and S.V. Smith, Seasonally hypersaline estuaries in mediterranean climatic regions, submitted *to Est. Coastal Shelf Sci.*, 1995b.

Nunes Vaz, R.A., and J.H. Simpson, Turbulence closure modeling of estuarine stratification, *J. Geophys. Res., 99*(C8), 16143-16160, 1994.

Olea, R.A., Optimal contour mapping using universal kriging, *J. Geophys. Res., 79*(5), 695-702, 1974.

Peeling, T., A proximate biological survey of San Diego Bay, California, Naval Undersea Research and Development Center, San Diego, Calif., *Tech. Rep. TP389*, 1974.

Rohr, J.J., and C.W. Van Atta, Mixing efficiency in stably stratified growing turbulence, *J. Geophys. Res., 92*, 5481-5488, 1987.

Safaie, B., Mixing of buoyant surface jet over sloping bottom, *J. Port, Waterway, Harbor, and Ocean Division, ASCE, 105*(WW4), 357-373, 1979.

Signell, R.P., and B. Butman, Modeling tidal exchange and dispersion in Boston Harbor, *J. Geophys. Res., 97*(C10), 15591-15606, 1992.

Simpson, J.H., J. Sharples, and T.P. Rippeth, A prescriptive model of stratification induced by freshwater run-off, *Est. Coastal Shelf Sci., 33*, 23-35, 1991.

Smith, E.L., Temperature fluctuations at a fixed position in San Diego Bay, Naval Undersea Research and Development Center, San Diego, Calif., *Tech. Rep. TP298*, 1972.

Stommel, H., and H.G. Farmer, On the nature of estuarine circulation, Part II, *Woods Hole Oceanogr. Inst. Tech. Rep. WHOI-52-51*, Woods Hole, Mass., 1952.

13

Buoyancy Phenomena in the Tweed Estuary

R. J. Uncles and J. A. Stephens

Abstract

Results are presented of buoyancy phenomena using data from a two week field program in the Tweed Estuary, UK. Main channel salinity distributions were strong functions of tidal range and freshwater inflow. During flooding spring tides and fairly low inflows, frontal systems were evident both inside the inlet at around mid-water and, somewhat later in the flood, in the main channel of the estuary. A tidal intrusion front sometimes occurred at the constricted neck of the inlet and an inflow Froude number (F_o^2) criterion, when applied to the neck, appeared to control the timing and shape of the 'plunge' line. Theoretically derived plunge times were within 0.3h of the observed frontal plunge times inside the inlet. The envelope of estimated, maximum inflow Froude numbers on flood tides at the inlet neck was such that $F_o^2>1$ during springs and $F_o^2<1$ during neaps. Inside the estuary main channel, the estimated maximum tidal current speeds at mid-water exceeded the buoyancy current speeds by about 0.18 to 0.25 m s^{-1} at mean spring tides and during low inflow, and the observed frontal migration speeds inside the estuary were consistent with those deduced from surface buoyancy current considerations. Vertical profiling measurements demonstrated that a gravitational circulation occurred in the inlet during the early flood period. Salinity distributions in the lower estuary were particularly complex. Large areas of high salinity waters abutted waters of much lower salinity with the formation of distinct frontal systems. These systems were largely controlled by the bathymetry of the region, especially the shoal morphology.

Introduction

The Tweed Estuary is one of the LOIS (Land Ocean Interaction Study) sites of the British Natural Environment Research Council. The riverine and coastal components of this study cover a catchment area and length of open coastline between the Yare Estuary and the Tweed (on the English-Scottish border, Figure 1(A)). Little information on circulation and salinity is available for the Tweed, despite some work on its chemistry [Gardner and Ravenscroft, 1991]. In this article we present results on salinity distributions and buoyancy phenomena (especially frontal phenomena) within the Tweed based on measurements made during a two-week field program between 14 and 29 September 1993.

The Tweed is particularly important as a fishery for trout and salmon [The Tweed Foundation, 1989 and 1992]. It is a fairly steeply rising and shallow estuary. The tidal limit is located approximately 13 km from the mouth [Mr. I. A. Fox, *pers. comm.*]. The mouth of the estuary is located between the coastal towns of Berwick-upon-Tweed (Berwick) and Tweedmouth. Depths at high water are typically a few meters in the upper estuary and several meters in the lower estuary. The estuary can be strongly stratified with respect to salinity and buoyancy effects are then important.

Buoyancy Effects on Coastal and Estuarine Dynamics
Coastal and Estuarine Studies Volume 53, Pages 175-193

The width at the tidal limit is about 100 m and the width near the old Berwick Bridge, at 2 km from the mouth, is about 300 m (Figure 1(B)). The width increases to about 700 m in the lower estuary, between the Old Bridge and the Harbor Pier. The mouth (a tidal inlet at low water levels) is confined between the Harbor Pier and a sand spit. The spit is dry at low water and is covered at high-water. The width at the narrowest part of the inlet (the neck) is about 50 m at low water.

In this article we show that the inlet neck appears to act as a control section for the inflow of high-salinity coastal waters during the flood portion of spring tides. The analysis is based on the internal hydraulic theory of Armi and Farmer [1986] and the review of estuarine tidal intrusion fronts by Largier [1992].

Observations

Longitudinal transects were undertaken throughout the saline reaches of the Tweed. Vertical profiles of salinity, temperature and turbidity were obtained from surface to bed during a flooding spring tide on 20 September 1993. The instruments were deployed from an inflatable boat that traveled between the inlet (Harbor Pier in Figure 1) and the freshwater-saltwater interface. Vertical profiles of salinity, temperature, turbidity and currents were measured in the neck of the inlet, close to the Harbor Pier wall, during the first 1.4h of a flooding spring tide on 28 September 1993.

Near-surface salinity, temperature and turbidity were measured during a flooding spring tide along three transverse tracks in the lower estuary on 21 September 1993 in order to delineate frontal features there. A self recording salinometer was strapped to the side of the boat which was then steered between navigational waypoints. During these measurements, a series of photographs was taken of the inlet region from 2h after low water to 2.2h before high water. Following this, photographs of the lower estuary were taken from an elevated site near the Lifeboat Station (Figure 1). Other near-surface salinity measurements were made in the lower estuary under similar tidal and freshwater inflow conditions during 20 September 1994.

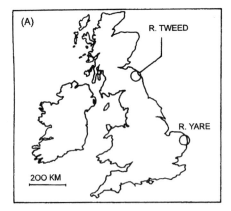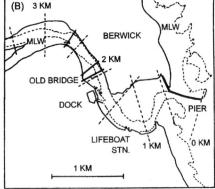

Figure 1. The Lower Tweed Estuary and its location within the British Isles. (A) The 'LOIS' area between the Tweed and Yare Estuaries, shown on a map of the British Isles. (B) The Lower Tweed. Distances along the estuary are marked off in 0.5 km intervals from the Pier Head at the seaward end of the Harbor Pier.

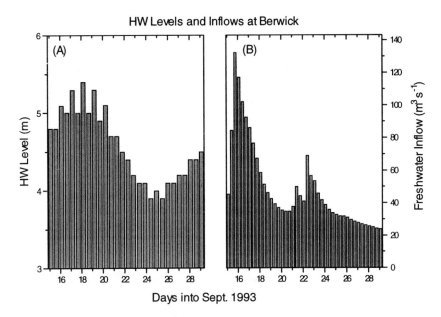

Figure 2. Environmental Data during 15 to 29 September 1993. (A) Predicted high-water levels at Berwick (m); levels are relative to local chart datum. (B) Freshwater inflow from the Tweed and Whiteadder Rivers $(m^3 s^{-1})$.

Environmental Data

Tides at Berwick are semidiurnal with mean spring and neap ranges of 4.1 and 2.5m, respectively. Mean high-water and low-water spring tide water levels are 4.7 and 0.6 m, relative to local chart datum [Hydrographic Office, 1993]. Mean high-water and low-water neap tide water levels are 3.8 and 1.3 m. High-water levels during the field work varied from 5.4 to 3.9 m (Figure 2(A)).

Freshwater inflow to the Tweed Estuary is the sum of runoff from the Tweed and Whiteadder Rivers. The long-term averaged inflow from the Whiteadder is about 8% of that from the Tweed. The long-term, monthly-averaged inflow to the estuary varies from about 140 $m^3 s^{-1}$ during January to about 30 $m^3 s^{-1}$ during July. The hourly-averaged freshwater inflows to the estuary during the field work varied from about 35 to 130 $m^3 s^{-1}$ (Figure 2(B)).

Results

Longitudinal Salinity Distributions

The main-channel salinity distributions were strong functions of tidal range and freshwater inflow. The flood behaviour observed on 20 September 1993 is relevant to the behaviour of tidal intrusion fronts (Figure 3). High-water (HW) level at Berwick was 4.7 m (mean springs) and the freshwater inflow was fairly low (35 $m^3 s^{-1}$). The salt wedge (salinity >30) had moved >1 km up-estuary by 2.6h after low-water (LW+2.6h). The halocline which separated upper and lower layers (salinities <5 and >30, respectively) was about 1 m thick

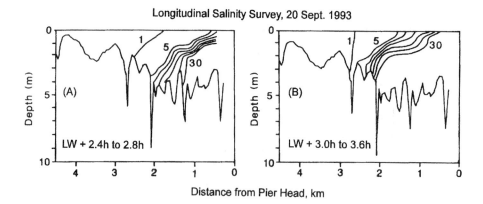

Figure 3. Salinity from longitudinal transects of the Tweed during a flooding spring tide on 20 September 1993. The HW level at Berwick was 4.7 m and the daily-averaged freshwater inflow was 35 m³ s⁻¹. (A) Transect between LW+2.4 and 2.8h. (B) Transect between LW+3.0 and 3.6h.

(Figure 3(A)). By LW+3.3h the salt wedge had moved >2 km into the estuary and the halocline thickness had increased to 1.5 m. In the lower reaches the halocline had mixed to the surface, so that the water column comprised a two layer system with high salinity waters overlain by a highly stratified upper layer of about 2 m thickness. Frontal systems were evident inside the inlet (< 0.5 km from the Pier Head) at LW+2.8h (Figure 3(A)) and in the main channel of the estuary (> 0.5 km) at LW+3.5h (Figure 3(B)).

Measurements in the neck of the inlet on 28 September 1993 illustrate the early flood dynamics of the flow there (Figure 4). Tides were small springs (4.4 m HW level at Berwick, Figure 2(A)) and freshwater inflows were low (25 m³ s⁻¹, Figure 2(B)). A vertical profiling station was occupied close to the Harbor Pier wall and in shallow water to the side of the main channel. At LW+0.2h the water column was homogeneous (salinity ~5 over a depth of 1.2 m) and the ebb-directed current profile was approximately linear. Speeds increased from bed to surface (Figures 4(A,B)). By LW+0.9h a near-bed, high salinity current was observed to be moving up-estuary. This current probably occurred earlier in the deep channel (~4 m) but required a greater depth of flooding water before spreading onto the shallower parts of the section. The upper 1 m still had low salinities (<7). By LW+1.4h the flooding, high salinity basal current had become thicker and the low salinity (<5) upper layer was <0.7 m thick (Figures 4(A,B)).

A classical gravitational circulation therefore occurred within the inlet during the early flood period of this low freshwater inflow, small spring tide. The surface layer thinned while ebbing and decelerating (Figure 4(B)) and the flooding basal current increased in speed. This thinning of the upper layer suggests either that (i) mixing between upper and lower layers was occurring, or (ii) the interface was advected into the section from down-estuary (seawards) or (iii) lateral spreading of the upper layer took place as intertidal areas flooded on the rising tide. A measure of the water column's stability is given by the gradient Richardson Number, Ri:

$$Ri = g\left(\partial\rho/\partial z\right)\Big/\left(\rho\left|\partial u/\partial z\right|^{2}\right)$$

where ρ is local density and z is depth co-ordinate [Turner, 1973]. During the early flood period $Ri>0.25$ except in the almost homogeneous upper layer, which suggests a stable layering. Dyer and New [1986] proposed the use of a bulk (or layer) Ri, Ri_{L}:

Inlet Neck on 28 Sept. 1993

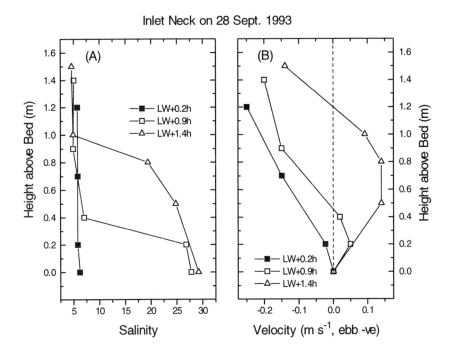

Figure 4. Vertical profiles of currents and salinity at the inlet neck on a flooding spring tide during 28 September 1993. Profiles are at LW+0.2, 0.9 and 1.4h. (A) Salinity. (B) Velocity.

$$Ri_L = gh\Delta\rho/(\rho U^2)$$

where h is water depth, $\Delta\rho$ is bed-to-surface density difference and U is depth-averaged velocity. According to them, when $Ri_L > 20$ a stable interface may exist between upper and lower layers, with internal waves but no significant tidal mixing; when $20 > Ri_L > 2$ the interface is modified by tidal mixing; and when $Ri_L < 2$ then strong tidal mixing occurs. At LW+1.4h, $Ri_L > 20$ and, again, the interface should be stable. However, this ignores possible advection of mixed waters from the deeper and somewhat faster currents in the main channel. The estimated deep channel tidal current speed at LW+1.4h is 0.27 m s^{-1} (deduced from a continuity model, see later) and the deep channel depth is 4.2 m, so that $Ri_L = 11$. This indicates that the interface may be modified by tidal mixing in, and advection from, the deep channel. Significant lateral spreading of the upper layer is unlikely because the inlet width at the neck increases relatively slightly, from 50 to 65 m between LW and LW+1.4h, while the neck section area increases from 160 to 190 m^2, and both increases are too small to explain the 50% reduction in upper layer thickness. An additional factor is the adjustment of the gravitational circulation to increasing tidal current speed. The isohaline interface that separates upper and lower layers must slope upwards towards the coastal sea. Up-estuary advection of this interface due to increasing flood currents therefore will lead to a reduced thickness of the upper layer at a fixed site.

If the fresher surface layer (of thickness h_l) behaved as a buoyancy current and were unopposed by the underlying layer (of thickness h-h_l) it would move down-estuary with a speed [Simpson and Britter, 1979; Simpson and Nunes, 1981]:

$$U_g = C(g'h_1)^{1/2} \qquad (1)$$

with:

$$C = \left[\frac{(2h - h_1)(h - h_1)}{h(h + h_1)} \right]^{1/2}$$

At LW+1.4h the interface lies 0.5 m beneath the surface (Figure 4(A)). In this straight reach of the inlet the interface's cross-channel tilt will be small and $h_l \approx 0.5$ m over the section. In the deep channel, h=4.2 m. The average upper and lower salinities are approximately 5 and 25, respectively (Figure 4(A)). Using Equation (1) and reducing U_g by 20% to account for mixing [Simpson and Britter, 1979; Simpson and Nunes, 1981] gives $U_g = 0.27$ m s^{-1}. The estimated flooding tidal current speed, U, in the deep channel at LW+1.4h (see later) is 0.27 m s^{-1}, which is faster than the observed, lower layer current speed to the side of the channel (0.14 m s^{-1}, Figure 4(B)). If the upper layer were pushed up-estuary by the maximum estimated tidal current in the deep channel of the inlet neck, then the speed of the buoyancy current over the ground would be essentially zero. To the side of the channel, the total depth at LW+1.4h is 1.5 m, and the estimated buoyancy current speed (when reduced by 20%) is approximately $U_g = 0.20$ m s^{-1}. The lower layer current is 0.14 m s^{-1} and $(U$ - $U_g) = $ -0.06 m s^{-1}, which is comparable with the observed, layer-averaged ebb-directed current speed measured in the upper layer at LW+1.4h (Figure 4(B)).

Transverse Salinity Distributions

In the lower estuary, within 2 km of the Pier Head (Figure 1), there were transverse variations of salinity due to the storage of lower salinity surface waters on the edges of the main-channel, where these abutted both the wide, flanking shoals and the sand spit. Strong transverse gradients were observed around mid to late flood. Transverse, topographically-induced salinity variations have been observed in other estuarine systems [e.g.: Huzzey, 1988].

On the early flood, high salinity coastal waters entered the inlet as a basal density current (Figure 4(B)), sometimes with the formation of a tidal intrusion front [Figure 3 and e.g.: Simpson and Nunes, 1981; Stigebrandt, 1988; Largier and Taljaard, 1991]. These flood currents were topographically steered into the lower estuary along the deep channel. Less saline waters were forced onto the edges of the flanking intertidal shoals, sometimes with the formation of long, buoyancy-induced frontal systems [e.g. Huzzey and Brubaker, 1988]. The sand-spit became submerged late in the flood and large volumes of high salinity coastal waters entered the lower estuary across the spit at spring tides. Lower salinity waters were further forced to the edges of the main-channel and sandwiched between high salinity waters on the flanking shoals. Deep channel waters then became effectively homogeneous in the lower 1 km of estuary.

Following vertical mixing in the lower estuary, other longitudinal frontal systems sometimes formed in the deep channel, possibly due to axial convergences [e.g.: Nunes and Simpson, 1985]. These could have been driven by transverse density variations resulting from tidal 'straining'.

Measurements 1 year later, during 20 September 1994, illustrate the flood behaviour of near-surface salinity (Figure 5) during a low freshwater inflow (24 m^3 s^{-1}), mean spring tide (4.7 m HW level at Berwick). During the first period of observations, from LW+1.5 to 2.0h, near-surface salinity within the inlet was low (<20) and <10 at the inlet neck (Figure 5(A)). A local salinity maximum (>15) occurred on the outside of the main channel bend immediately

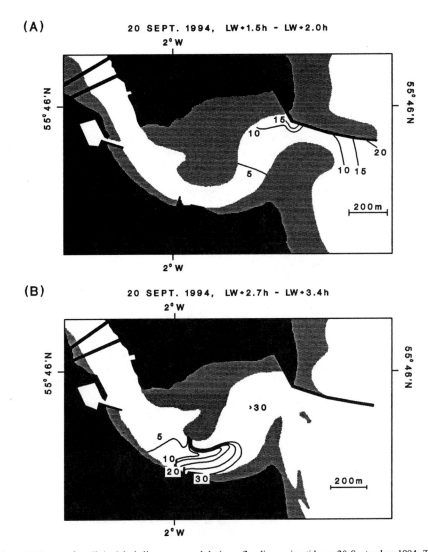

Figure 5. Near-surface (0.4 m) isohalines measured during a flooding spring tide on 20 September 1994. The track started at the Pier Head and ended at the Harbor Dock. (A) LW+1.5 to LW+2.0h. (B) LW+2.7 to LW+3.4h.

up-estuary of the inlet (Figure 5(A)). This was possibly due to centrifugal, cross-channel tilting of the isohalines. Proudman [1953, p. 77] has shown that the isopycnal tilt angle, θ, in a rotating water column water with an upper layer (1) and a lower layer (2), radius of curvature R and velocity of rotation U is:

$$g\tan\theta = R^{-1}\left(\rho_1 U_1^{\,2} - \rho_2 U_2^{\,2}\right)\Big/\left(\rho_1 - \rho_2\right) \tag{2}$$

Assuming a two layer flow with upper and lower layer salinities of 10 and 34, an upper layer velocity equal to zero, an estimated lower layer velocity of 0.37 m s^{-1} (see later) and a radius of curvature at the bend of 150 m, then Equation (2) gives a cross-channel isohaline tilt of:

$$\tan\theta = 0.5 \times 10^{-2}$$

Therefore, the interface elevation (set-up) is estimated to be 0.5 m over the 100 m width of the main, deep channel. This set-up is similar to the observed, upper layer thickness under similar tidal and inflow conditions (although at a slightly earlier time in the flood, Figure 4(A)). Therefore, the halocline possibly intersected the surface on the outside of the bend, which would have led to enhanced surface salinities there.

Observations made between LW+2.7 and 3.4h show that surface estuarine waters had been pushed up-estuary to the base of the spit (Figure 5(B)). An isohaline tilt was evident on the outside of the bend because of enhanced salinities there. A frontal system had developed between low salinity, surface waters in the main channel and high salinity waters flooding the shoals to the north of the channel (measured at about LW+3.1h). The fresher surface waters defined a buoyant plume, greatly modified by centrifugal effects and shoal topography, which was being forced up-estuary by the flooding tidal currents.

Frontal Features

Frontal phenomena were observed on several occasions during the September 1993 field work. Fronts were particularly noticeable after increased freshwater inflows, when buoyancy effects were strongest and floating storm litter was most abundant, and after strong winds, when surf-generated foam was carried into the estuary on the flood. Conditions for the visual manifestation of fronts were particularly favorable during the fine weather on 20 and 21 September 1993, which followed strong inflows (Figure 2(B)).

A 'V' shaped tidal intrusion front was observed at the neck of the inlet and pointing into the estuary (Plate 1(A)) at approximately LW+2.8h during the longitudinal, spring-tide transects on 20 September 1993 (deep-channel data inside the estuary are shown in Figure 3(A)). Tides were mean springs (4.7 m HW level at Berwick) and inflow was fairly low (35 m^3 s^{-1}, Figure 2). Salinity was highly stratified in the seaward reaches of the deeper channel during early and mid flood (Figure 3). The 'V' tip of a tidal intrusion front was observed in the vicinity of the Lifeboat Station (Figure 1) at LW+3.8h. The surface waters inside and outside the 'arms' of the 'V' had salinities of about 34 and 25, respectively. An accumulation of foam occurred along the arms (Plate 1(B)) and at the tip of the 'V'. Foam lines were still evident in this region at LW+4.4h. The location and timing of the front are consistent with observations made 1 year later under very similar conditions (Figure 5(B)).

Well defined, longitudinal foam lines were observed between the bridges (Figure 1) at about LW+3.2h during the transects on 20 September 1993. The salinities at a depth of 2 m across a transverse section located between the bridges (Figure 1) varied from about 5 over the shallower, near-bank regions to about 8 over the deep channel. Longitudinal foam lines also were observed in the upper reaches of the estuary. These occurred after intrusion of the salt wedge.

Frontal features in the lower estuary were investigated during the afternoon flood tide on 21 September 1993. Tides were small springs (4.5 m HW level at Berwick) and inflow was fairly low (42 m^3 s^{-1}, Figure 2). Photographs were taken from the Harbor Pier (Figure 1 and Plate 1(A)) of the tidal intrusion front between 2h after low-water and 2.2h before high-water and thereafter from an elevated site on the southern shore by the Lifeboat Station (Figure 1). A complex pattern of foam-highlighted fronts was observed, with some consistent features which were reproduced throughout the period. We will focus on two of these features: the inlet intrusion front and the across-shoals front, both of which were generated by the interaction of tidal currents with topography and buoyancy.

(A)   INLET TRANSECTS (LW+3.5h)

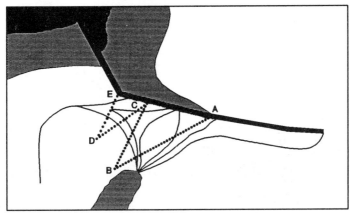

(B)   21 SEPT. 1993, LW+3.5h - LW+3.6h

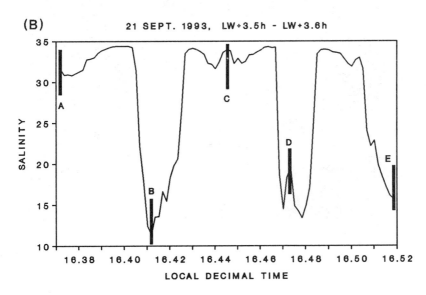

Figure 6. Tidal intrusion front and associated surface salinity in the inlet during the afternoon flood of 21 September 1993. (A) Schematic and qualitative representations of intrusion front plunge-line locations on six occasions (moving from the coastal waters into the estuary at times: LW+2.4, LW+2.7, LW+2.8, LW+3.0, LW+3.3, LW+3.5h, respectively). The zigzag track, A-E, was worked between LW+3.5 to 3.6h. (B) Near-surface (0.4 m) salinity as a function of time along the zigzag track, showing the marked drop in salinity crossing the arms of the 'V' shaped intrusion front.

(A)

(B)

Plate 1. Lower estuary fronts. (A) 'V' shaped tidal intrusion front at the neck of the inlet. The 'V' is pointing into the estuary. The photograph was taken from the Harbor Pier corner at LW+3.3h during the afternoon flooding spring tide on 21 September 1993. (B) Plunge line front and accumulations of foam in the main channel by the Lifeboat Station. The photograph was taken from the Lifeboat Station Pier at LW+4.5h, during the afternoon flooding spring tide on 21 September 1994. Darker estuarine waters are on the left and lighter coastal waters on the right. Anticlockwise rotating vortices occur on scales of cm to m. The spatial scale of the photograph is a few meters.

A tidal intrusion front was first apparent in the inlet at 2.4h after low-water on the 21 September 1993 (the sequence of events is illustrated in a qualitative and schematic way in Figure 6(A)).The buoyant surface waters of the upper layer were forced back towards the inlet

(Figure 6(A)). The front was located at the inlet neck and was slightly convex towards the sea by LW+2.8h. By LW+3.0h the surface front was slightly concave. By LW+3.3h the surface plume was 'V' shaped with the 'V' pointing inwards. By LW+3.5h the 'V' had progressed slightly further into the estuary (Plate 1(A)) with one frontal arm 'tied' to the end of the sand spit and the other moving inwards and along the Harbor Pier wall. At this time a 'ribbon' of differing surface water properties was apparent, moving away from the 'V' of the intrusion front into the estuary and following the curvature of the main channel (Figure 6(A)).

Near-surface salinity was measured between LW+3.5 to 3.6h along a zigzag track (Figure 6(A)) that passed through the lateral extent of the front between waypoints A-E. The near-surface salinity of waters adjacent to the Harbor Pier wall was ~30 at waypoint A (Figure 6(B)). Salinity increased to 34 towards the outer 'arm' of the intrusion and then rapidly fell to <15 as the frontal arm was traversed (approaching waypoint B, Figure 6(B)). The leg D-E traversed both arms of the 'V' shaped front.

The surface salinity field inside the estuary was complex following this time (Figure 7). The surface manifestations of various frontal systems were apparent as foam lines and color differences along the main channel and across the shoals (a schematic and qualitative representation of these demarcation lines, derived from ground-based photographs, is shown on the transect tracks in Figure 7(A)).

The intrusion of high salinity, near-surface coastal waters was evident between the inlet and the base of the spit, and across the shoals (Figure 7(B)). Relatively lower salinity waters (~25) 'hugged' the spit and the boundary between these and the intruding high salinity waters (>30) appears to have generated the observed longitudinal foam line (Figure 7(A)). Lower salinity waters (<20) were trapped on the shoals behind the inlet-estuary entrance. Relatively higher salinity waters (>25) occurred on the outside of the channel bend at the base of the spit, possibly due to centrifugal tilting of the isohalines there. These data imply that the low salinity upper layer was pushed up-estuary by the tidal currents with the formation of a frontal system across the shoals. A second frontal system developed in the main channel by the Lifeboat Station (Figures 1 and 7(A,B)). This front was often observed during the late flood of spring tides and frequently developed a strong foam line in which instabilities developed (Plate 1(B)). The isohalines (Figure 7(B)) resemble those observed a year later (Figure 5(B)) during higher tides and lower inflows and somewhat earlier in the flood.

At 1.0h before high-water the foam-line and surface color patterns implied a frontal system of the form illustrated qualitatively and schematically on Figure 8(A). Other frontal systems may have been present but were not evident from our photographs. The foam lines indicate convergences of surface waters which may have been generated either by axial convergences or juxtaposition of separate water bodies. Color differences are strongly suggestive of the latter. The foam-line which approximately followed the main channel had the form of an axial convergence (Figure 8(A)). The front which was oriented approximately northwest-southeast across the shoals, and which joined the main-channel foam-line, separated greenish coastal waters (to the east) from brownish estuarine waters (to the west).

Near-surface salinity was measured between LW+5.3 to 5.5h (0.8h before HW) across the main channel by the Lifeboat Station and across the shoals (waypoints 3-4-10-4-3 on Figure 8(A)). If the time axis also is viewed as an approximate distance axis, then two spatially separated frontal systems are evident from the record which also had clear visual manifestations (Figures 8(A)). Because the salinometer failed to switch on during the first two minutes of the transect, it is preferable to identify these fronts from the data recorded on the return transect (waypoints 10-4-3, Figure 8(A)). A salinity-temperature regression derived from the transect measurements indicate that the unrecorded salinity between waypoints 3-4 had the approximate form shown by the dashed line in Figure 8(B).

On the first leg of the return transect between waypoints 10 and 4 (Figure 8(B)) the salinity was initially high and varied between about 32 to 34. The salinity dropped from about 34 to 18 as the shoal front was traversed (drawn schematically on Figure 8(A)). The high-gradient zone between the water bodies was less than 40 m wide. On the remainder of the first leg, between the front and waypoint 4, the salinity was fairly low and increased

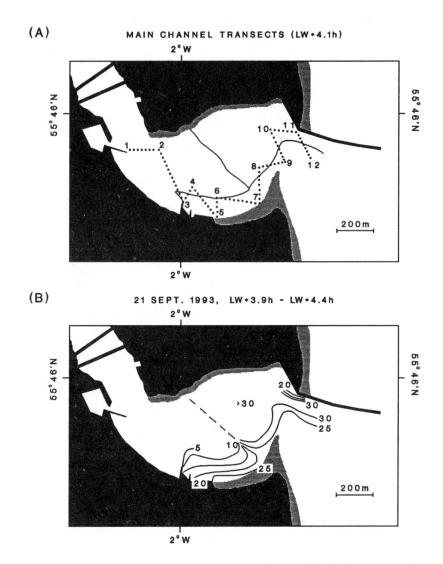

Figure 7. Main-channel transects and corresponding near-surface salinity in the lower estuary between LW+3.9 to 4.4h on 21 September 1993. (A) Survey track and waypoints 1 to 12. Also shown is a schematic and qualitative representation of the locations of some foam and color-highlighted demarcation lines derived from ground-based photographs. (B) Near-surface (0.4 m) isohalines.

approaching the edge of the main channel (18 to 24) although a region of lower salinity surface waters (24 to 18 with a width of about 40 m centered on waypoint 4) was crossed which had no obvious visual manifestation.

On the second leg of the return transect, between waypoints 4 to 3 (Figure 8(B)) the salinity increased from 19 on the north side of the channel to a mid-channel maximum of 33.

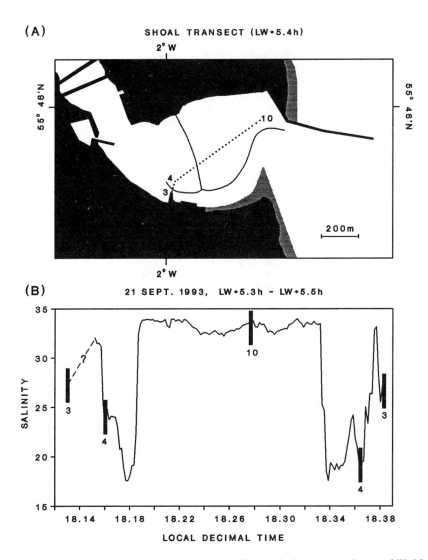

Figure 8. Shoal transects and corresponding near-surface salinity in the lower estuary between LW+5.3 to 5.5h on 21 September 1993. (A) Survey track and waypoints 3, 4 and 10. Also shown is a schematic and qualitative representation of the locations of some foam and color-highlighted demarcation lines derived from ground-based photographs. (B) Near-surface (0.4 m) salinity along the track 3-4-10-4-3.

Surface salinity subsequently fell to about 26 near the Lifeboat Station on the south side of the channel (waypoint 3). The mid-channel salinity maximum was located about 10 m to the south of the foam line (Figure 8(A)). Recorded salinity on the outward legs of the transect showed similar behaviour.

Inflow Froude Number and Buoyancy Currents

The dynamics of the inflowing coastal waters during the flood tides on 20 and 21 September 1993, particularly the formation of a 'V' shaped intrusion front (Figure 6(A)), appeared to depend on topographical and velocity conditions at the neck of the inlet [e.g.: Stigebrandt, 1988 and Largier and Taljaard, 1991]. Where a feature such as a sill or width constriction acts as a control on the flow, then an inflow Froude number can be invoked, F_0, which governs the flow and which is specified in terms of the flow and topographical properties at the control section [Armi and Farmer, 1986].

The square of the inflow Froude number, $F_0{}^2$, is defined by:

$$F_0{}^2 = Q_0{}^2 \Big/ \left(g' h_0{}^3 b_0{}^2 \right) = U_0^2 / g' h_0$$

where Q_0, U_0, b_0, h_0 and g' are rate of discharge, cross-sectionally averaged velocity, width and depth at the sill or constricted mouth and reduced gravity, respectively. For the Tweed on 21 September 1993, measurements gave coastal salinity as 34 and near-surface salinity immediately outside the 'V' of the tidal intrusion front as 15 (Figure 6(B)), so that reduced gravity was $g'=0.15$ m s^{-2}.

A numerical model of tidal flow through the inlet neck was constructed based on continuity of water volume. This is realistic in view of the short length of the tidal Tweed (~13 km) and the fact that maximum flood tidal speeds are reached at mid-water on the rising tide. The surface area of the tidal Tweed, A_s, and the cross-sectional area at the neck, A_0, were tabulated as functions of water level, ς, to yield the flow Q_0 and velocity U_0 at the neck (flood positive) in terms of the rate of change of water level and the freshwater inflow:

$$Q_0 = A_0 U_0 = A_s \partial\varsigma / \partial t - Q_f$$

$F_0{}^2$ was then derived for each flood tide of the fieldwork period, assuming a constant value of $g'=0.15$ m s^{-2} throughout.

The estimated inflow Froude numbers during the afternoon flood tide of 21 September 1993 exceeded unity at the neck (Figure 9(A)). The inflow Froude number at the neck, $F_0{}^2$ exceeded 0.3 at LW+2.1h (Figure 9(B)). At this time the surface layers were expected to be 'blocked' down-estuary of the neck. At LW+2.6h, $F_0{}^2=1$ at the neck and plunging would be expected to occur there. Maximum $F_0{}^2$ occurred at LW+3.7h. If a surface salinity of 5 is assumed rather than 15, then $F_0{}^2$ exceeds 0.3 at LW+2.3h and $F_0{}^2=1$ at LW+3.0h. Therefore, these times are not overly sensitive (<0.4h) to the observed upper layer salinity.

For strong inflows ($F_0{}^2>1$) the channel width at the point of plunging, relative to the inlet neck width, b_p/b_0, is given theoretically by [Largier, 1992]:

$$b_p / b_0 = F_0$$

When $F_0{}^2>1$, tidal forcing was strong and the estimated maximum ratio of channel to inlet neck widths at plunging was approximately 1.4. This indicates that plunging (and thus the location of the 'V' shaped front) occurred up-estuary of the inlet neck after about LW+2.6h.

The envelope of estimated maximum inflow Froude numbers ($F_0{}^2$, Figure 9(B)) on flood tides at the inlet neck was such that $F_0{}^2>1$ during springs (>4.4 m HW level at Berwick).

This suggests that an intrusion front generally occurred at the inlet neck during springs when $F_0{}^2=1$, and then moved into the estuary as flood current speeds increased and $F_0{}^2>1$. An intrusion front was not observed in the inlet during the afternoon flood of 22 September 1993, the day following the period of observations considered here (Figures 6 to 8) and calculations indicate that $F_0{}^2<1$ then (Figure 9(B)).

Inflow Froude Number on Flood

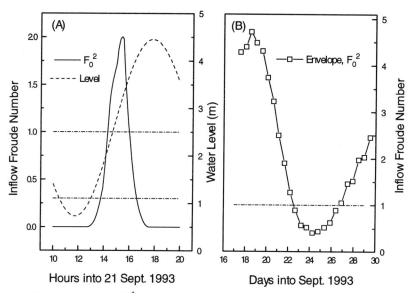

Figure 9. Inflow Froude number ($F_0{}^2$) at the inlet neck during the afternoon flood tide of 21 September 1993, as estimated from a water-volume continuity model of the flow there. (A) $F_0{}^2$ and water level variations during the period, relative to local chart datum. (B) Envelope of maximum $F_0{}^2$ on flood tides during 17-29 September 1993, assuming a constant value of reduced gravity (0.15 m s^{-2}). The critical value for plunging at the inlet ($F_0{}^2 = 1$) is exceeded on every spring tide, irrespective of freshwater inflow during this period.

The intrusion front was observed to occur at the inlet neck around mid-water. At this time the neck still acted as a constriction to the flow, thereby satisfying (in a qualitative sense) the requirements of the Armi and Farmer [1986] model. The cross-sectional areas and inlet widths from the Pier Head to just inside the estuary's main channel showed pronounced minima at the neck (300 m from the Pier Head in Figure 10(A)). According to hydrographic charts [Hydrographic Office, 1992], depths tended to increase moving up-estuary through the inlet, reaching a maximum of 6.5 m at mid-water, about 70 m up-estuary of the neck (Figure 10(A)).

The continuity model has been used to compute section-averaged current speeds through the inlet at mid-water (U_b in Figure 10(B)). At this time the tidal current speeds are maximal and accelerations are small. Therefore, the quadratic frictional drag at the bed is approximately balanced by a surface slope force of magnitude, P, for these barotropic tidal currents. An approximate estimate for the tidal speed at any site of depth h on a section is then:

$$U \approx U_b \left(h/h_b \right)^{1/2} \approx \left(Ph/C_D \right)^{1/2}$$

where C_D is the quadratic drag coefficient and h_b the mean depth over the section (section area divided by surface width). In particular, at mid-water an estimate for the maximum current speed is:

$$U_m \approx U_b \left(h_m/h_b \right)^{1/2}$$

where h_m is the depth of the deepest part of the section. Both section-averaged and estimated maximum current speeds at mid-water have maximum values in the vicinity of the inlet neck (Figure 10(B)).

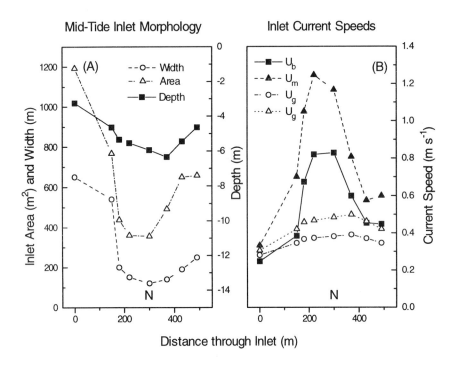

Figure 10. Inlet morphology at mid-tide and estimated tidal and buoyancy current speeds. (A) Inlet area, width and depth at mid-tide as functions of distance from the Pier Head, showing the constriction at the neck (N) of the inlet. (B) Estimated section-averaged and section-maximum tidal current speeds through the inlet at mid-water, together with upper and lower estimates of the buoyancy current speeds.

Using an upper layer thickness in the buoyancy current equation, Equation (1), of 1.5 m in a total depth of h_m, a lower layer salinity of 34 and an average salinity in the upper layer of 15, gives an estimate of the buoyancy current speed in the deepest part of the inlet (Figure 10(B)). Reducing U_g by 20% to account for mixing [Simpson and Britter, 1979; Simpson and Nunes, 1981] leads to $0.4 > U_g > 0.3$ m s^{-1}. Repeating the calculation for an upper layer salinity of 5 and a thickness of 2 m gives an upper limit to the buoyancy current speed of $0.5 > U_g > 0.3$ m s^{-1} (Figure 10(B)). Thinner upper layers (e.g. Figure 4(A)) lead to slower buoyancy current speeds. The estimated, maximum flood tidal current speed in the vicinity of the inlet neck at mid-water, U_m, is 1.2 m s^{-1} and the section-averaged speed, U_b, is 0.8 m s^{-1}.

If the upper layer is pushed up-estuary by the maximum estimated tidal current in the deep channel of the inlet neck, then the migration speed of the buoyancy current front would be in the range of approximately $0.9 > (U_m - U_g) > 0.7$ m s^{-1} and $0.45 > (U_b - U_g) > 0.35$ m s^{-1}. Either side of the neck the frontal speed would be < 0.3 m s^{-1}. The observed apparent migration speed of the front through the inlet is < 0.1 m s^{-1}. Therefore, the apparent frontal migration speed, deduced by treating the upper layer as a buoyancy current, appears to be too fast in the vicinity of the neck.

Inside the estuary main channel ($x > 0.5$ km) the estimated maximum tidal current speeds exceed the buoyancy current speeds by about 0.18 to 0.25 m s^{-1} (Figure 10(B)). The main channel salinity front observed at LW+4.2h (Figure 7(B)) had an apparent migration speed between the inlet-estuary entrance and its observed location of about 0.19 m s^{-1}, which is not unduly slow if the decreasing tidal current speeds following mid-water are taken into account. The estimated up-estuary migration speed during the fastest flood tidal currents on 20 September 1994 (Figure 5(B)) was 0.27 m s^{-1}. Therefore, frontal migration speeds inside the estuary are consistent with those deduced from surface buoyancy current considerations.

Summary

As expected from theoretical considerations, the main-channel salinity distributions were strong functions of tidal range and freshwater inflow [Uncles and Stephens, in press; Fischer et al., 1979; Turner, 1973]. During flooding spring tides and fairly low inflows, frontal systems were evident both inside the inlet at around mid-water and in the main channel of the estuary later into the flood.

The 'V' shaped front which formed in the neck of the inlet was particularly well developed. Vertical profiling measurements illustrated the early flood dynamics of the flow there. A gravitational circulation occurred during the early flood period. The flooding basal current increased in speed and the upper layer thinned while it ebbed and decelerated. Thinning of the upper layer most likely was due to inward advection of the interface as the gravitational circulation adjusted to increasing barotropic flood currents. Treating the upper layer as a buoyancy current indicated that the main channel upper layer should have been 'blocked' at LW+1.4h, which is 0.5h earlier than that deduced from an inflow Froude Number criterion.

The tidal intrusion front that was observed at the constricted neck of the inlet was similar to those described and reviewed by Largier [1992]. In particular, an inflow Froude number criterion, when applied to the neck, appeared to control the timing and shape of the plunge line within the inlet (where high salinity waters plunged beneath lower salinity estuarine waters) in a more satisfactory manner than criteria based on the surface buoyancy current hypothesis. The intrusion front was observed to occur at the inlet neck around mid-water. At that time, and earlier in the flood, the neck acted as a constriction to the flow. A water-volume continuity model was used to compute section-averaged current speeds through the inlet at mid-water. If the upper layer were pushed up-estuary by the estimated maximum tidal current in the deep channel of the inlet neck, then the up-estuary migration speed of the buoyancy current front would be in the range of about 0.9 to 0.7 m s^{-1} (if subjected to maximum tidal speeds over the neck section) and 0.45 to 0.35 m s^{-1} (if subjected to section-averaged tidal speeds). Both speeds appear to be too fast in the vicinity of the neck. The observed apparent migration speed of the front through the inlet was ~0.1 m s^{-1}.

According to the review by Largier [1992], on the early flood tide the outflowing surface waters from the estuary to the coastal zone are blocked when the inflow Froude number at the inlet neck is such that $1 > F_0^2 > 0.3$. These surface waters are then pushed back towards the neck as the flood currents increase. Plunging occurs at the neck when $F_0^2 = 1$. The estimated inflow Froude number on the afternoon flood of 21 September 1993 exceeded unity during strongest flood currents and F_0^2 exceeded 0.3 at LW+2.1h. At that time the surface layers were expected to be 'blocked'. At LW+2.6h, $F_0^2 = 1$ and plunging would be expected to occur at the neck. F_0^2 reached its maximum at LW+3.7h and plunging would have been expected to occur up-estuary of the neck following LW+2.6h. These computed theoretical times are within 0.3h of the observed frontal plunge times.

The envelope of estimated, maximum inflow Froude numbers on flood tides at the inlet neck was such that $F_0^2 > 1$ during springs and $F_0^2 < 1$ during neaps. This suggests that an intrusion front generally occurred inside the inlet during spring tides and then moved into the estuary as flood currents increased in speed approaching mid-water. An intrusion front was not observed in the inlet during the day which followed observations of a strong intrusion front (when $F_0^2 > 1$) and calculations indicated that $F_0^2 < 1$ then.

Inside the estuary main channel, the estimated maximum tidal current speeds at mid-water exceeded the buoyancy current speeds by about 0.18 to 0.25 m s^{-1} at mean spring tides and during low inflow. The main channel salinity fronts had apparent migration speeds of about 0.19 m s^{-1} to 0.27 m s^{-1}. Therefore, frontal migration speeds inside the estuary were consistent with those deduced from surface buoyancy current theory.

On the early flood, vertical profiling in the inlet showed that high salinity coastal waters entered the inlet as a basal density current. This current was topographically steered into the lower estuary along the deep channel and transverse salinity gradients occurred due to centrifugal, cross-channel tilting of the isohalines. Less saline waters were forced onto the edges of the flanking intertidal shoals, sometimes with the formation of long, buoyancy-induced and shear-induced frontal systems. When an intrusion front occurred inside the inlet, the high salinity waters plunged beneath estuarine waters. With rising water levels, the basal current bifurcated as it left the inlet and entered the estuary. One branch turned into the main channel while the other branch retained its course and flooded the shoals. The latter abutted fresher waters which had been pushed to the edges of the main channel by the advancing, high salinity waters of the main-channel branch of inflowing waters. The abutment generated a long, northwest-southeast oriented front that approximately followed the highest ridge of the shoals. The fresher surface waters in the main channel defined a buoyant plume, greatly modified by centrifugal effects and shoal topography, which was forced up-estuary by the flooding tidal currents.

In addition to the strongly plunging intrusion front which developed inside the inlet, a second, strong 'V' shaped frontal system developed in the main channel by the Lifeboat Station. This front was often observed during the late flood of spring tides and frequently developed a pronounced foam line in which instabilities developed. It appeared to form as the main channel salinity front, pushed up-estuary by high salinity surface waters, reached the vicinity of that section.

Acknowledgments. We are grateful to Mr. I. A. Fox, Tweed River Purification Board, for supplying hourly freshwater inflow data for the Tweed and Whiteadder Rivers. We appreciate the kind assistance and hospitality extended to us by Capt. Jenkinson, Queen's Harbormaster, Berwick-Upon-Tweed, and Mr. A. Veitch, Tweed River Superintendent. We thank Mr. Norman Bowley, PML, for his assistance with the field work. This work forms part of the Plymouth Marine Laboratory's contribution to the Land Ocean Interaction Study, Community Research Project of the NERC (LOIS contribution No. 22).

References

Armi, L., and D. M. Farmer, Maximal two-layer exchange through a contraction with barotropic net flow, *Journal of Fluid Mechanics, 164*, 27-51, 1986.

Dyer, K. R., and A. L. New, Intermittency in estuarine mixing, in *Estuarine Variability,* edited by D. A. Wolfe, pp. 321-339, Academic Press, Orlando, Florida, USA, 1986.

Fischer, H. B., E. J. List, R. C. Y. Koh, J. Imberger, and N. H. Brooks, *Mixing in inland and coastal waters,* Academic Press, 483 pp., 1979.

Fox, I. A., *personal communication,* Tweed River Purification Board, Burnbrae, Mossilee Road, Galashiels TD1 1NF, UK.

Gardner, M. J., and J. E. Ravenscroft, The range of copper-complexing ligands in the Tweed Estuary. *Chemical Speciation and Bioavailability, 3*, 22-29, 1991.

Huzzey, L. M., The lateral density distribution in a partially mixed estuary. *Estuarine, Coastal and Shelf Science, 9*, 351-358, 1988.

Huzzey, L. M., and J. M. Brubaker, The formation of longitudinal fronts in a coastal plain estuary, *Journal of Geophysical Research, 93*, 1329-1334, 1988.

Hydrographic Office, Admiralty Chart no. 1612, Hydrographer of the Navy, Taunton, Somerset, UK, 1992.

Hydrographic Office, Admiralty Tide Tables, Vol. 1, Hydrographer of the Navy, Taunton, Somerset, UK, pages 65 and 307, 1993.

Largier, J. L., Tidal intrusion fronts, *Estuaries, 15,* 26-39, 1992.

Largier, J. L., and S. Taljaard, The dynamics of tidal intrusion, retention, and removal of seawater in a bar-built estuary, *Estuarine, Coastal and Shelf Science, 33,* 325-338, 1991.

Nunes, R. A., and J. H. Simpson, Axial convergence in a well-mixed estuary, *Estuarine, Coastal and Shelf Science, 20,* 637-649, 1985.

Proudman, J., *Dynamical Oceanography,* Methuen, London, 409 pp., 1953.

Simpson, J. H., and R. A. Nunes, The tidal intrusion front: an estuarine convergence zone, *Estuarine, Coastal and Shelf Science, 13,* 257-266, 1981.

Simpson, J. E.., and R. E. Britter, The dynamics of the head of a gravity current advancing over a horizontal surface, *Journal of Fluid Mechanics, 94,* 477-495, 1979.

Stigebrandt, A., Dynamic Control by Topography in Estuaries, in *Hydrodynamics of Estuaries, Volume 1 (Estuarine Physics),* edited by B. Kjerfve, pp. 17-25, CRC Press, Boca Raton, Florida, USA, 1988.

Turner, J. S., *Buoyancy Effects in Fluids,* Cambridge Monographs on Mechanics and Applied Mathematics, Cambridge University Press, 368 pp., 1973.

Tweed Foundation, Tweed towards 2000, in *Tweed towards 2000 (A Symposium on the Future Management of the Tweed Fisheries),* edited by D. Mills, Published by the Tweed Foundation, 27 Main Street, Tweedmouth, Berwick-upon-Tweed TD15 2AB, UK, 128 pp., 1989.

Tweed Foundation, *Review and Progress Report,* Published by the Tweed Foundation, 27 Main Street, Tweedmouth, Berwick-upon-Tweed TD15 2AB, UK., 30 pp., 1992.

Uncles, R. J., and J. A. Stephens, Salt intrusion in the Tweed Estuary, *Estuarine, Coastal and Shelf Science,* in press, 1996.

14

Salt Transport Calculations from Acoustic Doppler Current Profiler (ADCP) and Conductivity-Temperature-Depth (CTD) Data: A Methodological Study

David J. Kay, David A. Jay and Jeffery D. Musiak

Abstract

Definition of the mechanisms for scalar transport is important in understanding of biological, chemical, and geological processes in estuaries. The purposes of this study are to 1) understand the meaning of two distinct salt balance expressions based on a tidally varying estuarine volume and a non-varying mean volume, 2) test various simplifications to the salt balance expression based on wave theory, and 3) test the method of using a 1.2 MHz acoustic Doppler current profiler (ADCP) and a conductivity-temperature-depth profiles (CTD) for studying the important transport processes. Velocity measurements made with a 1.2 MHz ADCP and salinity measurements made with a CTD profiler at a cross-section in the Columbia River estuary were used to examine the flux of salt in the estuary. Two forms of the salt balance are examined. The 'full balance' results from a) direct integration of the temporal correlations of velocity and salinity over the time varying cross-sectional area, and b) averaging of this product over a tidal cycle. This expression gives the net change of estuarine salt content over a tidal cycle. The 'mean balance' results from spatial integration of the mean salt conservation equation over the mean cross-sectional area. This removes the necessity of calculating numerous terms involving correlations that are, at least for micro and mesotidal estuaries, in sum small. This expression represents the change in the time-mean salinity in the time-mean upstream estuarine volume. Possible simplifications to these salt balances based on results from wave theory are also examined and tested. Decomposition of the salt flux expression into lateral and vertical "shear terms" was employed with both balances to examine the effects of lateral and vertical variations in the velocity and salinity fields on the net salt flux. Vertical shear terms were found to be significant while lateral shear terms were small. The salt flux was dominated by a balance between landward tidal advection of the tidal salinity field and the seaward mean advection of the mean salinity field. Results suggest that ADCP technology, especially the higher resolution broad-band ADCP, can be used in combination with scalar measurements to make high quality flux measurements in estuaries.

Introduction

Urgent management questions posed by extensive alteration of coastal systems require measurement, comprehension, and prediction of nutrient, carbon and total suspended matter transports. Any ability to predict long-term changes in fluxes must be based on a knowledge of flux mechanisms and their role in and response to changes in estuarine circulation. A knowledge of the salinity dynamics is a necessary precursor to determination of the fate of

Buoyancy Effects on Coastal and Estuarine Dynamics
Coastal and Estuarine Studies Volume 53, Pages 195-212
Copyright 1996 by the American Geophysical Union

other scalars that are harder to measure and whose concentrations are affected by, but more weakly coupled to, the velocity field. The use of ship-board acoustic Doppler current profiler (ADCP) time series and salinity profile time series covering an estuarine cross-section allows accurate estimates of salt transport that only recently are being attempted. In this paper we will present and discuss results of salt flux calculations using harmonic decomposition into tidal species of measurements in the Columbia River estuary using a 1.2 MHz ADCP to measure velocities and a conductivity-temperature-depth sensor (CTD) to measure salinity. The purposes are to a) compare two different approaches to the salt balance based on 1) a tidally varying estuarine volume (henceforth referred to as the 'full salt balance') and 2) a non-varying estuarine volume (henceforth referred to as the 'mean salt balance'), and b) test the applicability of simplifications to the salt balance expression (based on mass conservation and Generalized Lagrangian Mean or GLM theory; Andrews and MacIntyre 1978) in the Columbia River estuary, and c) examine the contributions of lateral and vertical effects to the salt balance by decomposing the salt balance expression into terms involving correlations of deviations from vertical and lateral averages of velocity and salinity.

The estuarine salt balance has been of interest to estuarine scientists for more than three decades.   Numerous measurements of fluxes of salt and suspended particulate matter, investigating both vertical and lateral effects, have been made [e.g. Dyer, et al. 1992; Uncles, et al. 1985; Hughes and Rattray 1980; Kjerfve and Proehl 1979; Murray and Siripong 1978; Hansen 1965]. These studies have involved the use of relatively sparse data in the cross-section and some form of interpolation in both the vertical and the lateral directions.   The development of the Acoustic Doppler Current Profiler (ADCP) and its application to estuaries offers a potential for another look at the salt balance in estuaries.   The use of a boat-mounted ADCP offers a potential advantage over other techniques in measuring estuarine velocity fields because, although some interpolation is still necessary, the velocity fields can be measured with a single boat and with higher resolution than other techniques.   The 1.2 MHz narrowband ADCP (as employed here) allows for 1 m spatial resolution in the vertical and horizontal resolution of O(100m) or less, depending on boat speed and averaging.   The more recently developed broad-band ADCP can improve horizontal resolution by 3-5 fold and vertical resolution by a factor of 2 to 4. ADCPs have a weakness, however, which is their inability to reliably measure velocity near the seabed and near the free surface.

It is important to establish a consistent nomenclature for the numerous transport mechanisms that occur in stratified tidal flow. The net salt flux through an estuarine cross section can be broken down into a) an advection of the cross-sectional mean salinity by the cross-sectional mean velocity (both time mean and tidal components), b) correlations between the vertical and lateral deviations of both time mean and tidal salinity and velocity fields (termed the 'shear effect', Taylor, 1954), and c) the correlations of the salinity and velocity fields with cross sectional area. Tidal advection of the cross-sectional mean salinity is referred to as tidal pumping. Tidal pumping occurs when saline water entering the estuary on flood tide is mixed with fresher water in the inner estuary resulting in fresher water being exported on ebb than was imported on flood [Uncles et al. 1985]. In river estuaries, the net effect of the tidal pumping alone is an increase in estuarine salinity. Shear dispersion, associated with vertical and transverse variations in velocity and salinity fields, may also cause net transport. Salt transport caused by transverse variations may result from many factors, including topography, channel curvature, or the earth's rotation.   Vertical variations are sensitive to topography, tidal range, riverflow, and the level of stratification. An early description of the correlations of vertical variations in the salinity and velocity was made by Bowden [1963, 1965]. Hansen [1965] recognized the importance of variations over the entire cross-section and applied a method that included such variations, but did not separate them into lateral and vertical components. Fisher [1972] explicitly included both lateral and vertical variations and found that lateral effects can be important. Rattray and Dworski [1980] compared different lateral and vertical transport decompositions and showed that the relative contributions of lateral and vertical effects were sensitive to the decomposition employed.

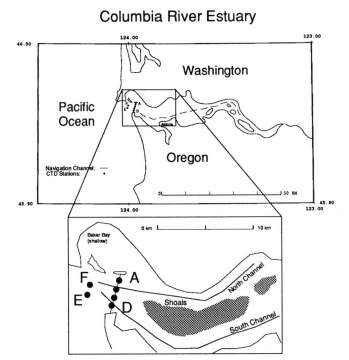

Figure 1. Columbia River Estuary. The position of the cross-section is marked by black dots, with four CTD stations A through D.

The complexity and large number of terms that result from most decompositions of scalar transport render simplification beneficial. We test here three results from wave theory that can be used to simplify the salt balance expression by exclusion of large terms (that may have large measurement uncertainties) [Jay 1991]. First, the measured Eulerian cross-sectionally mean transport may be equated to the sum of the Stokes drift compensation flow and the riverflow . Second, a 'mean salt balance'; an estimate of the flux through the mean (tidal variations removed) cross-section results from the use of standard wave theory and a kinematic boundary conditions. Lastly, a further substitution based on the results of Generalized Lagrangian Mean theory is possible if tidal variations in salinity result only from the time variations in the mean field and mixing [Andrews and MacIntyre 1978]. These possible simplifications will be discusses in this paper.

The Prototype System

The Columbia River estuary (Figure 1) is located in the northwestern United States, has a salinity intrusion of about 45 km, varies in width from about 5 to 15 km, and has a mean channel depth of about 13 m. The main part of the estuary is characterized by two channels; the south channel is the main navigation channel, in which the major discharge of freshwater occurs, and the north channel is where the strongest influx of salt occurs [Jay and Smith 1990]. The section where the study was conducted is seaward of the downstream end of the mid-estuary sands that divide the two channels; nonetheless, the effects of the two channels can be seen in the bottom topography and in water properties as well. The Columbia River discharge varies seasonally, with highest flow in the late spring and lowest flow in the late summer and early fall. In the past the natural discharge ranged from 1000 m^3 s^{-1} to about 35,000 m^3 s^{-1}, but present flow regulations and water withdrawal have resulted in a monthly average discharge in the range of 2800 - 13,000 m^3 s^{-1} [Sherwood et al., 1990]. The density field varies seasonally, being quite variable during low flow periods and partially mixed to highly stratified under high flow conditions. Significant variations in stratification occur on a bi-monthly cycle also, with higher stratification during neap tides than during spring tides. Tides in the estuary are mixed semidiurnal, with a mean range of 2.6 m.

Data Collection

This study was conducted at a cross section 8 km landward from the mouth of the Columbia River estuary (Figure 1). Salinity and velocity data used were collected from a ship over a 37 hr period from 23:00 PST, July 27 to 12:00 PST, July 29, 1991. Riverflow was approximately 4400 m^3 s^{-1}, the greater diurnal range was 2.6 m and lesser diurnal range was 1.7 m. The collection strategy was to visit as many CTD stations as possible, consistent with covering the entire transect in under 60 minutes, providing an optimal compromise between spatial and temporal resolutions. The vessel steamed in a rectangular pattern visiting 6 CTD stations approximately once every 50 minutes while continuously collecting velocity data using a direct-reading 1.2 MHz ADCP. Four of the stations (stations A through D, Figure 1) were positioned so as to cover a cross section of the estuary and the remaining two (stations E and F, Figure 1) were situated 1 km seaward of the cross section to measure along channel gradients. The current profiles were recorded continuously while the ship made the circuit of CTD stations. Recorded ADCP profiles were averages taken over ~70 ADCP 'pings' in ~20 seconds. The profiles of velocity were averaged over 1 m bins in the vertical and ship velocities were subtracted out using the bottom tracking capabilities of the ADCP. Direction and position information were obtained using a gyrocompass and GPS navigation. ADCP measurement errors are discussed in Jay and Musiak [1995].

The velocity field at the cross section was obtained by extracting from the database those profiles measured inside a box, centered at the CTD stations, of width 250 m (across estuary) and length 135 m (along estuary). The box sizes were designed to maximize the number of ADCP profile measurements while not covering an area so large that there were significant bathymetry changes in the box or large systematic variations in velocity. Over the 37 hours that measurements were made at the cross section, an average of 220 velocity profiles and 42 CTD profiles were measured in each box. After corrections were made for tidal height, these measurements gave time series of velocity and salinity in 1 m increments at each station for 75-80% (velocity) and 100% (salinity) of the water column. Because the salinity and velocity times series data points were not commensurate in time and the data collection period was not an exact multiple of a tidal period, each of these time series was subject to harmonic analysis to deconvolve the signal into a mean flow (Z_0) and magnitudes and phases for eight major tidal species represented by tidal constituents K_1, M_2, M_3, M_4, $2MK_5$, M_6, $3MK_7$, and M_8. The harmonic analysis routine used was that developed by Foreman [1978] and Foreman and Henry [1979].

Velocity Profile

CTD density data are valid all the way to the seabed for all stations but ADCP velocity data from the bottom 15% of the water column are unreliable because the acoustic signal is contaminated by reflections of the sound from the seabed. It is also necessary to extrapolate ~ 2 m of velocity data near the free surface. To determine near bed velocities and examine the errors in the flux measurements due to contamination of near-bed data, a velocity profile model was used to more accurately estimate the near-bed velocities. This method, a generalization of a technique used for steady, neutrally-stratified flow by Gelfenbaum and Smith [1986], was developed as part of the analysis of the estuarine momentum balance. The model calculates a complete velocity profile, however, since good velocity data exist in the top 85% of the water column, only the portion of the calculated velocity profile for the bottom 15% is used in the flux calculations. The model involves integrating a linearized momentum balance from the bed to the free surface ζ (b. c. $\delta u/\delta z|_{\zeta}=0$ was used for extrapolation to free surface) and subtracting this from the momentum balance integrated from a point z in the vertical to ζ. An iterative integration is performed to find the bedstress that minimizes the discrepancy between the observed and the calculated profile. Defining z as the vertical coordinate (positive upward) and s as the local stream-wise coordinate (positive landward), the linearized equation of motion in the direction of flow is:

$$\frac{\partial u}{\partial t} + g\frac{\partial \zeta}{\partial s} + \frac{g}{\rho}\int_{z}^{\zeta}\frac{\partial p}{\partial s}dz' - \frac{\partial}{\partial z}\left(K_m\frac{\partial u}{\partial z}\right) = 0 \tag{1}$$

The convective accelerations ignored here are, when integrated over the water column, small for low angle beds. Their neglect may, however, cause errors at some depths, particularly in interfacial layers. Equation (1) is integrated from a point z in the flow, to the free surface ζ.

$$\int_{z}^{\zeta}\frac{\partial u}{\partial t}dz + g(\zeta - z)\frac{\partial \zeta}{\partial s} + \frac{g}{\rho_o}\int_{z}^{\zeta}\int_{z'}^{\zeta}\frac{\partial p}{\partial s}dz''dz' + \tau_{zs}|_{z} = 0 \tag{1a}$$

Equation 1, when integrated from the roughness height z_o to the surface ζ, gives:

$$\int_{z_o}^{\zeta}\frac{\partial u}{\partial t}dz + g(\zeta - z_o)\frac{\partial \zeta}{\partial s} + \frac{g}{\rho_o}\int_{z_o}^{\zeta}\int_{z'}^{\zeta}\frac{\partial p}{\partial s}dz''dz' + \tau_b = 0 \tag{1b}$$

Then, dividing (1a) by H and dividing (1b) by (ζ-z) and subtracting, the surface-slope term is removed and, after rearrangement, the following equation results:

$$\frac{\partial u}{\partial z} = \frac{1}{Km}\left[\underbrace{\frac{(\zeta-z)}{H}\int_{z_o}^{\zeta}\frac{\partial u}{\partial t}dz'}_{(i)} - \underbrace{\int_{z}^{\zeta}\frac{\partial u}{\partial t}dz'}_{(ii)} + \underbrace{\frac{g}{\rho_o}\frac{(\zeta-z)}{H}\int_{z_o}^{\zeta}\int_{z'}^{\zeta}\frac{\partial p}{\partial s}dz'dz'}_{(iii)} - \underbrace{\frac{g}{\rho_o}\int_{z}^{\zeta}\int_{z'}^{\zeta}\frac{\partial p}{\partial s}dz'dz'}_{(iv)} + \underbrace{\frac{(\zeta-z)}{H}\tau_b}_{(v)}\right] \tag{2}$$

where $Km = u_*(kze^{(\frac{-z}{H})})S(Rig)$ and $S(Rig) = (1+3Rig)^{-1}$

R_{ig}, the gradient Richardson number, is the ratio of gravitational to inertial forces,

$$R_{ig} = \frac{\frac{-g}{\rho_o}\frac{\partial p}{\partial z}}{\left(\frac{\partial u}{\partial z}\right)^2}$$

The friction velocity u_* appears in (2) in two places. The first is in term (v), $\tau_b = u_*^2/\rho$. The second is in the eddy diffusivity K_m that multiplies the entire equation. The terms in (2) are evaluated from data and u_* is iterated so that the calculated profile matches as well as possible (in a least squared sense) the measured profile in the outer 85% of the flow. Terms (i), (ii), and (v) on the right hand side of (2) are evaluated for the top 85% of the water column, which in our case is above about 3 m from the bed. Terms (iii) and (iv) are evaluated for the entire water column as density data are good all the way to the bottom for all stations. The gradient $dp(z,t)/ds$ is obtained from density differences between cross-section stations (A through D) and the two stations seaward of the cross-section (stations E and F). Although the deepest valid velocity data are ~3 m above the bed, estimates of the Richardson number and the acceleration are necessary for the bottom 3 m. These are brought down to the bed from the first good data point (3 m above the bed) as quadratic functions of depth, allowing for an estimate of the stratification correction and, therefore, K_m. The K_m stratification correction used is that of Lehfeldt and Bloss [1988]. The model appears to fit the data in the outer 85% of the flow well at maximum flood and ebb tides, but not as well around slack water when errors in acceleration may be large and the eddy-diffusivity turbulence modeling may not adequately represent the vertical momentum exchange. Examples of the model output and their comparison to the data are shown in Figure 2. A comparison of flux calculations using modeled near-bed velocities and those using raw ADCP near-bed velocity data is made in a section on errors below.

The use of this boundary calculation method of extrapolating to the bed should work better with a) newer ADCP that samples to within 8% of depth to bed. and b) a more sophisticated turbulence closure. The method is well suited for ADCP studies where determination of bed-stress is important.

The Salt Balance

In our examination of the longitudinal salt flux we are interested in three primary issues; a) determining the relative importance of tidal transport versus mean transport in the Columbia River, b) comparing the role of lateral and vertical shear dispersion in transport processes, and c) examining various simplifications to the salt balance expression. The velocity and salinity are decomposed into a time-mean and tidally varying parts, explicitly separating the first 8 tidal species. Higher frequency turbulent fluctuations were neglected as contributing negligibly to horizontal scalar fluxes. Thus:

$$u = \overline{u} + \Sigma u'_j \qquad\qquad S = \overline{S} + \Sigma S'_j \qquad\qquad A = \overline{A} + \Sigma A'_j$$

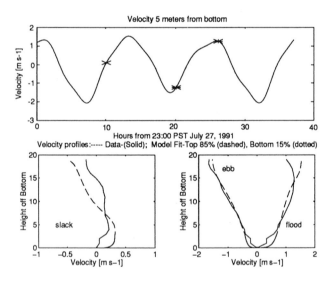

Figure 2. Top: Time Series of velocity at 5 m above the bottom, reconstructed from harmonic constituents. The marks (* - ebb and flood profiles, x - slack profile) show the times of the profiles pictured in the lower panels. Bottom: Reconstructed low-current, ebb, and flood current profiles collected with the ADCP (solid)) and corresponding momentum balance model output profile (top 85% of flow - dashed. bottom 15% of flow - solid).

where the overbar signifies a time mean, the prime signifies the tidal part, and the index j spans the first eight tidal species. U is the velocity, S is the salinity and A is the cross-sectional area. These definitions assume that temporal changes in u, S, and A are either tidal (accounted for in the primed variables) or occur slowly relative to a tidal day (a "two-timing" assumption, Middleton and Loder, 1989). Velocity and salinity fields were then divided into cross sectional averages, deviations from vertical averages, and deviations from lateral averages [Dyer 1974]; denoted as follows:

$$\overline{u}_t = \overline{u} - \overline{\{u\}} \qquad \overline{u}_v = \overline{\{u\}} - \overline{<\{u\}>}$$

$$u'_{t_j} = u'_j - \{u'_j\} \qquad u'_{v_j} = \{u'_j\} - <\{u'_j\}>$$

Therefore,

$$\overline{u}_t + \overline{u}_v = \overline{u} - \overline{\{<u>\}} \qquad u'_{t_j} + u'_{v_j} = u_j - \{<u'_j>\}$$

where { } signifies a lateral average, ⟨ ⟩ signifies a vertical average, and summation over j is implied. The subscript t signifies the deviation from the lateral average and subscript v signifies the deviations of the lateral average at a given depth from the cross-sectional average. This decomposition allows separation of the flux into contributions from the cross-sectional mean, and the lateral and vertical deviations from the mean, the so called 'shear terms'. Once the velocities and salinities are decomposed as above, lateral and vertical dispersion effects on the total salt flux can be examined if velocities and salinities have been adequately sampled.

Salt Conservation in Time Varying Volume (Full Salt Balance)

An examination of the 'full salt balance' begins directly from conservation of salt into and out of a temporally varying control volume [Dyer 1974]. It involves the evaluation of the flux (the product $u(x_o,y,z,t)A(x_o,t)S(x_o,y,z,t)$) everywhere in the cross-section and throughout a tidal cycle, or equivalently, the change in average (over a tidal cycle) amount of salt in the time varying upstream estuarine volume. Defining b as the estuary width, the control volume expression equating influx with change in inventory is:

$$\overline{F} = \overline{\int_{-h}^{\zeta} b(x_o,z)\{u(x_o,y,z,t)S(x_o,y,z,t)\}dz} = \frac{\partial}{\partial t}\int_{x_o-h}^{x_h}\int b(x,z)\{S(x,z,t)\}dzdx \qquad (3)$$

Equation 3 is simply the integral salt conservation equation for a control volume that extends from a cross-section at (x_o) to the head of the tide at the landward end of the estuary (x_h) as shown in Figure 3. The width b is assumed to be independent of time; acceptable in this case because large tidal flats do not border the cross-section. The time average is over the entire integral.

Schematic Estuary

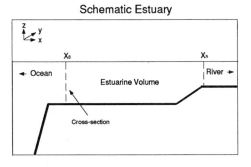

Figure 3. Schematic Estuary.

When velocity and salinity fields are decomposed as above, the left hand integral in Equation 3 can be written as [Dyer 1974]:

$$\overline{F} = \overline{\overline{A} < \{u\} > < \{S\} >} + \overline{\overline{A} < \{u_j'\} > < \{S_j'\} >} + \overline{\overline{A_j'} < \{u_j'\} > < \{S\} >} +$$
$$\quad\quad\quad 1 \quad\quad\quad\quad\quad\quad 2 \quad\quad\quad\quad\quad\quad 3$$

$$\overline{A_j' < \{S_j'\} > < \{u\} >} + \overline{A_j' < \{S_j'\} > < \{u_j'\} >} + \overline{< \{A_j'u_{tj}'S_{tj}'\} >} + \overline{< \{A_j'u_{vj}'S_{vj}'\} >}$$
$$\quad\quad 4 \quad\quad\quad\quad\quad\quad 5 \quad\quad\quad\quad\quad\quad 6 \quad\quad\quad\quad\quad 7$$

$$+\overline{\overline{A} < \{\overline{u}_t\overline{S}_t\} >} + \overline{\overline{A} < \{\overline{u}_v\overline{S}_v\} >} + \overline{\overline{A} < \{u_{tj}'S_{tj}'\} >} + \overline{\overline{A} < \{u_{vj}'S_{tvj}'\} >}$$
$$\quad\quad 8 \quad\quad\quad\quad\quad\quad 9 \quad\quad\quad\quad\quad\quad 10 \quad\quad\quad\quad\quad 11 \quad\quad\quad (4)$$

Term 1 is the product of mean (cross-sectional and time) advection of the mean salinity field. Term 2, involving cross-sectional means also, is the tidal advection of the tidal salinity field, also known as tidal pumping. Terms 3-7 are those associated with correlations with tidal changes in cross-sectional area. Terms 8-11 are shear terms, involving correlations of lateral and vertical deviations in velocity and salinity. The numerous triple correlations terms, which have been found small in the Columbia river estuary [Hughs and Rattray 1980, also referred to as HR80] are the most difficult to measure and are neglected as small in sum.

It is vital to determine the variation of the magnitude of F and the relative sizes of each of the terms over the neap-spring cycle and with changes in riverflow because these a) indicate how transport mechanisms vary with external forcing, b) are vital to validating numerical transport models, and c) assist in understanding effects of anthropogenic changes.. In addition, their difference from system to system is potentially a criterion for useful classification of estuaries. In salt balance studies where the magnitude of the flux is desired, but measurements are sparse, substitutions based on mass conservation and resulting from Generalized Lagrangian mean theory can be made that might allow accurate estimates.

Water Conservation

Because correct salt flux calculations require accurate measurements of mass flux, and because the water balance offers a potential simplification to the salt balance, a test of water conservation is a crucial precursor to evaluating a salt balance. Tidal propagation in estuaries is known, from wave theory, to cause a mean non-Eulerian mass flux in the direction of propagation called the Stokes drift [Uncles et al 1985]. The Stokes drift drives a measurable Eulerian return flow, referred to as the Stokes drift compensation flow. The total measured mean Eulerian flow is then the sum of the river flow and the Strokes drift compensation flow [Jay 1991]:

$$-\overline{Q}_R - \overline{Q}_S = \overline{Q}_E \quad\quad\quad (5)$$

where Q_R is the river transport, Q_S is the Stokes drift, and Q_E is the measured Eulerian mass flux. Q_R and Q_S are positive definite quantities and landward transport is positive in sign. Thus there is a mean Eulerian flux, Q_E, driven by a mean surface slope generated by both the riverflow and the Stokes drift. Estuaries often show systematically steeper surface slopes on spring tides [Allen et al. 1980], reflecting the enhanced Stokes transport, which varies with the square of the tidal amplitude. Term 1 in Equation 4 is simply the product of the Eulerian transport and the cross-sectionally averaged salinity. If riverflow transport is known and the Stokes drift can be reliably estimated from measured surface velocities and tidal heights, the mass conservation expression can be substituted into the salt flux expression. This substitution may provide a means of avoiding large errors that often accompany measurements of the time mean cross-sectionally averaged velocities at an estuarine cross section.

GLM Simplification

A further substitution based on the results of Generalized Lagrangian Mean theory is possible if tidal variations in salinity result only from the time variations in the mean field and mixing [Andrews and MacIntyre 1978, Jay 1991]. For small-amplitude tides (relative to mean depth) and channelized estuaries, there is an equivalence of the tidal advection of the tidal salinity field and the product of the Stokes transport of the mean salinity:

$$\overline{A\{< u'_j >\} < \{S'_j\} >} = \overline{Q}_S < \{\overline{S}\} > \tag{6}$$

The left hand side of Equation 6 is identical to the tidal pumping term in the salt balance expression (term 2 of Equation 4). When this substitution is made in Equation 4, terms 1 and 2 are replaced by $Q_R \overline{\{[S]\}}$. This may be an improved starting point if riverflow is a better constrained quantity than the cross-sectional mean velocity field.

Salt Conservation in Non-Varying Volume (Mean Salt Balance)

An alternate expression of a 'mean salt balance' has been proposed by Jay [1991]. Salt is conserved by spatially integrating the tidally averaged salt conservation equation over the time average volume. The salt balance equation that results, however, has a slightly different interpretation than the 'full salt balance' (Equation 4). The approach, based on standard assumptions of small amplitude wave theory, spatially integrates the tidally averaged salt conservation equation over the time-average volume [Jay 1991]. This differs from Equation 4, which results from spatially integrating the instantaneous salt conservation equation, and subsequently taking the time average. This balance expresses a measure of the average flux through a mean cross-section, or equivalently, the change (over a tidal cycle) in average amount of salt in the mean upstream estuarine volume:

$$\overline{F'} = \int_{-h}^{\overline{\zeta}} \overline{b(x_o,z)\{uS\}}dz = \tfrac{\partial}{\partial t}\int_{x_0-h}^{x_b}\int^{\overline{\zeta}} \overline{b(x,z)\{S(x,y,z,t)\}}dzdx \tag{7}$$

The derivation of (7) is lengthy, and the reader is referred to Jay [1991] for details. When the flux on the left hand side of (7) is broken down into averages and deviations from averages, the following expression results:

$$\overline{F'} = \overline{A< \{u\} >< \{S\} >} + \overline{A< \{u'_j\} >< \{S'_j\} >} + \overline{A < \{\overline{u}_t\overline{S}_t\} >} +$$
$$\overline{A < \{\overline{u}_v\overline{S}_v\} >} + \overline{A< \{u'_{tj}S'_{tj}\} >} + \overline{A< \{u'_{vj}S'_{vj}\} >} \tag{8}$$

The will henceforth be referred to as the 'mean salt balance'. Equation (8) does not involve time variations in area because they are eliminated using a small amplitude assumption and a kinematic boundary condition at the free surface. Jay [1991] proposed to use the mass conservation and GLM to further simplify this expression, replacing the first two terms as was discussed above.

It is emphasized that (4) and (8) actually calculate slightly different salt inventories and F and F' are distinct fluxes. Examination of the right side of (3) and (7) illustrates this fact. Equations (3) represents a measure of the total change in salt upstream of the cross-section integrated over a tidal cycle, while (7) represents a measure the change in the amount of salt in a tidally mean upstream estuarine volume. To illustrate the difference, it is noted that, writing the upstream volume as $V = \overline{V} + V'$ and the upstream volume-averaged salinity as $[S] = [\overline{S}] + [S]'$; the rate of change of the salt inventory can be decomposed as:

$$\tfrac{\partial}{\partial t}\overline{(V[S])} = \tfrac{\partial}{\partial t}\overline{(\overline{V}[\overline{S}]} + \overline{V'[S]'})\tag{9}$$

$$\underset{1}{} \qquad \underset{2}{} \qquad \underset{3}{}$$

The left hand side of (9), term 1, is the change in the tidal cycle average of the total estuarine salt content, as calculated by the 'full salt balance' (4). The first term on the right hand side of (9), term 2, is the change in the product of the mean salinity and the mean upstream estuarine volume, as calculated by the 'mean salt balance' (8). Models of wave processes often set boundary conditions on mean quantities so the value of term 2 may be more interesting from a wave flux model point of view. Term 2 is the rate of change in salt content of the landward volume at mean sea level, from tidal cycle to tidal cycle. If V' and S' are small in magnitude or uncorrelated, term 3 is small and term 2 dominates. In most cases where tidal transport of salt occurs, however, V' and S' are correlated. Only if the magnitudes of V' and S' are small will the two balances give similar results. This may be the case in a micro-tidal estuary or meso-tidal system such as the Columbia River.

Results and Discussion

Velocity and Salinity Fields

For the computation of the velocity and salt fields and the salt flux, the cross-section studied was decomposed into sub-areas of vertical extent of 1 m and variable width as shown in Figure 4. Time series of velocity and salinity from each one meter depth interval at each of the four stations were assumed to represent average values over each of the sub-areas shown in Figure 4.

The residual (Z_0) and semidiurnal (M_2) velocity and salinity structure at the cross-section is shown in Figure 5. The residual velocity structure is characterized by outward net flow in the south, navigation channel and a small inward net flow in the north channel near the bed. The semidiurnal velocity structure shows a $>20°$ phase change over the water column, with the tide at the bed changing direction earlier than at the surface because of the greater inertia of the faster moving surface water. These patterns are consistent with those derived from current meter records for this section by Jay and Smith [1990]. The cross section is situated downstream of a channel curvature with center of curvature to the north and upstream of a bend with the opposite curvature. The Z_0 salinity shows sloping isohalines consistent with control of the salinity field by the centrifugal force due to channel curvature acting on the net seaward velocity on the more landward of the two bends. This is to be expected because the net flow is outward. M_2 salinity structure at the cross section shows a maximum amplitude at mid-depth in the North Channel, consistent with large vertical excursions of a mid-depth interfacial layer in salinity and filling of most of the estuary's tidal prism from the North Channel.

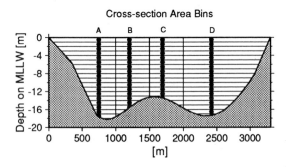

Figure 4.   Cross-section of Columbia River estuary studied, showing distribution of sub-areas used for applying harmonic decomposition of velocity and salinity time series. Sub-areas measure 1 m in the vertical and have variable width. Solid circles show positions of CTD salinity data.

Mass Fluxes

The measured Eulerian water transport through the mean cross-section was measured to be
-5900 $m^3 s^{-1}$. Since riverflow transport is known and the Stokes drift can be reliably estimated
from measured surface velocities and tidal heights, the mass conservation expression (Equation
5) is used to check the consistency of the measured Eulerian transport. Correlations between
measured tidal heights at km 30 (corrected to the study cross-section using known phase lags,
Giese and Jay 1987) and surface velocities obtained with the ADCP give a value of -2100 $m^3 s^{-1}$
for the Stokes drift. Riverflow at the upstream end of the estuary was measured by U.S.
Geological Survey to average -4400 $m^3 s^{-1}$ over the measurement period. The sum of the river
transport and the Stokes drift compensation transport (-6300 $m^3 s^{-1}$) matched reasonably well
the measured Eulerian transport (-5900 $m^3 s^{-1}$); the difference is about 7%. A more accurate
result might have been obtained with a pressure gauge located on the cross-section. The
results are summarized in Table 1.

Salt Fluxes

The role of shear dispersion, especially lateral shear dispersion, is important in
understanding whether the common approach of treating estuaries as two-dimensional systems
in the along channel and vertical directions, neglecting lateral currents and lateral variations
in longitudinal currents, is a useful characterization of transport processes. In macro-tidal
systems with large variations in width over the tidal cycle and

Table 1. Mass Flux

TERM	Volume Flux [$m^3 s^{-1}$]
$-\overline{Q}_R$	-4400
$-\overline{Q}_S$	-2100
$-\overline{Q}_R - \overline{Q}_S$	-6300
$\overline{Q}_E$	-5900

in wide estuaries, the lateral shear dispersion terms are often important [Uncles et al., 1985].
The Columbia River doesn't fall into either of these classes, so lateral shear   terms might be
expected to contribute only slightly to the salt flux. The calculations presented below suggest
that, at the cross section studied, the lateral shear terms are in fact small despite its proximity to
the division between the north and south channels.

The total flux, based on the 'full salt balance' (Equation 3), in the cross-section, before
breakdown into mean and shear terms, is shown in Figure 6. The residual flux is seaward
everywhere except near the bed in the north channel. Consistent with the results of Jay and
Smith [1990], salt enters the system in the north channel and leaves it in the south channel,
requiring salt transport from north to south across the region of sand shoals in the center of the
estuary landward of the cross-section (see Figure 1). The tidal flux is landward everywhere in
the cross-section. The vertical structure of the shear terms is shown in Figure 7a. The vertical
shear dispersion terms (panels A & B) are positive everywhere in the vertical, with maximums
at the bed and surface in both the residual and the tidal. This is consistent with the

predominance of mean two-layer circulation in the Columbia River, i.e. higher than average salinity water flowing in at the bottom and lower than average salinity water flowing outward at the top. The vertical distribution of the lateral shear terms (panels C & D) are also shown for comparison, although their contribution was found to be small. The lateral distribution of the lateral shear terms are shown in Figure 7b. The lateral terms, though small, have a distribution consistent with the channeling of river flow down the south channel, at stations C and D; i.e. a strong outward flow of lower than average salinity water in the south channel.

The relative size of the terms in the flux expression suggests the relative importance of the flux mechanisms in the Columbia River. Although the terms in the 'mean salt balance' expression are a subset of the terms in the 'full salt balance', the results of the two expressions are presented separately for comparison. The terms in the full salt balance (Equation 4) are listed in column 1 of Table 2, and column 2 is the same balance but with the mass substitution made. For comparison, the results of analysis by Hughes and Rattray [1980], conducted at essentially the same cross section, are included (column 3). River discharge during the low flow segment of Hughes and Rattray's data was 4600 m^3 s^{-1}, very similar to that in our study. The terms in the mean salt balance (Equation 8) are listed column 4 of Table 2.

Tidal pumping (row two of Table 2) was found to be the biggest contributor to the up-estuary salt flux in both balances (about 63% of the total up-estuary contributions), followed by the vertical shear terms (about 20% of the total), with contributions due to lateral shear effects and area correlations being the least important terms (around 6% of the total for the lateral shear terms and 9% of the total for area-correlation terms). In comparison, Hughes and Rattray [1980] found contributions from the area-velocity correlation term to be about 30% of the total up-estuary flux. In this study, a phase difference between sectional mean velocity and tidal height was found to be 56^0, reducing the magnitude of terms involving correlations of A and u. Our measurement of the total flux of salt gives a net rate of change in salt content in the estuary (column 1 of Table 2) of $+4.8 \times 10^4$ ppt m^3 s^{-1}; reduced to $+3.8$ ppt m^3 s^{-1} (col. 2) if the Stokes drift and riverflow terms (rows 9 & 10) are substituted for the mean Eulerian term (row 1).

Table 2. Components of Salt Fluxes

TERM	Total Balance: [10^4 ppt m^3 s^{-1}]		Total Balance: HR80:	Mean Balance:
	not using (5)	using (5)	Low Flow	[10^4 ppt m^3 s^{-1}]
1: $\overline{A}\langle\{u\}\rangle\langle\{S\}\rangle$	-10.8		-15.7	
2: $\overline{A}\langle\{u'_i\}\rangle\langle\{S'_i\}\rangle$	10.1	10.1	4.4	10.1
3: $\overline{A'_j\langle\{u_j\}\rangle\langle\{S\}\rangle}$	1.7	1.7	4.7	
4: $\overline{A'_j\langle\{S'_j\}\rangle\langle\{u\}\rangle}$	-0.2	-0.2	-0.5	
5: $\overline{A}\langle\{\overline{u_t}\,\overline{S}_t\}\rangle$	0.4	0.4	0.1	0.4
6: $\overline{A}\langle\{\overline{u_v}\,\overline{S}_v\}\rangle$	2.1	2.1	4.4	2.1
7: $\overline{A}\langle\overline{\{u'_{tj}S'_{tj}\}}\rangle$	0.6	0.6	0.7	0.6
8: $\overline{A}\langle\overline{\{u'_{vj}S'_{vj}\}}\rangle$	1.1	1.1	0.3	1.1
9: $-\overline{Q}_R\langle\{S\}\rangle$		-8.0		-8.0
10: $-\overline{Q}_S\langle\{S\}\rangle$		-3.8		-3.8
TOTAL	+4.8	+3.8	-1.6	+2.5

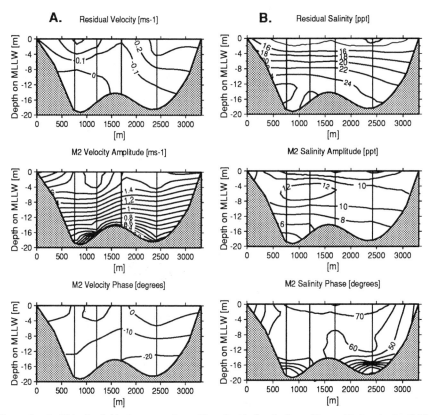

Figure 5. A. Velocity field at cross-section. Top: Residual velocity. The residual velocity field is characterized by outward net flow in the south, navigation channel and a small inward net flow in the north channel near the bed. Middle: Semidiurnal velocity amplitude. Bottom: semidiurnal velocity phase. B. Salinity field at cross-section. Top: Residual salinity. The residual salinity structure shows sloping isohalines consistent with the centrifugal forces due to channel curvature acting on the net seaward velocity. Middle: Semidiurnal salinity amplitude. Bottom: semidiurnal salinity phase.

The uncertainty in these results, which may be as high as 30%, is governed by a number of factors discussed below.

Mean Balance

The terms in the fourth column of Table 2 comprise the 'mean salt balance' with the mass conservation substitution made. The total represents the change in the average salinity in the average upstream estuarine volume and doesn't include the contributions involving variable cross-sectional area (rows 3 & 4). A comparison of the totals of columns 1 and 4 suggest that the rate of change in the mean inventory in the mean estuarine volume (evaluated in Equation 8) differs from the rate of change in the total mean inventory (evaluated in Equation 4) by a factor of about 2, so that changes in the correlation of tidal variations of upstream estuarine volume and upstream salinity is not negligible in the Columbia River.

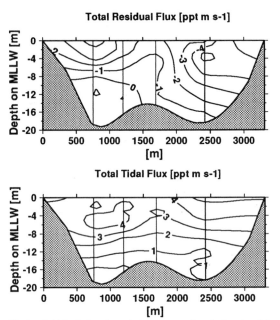

Figure 6. Total residual and tidal salt flux through the cross-section, before breakdown into mean and shear terms. The residual sakt flux is seaward everywhere except near the bed in the north channel. The tidal salt flux is landward everywhere in the cross-section.

Our measurements are not consistent with the Generalized Lagrangian Mean theorem (Equation 6). The product of the tidal advection of the tidal salinity field (LHS of Equation 6) is calculated to be 10.1×10^4 ppt m^3 s^{-1}, while product of the Stokes drift and the mean salinity, (RHS of Equation 6), was only 3.8×10^4 ppt m^3 s^{-1}, about a factor of 2.7 difference. Interestingly, the results of Hughes and Rattray are not inconsistent with the GLM theorem within measurement error; their results show only a 7% difference between the $\overline{A < \{u'\} > < \{S'\} >}$ and $\overline{A'_j < \{u'_j\} > } < \{\overline{S}\} >$. Though $\overline{A'_j < \{u'_j\} > } < \{\overline{S}\} >$ is not exactly the Stokes drift transport:

$$b\overline{u'\xi'}\big|_\xi < \{\overline{S}\} >$$

the difference between the two is very small. The meaning of these very divergent results is unclear.

The landward salt transport suggested by Table 2 may or may not be real, but evidence exists that there may have been a real increase in salt in the estuary during the period over which this study was conducted. An increase in tidal cycle averaged elevation of 0.17 m was found during the data collection period. This may be due to atmospheric effects or to the change in the diurnal inequality. Such an increase would, if it were assumed that this incoming water was at the observed cross-sectional mean salinity, contribute a rate of increase of salt in the estuary of 1.8×10^4 ppt m^3 s^{-1}, a significant fraction of net salt increase measured. This, of course, is only a rough order of magnitude because the average salinity of this "extra" water is unknown and the assumption was made that this elevation gain took place over the whole surface area of the estuary. Next, examination of ebb tide CTD casts separated by 25.3 hours (about 30 minutes longer than twice the semidiurnal period and an hour longer than the diurnal period) show an increase of about 2% in the cross-sectionally averaged salinity (Figure 8). Two factors may have contributed to a net increase in average salinity in the

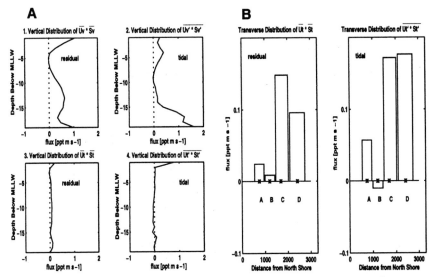

Figure 7. A: 1) Vertical distribution of vertical shear term (residual). 2) Vertical distribution of vertical shear term (tidal). 3) Vertical distribution of lateral shear term (residual). 4) Vertical distribution of lateral shear term (tidal). The vertical shear dispersion terms (panels A & B) are positive everywhere in the vertical, with maximums at the bed and surface in both the residual and the tidal. S_t, S_vm u_t, abd u_v are defined in the Salt Balance section. B: 1) Transverse distribution of lateral shear term (residual). 2) Transverse distribution of lateral shear term (tidal). These distributions, though small, are consistent with the channeling of river flow down the south channel, near station D; i.e. a strong outward flow of lower than average salinity water in the south channel.

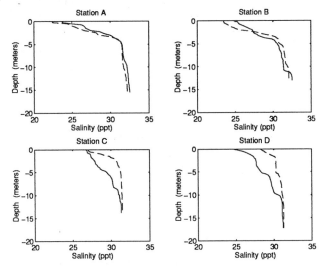

Figure 8. Ebb salinity casts at each of the four stations at the cross section showing difference in salinity. Earlier casts (solid) were conducted 25.4 hours before the later casts (dotted). The 2% increase in salinity over this period suggests that the salt inventory of the estuary was increasing.

estuary. First, the tide during this period was in a transition from spring to neap. During neaps the system becomes more stratified and the salinity intrusion increases, bringing more salt into the estuary. Second, mean salt inventory is also sensitive to river flow conditions, with salinity varying (for steady riverflow) with the square of the riverflow. Q_R during the data collection period was approximately +4400 m^3 s^{-1} at the mouth of the river. Large fluctuations in discharge occurred during the days before the data collection period, falling from about 6000 m^3 s^{-1} on July 25th, 3 days before the data collection commenced.

Error Analysis

Evaluation of errors associated with these flux measurements is complex, but the principle errors in the flux calculation input can be broken down into measurement errors associated with the velocity and salinity measurements, errors related to the fit of the harmonic analysis of u and S, and errors caused by extrapolation of velocity and salinity to the peripheries of the estuary.

The principal errors associated with measurement are those caused by spatial averaging as the boat is underway, the ADCP's limited ability to track high shear, and the perturbations of the gyrocompass by vessel turns. The velocity profile sampling design was to record an average of ~ 70 pings over a period of ~20 seconds. This scheme, while minimizing random error, causes a 'spatial averaging' of about 40 - 80 m at a typical boat speed of 2 - 4 m s⁻¹. The profiles used in this calculation, however, are those taken at or near the CTD station so boat speeds are likely smaller in these areas, with many taken at velocities of only ~ 0.5 - 2 m s⁻¹. Failure of the ADCP to track shear likely causes errors in velocities measured at times or positions of high shear. The tracking system in the ADCP has a non-zero response time that for a constant shear, results in an error of about 3.5 times the velocity change over 1 m. For a typical strongly sheared ebb flow (e.g. as in Figure 2) the error resulting from the shear is < 3.5 cm s⁻¹, less than 5% of the vertically integrated velocity. In locations of higher shear where salinity and velocity are correlated, this error may result in a bias in the flux magnitude. High shears exist near the bed, but in this region the ADCP fails because of interference by the acoustic signal reflected from the bed, so this ADCP data were not used. The momentum balance model described earlier was used to estimate velocities near the bed. Examination of other simpler near-bed extrapolation schemes showed that the flux measurements are not, in fact, very sensitive to the scheme used. For example, the calculation was carried out using ADCP data all the way to the bed (i.e. without use of the profile model) and flux magnitudes changed by less than 5%.

The errors associated with harmonic analysis may be estimated by evaluating the rms residual error of the reconstructed time series relative to the raw data. For velocity residual error is about 7%; for salinity it is about 10%. These errors result primarily from an inability of the harmonic analysis routine to accurately determine the phase of each contributing constituent and secondarily from the net increase in salinity during the observation period, which cannot be "fit" by the harmonic analysis. To examine the contribution from phase errors, consider the error in the flux resulting from an error in phase of the velocity and salinity constituents for a single bin. The expression for tidally averaged flux:

$$F = \tfrac{1}{2} * A v * A s * \cos(\delta s - \delta v)$$

would have a relative error of:

$$\tfrac{\Delta F}{F} = [\tan(\delta s - \delta v)] * \Delta(\delta s - \delta v)$$

where $A v$ and $A s$ are the velocity and salinity amplitudes, δs and δv are their phases, and $\Delta(\delta s - \delta v)$ is the error in the phase difference. If the error in phase of the velocity and salinity constituents for the bin were 5° (about ten minutes of M_2 period), and this error were of opposite sign for salinity and velocity (a worst case scenario, making $\Delta(\delta s - \delta v)$ equal to 10°), the relative error due to uncertainty in phases is a ΔF/F of 30%. (A typical value of 60° was used for (δs-δv), the phase difference between salinity and velocity.) This single-bin error is greater than the error in the total flux because the terms evaluated in the flux calculation are correlations between cross-sectional averages or deviations of salinity and velocity. This integration averages errors from individual stations. For example, even if it assumed that the phase errors are correlated in the vertical because velocities are collected simultaneously in the vertical (again a worse case scenario) but not correlated across stations, the above estimated single bin error of 30%, when treated as a random error over 4 stations, is reduced to 15%. Because the amplitudes of the velocity and the salinity appear in the flux expression as simple products and not as part of a trigonometric function (as is the case with the phases), the error of 7-10% in these for each depth bin results in relative errors in the flux of the same percentage, and are reduced significantly during the averaging making their added contribution smaller than that of the phase. Another source of error in our calculations is the lateral extrapolation of salinity and velocity at stations A and D to the north and south shores of the estuary. These errors, while impossible to quantify, are likely as big or bigger than errors associated with velocity measurement, or harmonic analysis.

Summary and Conclusion

Water and salt fluxes at km 8 of the Columbia River estuary were calculated from Acoustic Doppler Current Profiler (ADCP) and CTD data. Mass (water) was conserved within < 7%. Two approaches to the examination of the salt flux and estuarine salt inventory were employed. The first is the 'full salt balance', a measure of the change in salt inventory in a time-varying upstream of the cross-section. The second is the 'mean salt balance', which calculates a change in the salt inventory in a time-average upstream estuarine volume. Both the 'full' and 'mean' balances showed an increase in salt over the period studied. The difference in results between the two balances suggests that changes in the correlation between estuarine volume and salinity are not negligible in the Columbia River estuary.

The salt balance was found to be primarily between the mean advection of the mean salinity field (100% of seaward transport) and the tidal advection of the tidal salinity field (63% of total landward transport). Vertical shear dispersion effects were found to be significant (about 20% of total landward transport) consistent with the predominance of gravitational circulation in the Columbia River. Lateral shear dispersion was found to be small (about 6% of total landward transport). Although substantial lateral variations in the flux field exist, they make only a small contribution to the net salt transport. The tidal cycle averaged salt transport at the mouth of the Columbia River is governed primarily by cross-sectional mean and vertical shear processes.

Acknowledgments. The field work and the bulk of the analysis was conducted under the support of NSF grant OCE-8907118, the Columbia River Estuary Land Margin Ecosystem Research Program. A portion of the data analysis conducted by Jeff Musiak was supported by ONR grant N00014-94-1-0009.

References

Allen, G. P., J. C. Solomon, P. Bassoulet, Y. Du Penhoat, and C. DeGrandpe, Effects of tides on mixing and suspended sediment transport in macrotidal estuaries, *Sedimentary Geology, 26*, 69-90, 1980.

Andrews, D. G. and M. E. MacIntyre, An exact theory of nonlinear waves on a Lagrangian-mean flow, *Journal of Fluid Mechanics, 89*, 609-646, 1978.

Baird, D. P., E. D. Winter and G. Wendt, The flux of particulate material through a well mixed estuary, *Continental Shelf Research, 7*, 1399-1403, 1987.

Dyer, K. R., *Estuaries: A Physical Introduction*, J. Wiley and Sons, London, pp. 64-70, 1973.

Dyer, K. R., The salt balance in stratified estuaries, *Estuarine Coastal Marine Science, 2*, 275-281, 1974.

Dyer, K. R., W. K. Gong and J. E. Ong, The cross sectional salt balance in a tropical estuary during a lunar tide and a discharge event, *Estuarine, Coastal and Shelf Science, 34,* 579-591, 1992.

Fischer, H. B., Mass transport mechanisms in partially mixed estuaries, *Journal of Fluid Mechanics,* 671-687, 1972.

Foreman, M. G. G., Manual for tidal currents analysis and prediction, *Institute of Ocean Sciences report 78-6,* Sidney, B.C., 101 pp., 1978.

Foreman, M. G. G. and R. F. Henry, Tidal analysis based on high and low water observations, *Institute of Ocean Sciences report 79-15,* Sidney, B.C., 39 pp., 1979.

Gelfenbaum, G. and J. D. Smith, Experimental evaluation of a generalized suspended sediment transport theory, in Knight, R. J. and J. R. Mclean, eds., *Shelf Sands and Sandstones, Canadian Soc. Of Petr. Geol.,* Men II, pp.133-144, 1986.

Hansen, D. V., Salt balance and circulation in partially mixed estuaries, pp. 45-51, in *Estuaries,* Publication No. 83, American Association for the Advancement of Science, Washington, D. C., 1965.

Hansen, D. V. and M. Rattray, Jr., Gravitational circulation in straits and estuaries, *Journal of Marine Research, 23,* 104-122, 1965.

Hansen, D. V., and M. Rattray, Jr., New dimensions in estuary classification, *Limnology and Oceanography* 11, 319-326, 1966.

Hughes, F. W. and Rattray M. Salt flux and mixing in the Columbia River estuary, *Estuarine and Coastal Marine Science 10,* 479-493, 1980.

Jay, D. A., Estuarine salt conservation: a Lagrangian approach, *Estuarine, Coastal and Shelf Science, 32,* 547-565, 1991.

Jay, D. A. and J. D. Smith, Circulation, density distribution and neap-spring transitions in the Columbia River Estuary, *Progress in Oceanography, 25,* 81-112, 1990

Kjerfve, B. and J. A. Proehl, Velocity variability in a cross-section of a well-mixed estuary, *Journal of Marine Research, 37,* 409-418, 1979.

Lewis, R. E., Circulation and mixing in estuary outflows, *Continental Shelf Research, 3,* 201-214, 1984.

Lewis, R. E., and J. O. Lewis, The principal factors contributing to the salt flux in a narrow, partially stratified estuary, *Estuarine Coastal Marine Science, 16,* 599-626, 1983.

Middleton, J. F., and J. W. Loder, Skew fluxes in polarized wave fields, *Journal of Physical Oceanography, 19,* 68-76, 1989.

Murray, S. P. and A. Siripong, Role of lateral gradients and longitudinal dispersion in the salt balance of a shallow, well mixed estuary, in (ed. B. Kjerfve) *Estuarine Transport Mechanisms,* University of South Carolina Press, 113-124, 1978.

Rattray, M, and J. G. Dworski, Comparison of methods for analysis of the transverse and vertical circulation contributions to the longitudinal advective salt flux in estuaries, *Estuarine and Coastal Marine Science, 11,* 515-536, 1980.

Sherwood, C. R., D. A. Jay, R. B. Harvey, P. Hamilton, and C. A. Simenstad, Historical changes in the Columbia River estuary, *Prog. Oceanogr., 25,* 299-352, 1990.

Uncles, R. J. Elliott, R. C. A. and S. A. Weston, Dispersion of salt and suspended sediment in a partly mixed estuary, *Estuaries, 8,* 256-269, 1985.

15

Tidal Pumping of Salt in a Moderately Stratified Estuary

W. Rockwell Geyer and Heidi Nepf

Abstract

Observations during high discharge conditions in a 3-km reach of the Hudson River estuary indicate a significant contribution of tidal pumping, i.e., correlation between tidal variations in salinity and velocity. Based on measurements of velocity and salinity at several cross-sections, tidal pumping is found to be larger than the upstream salt flux due to the estuarine circulation. Similar measurements, obtained in the same reach of the estuary during low discharge conditions, indicate that tidal pumping is insignificant relative to the salt flux due to estuarine circulation. The tidal pumping during high discharge conditions is caused by large vertical excursions of the halocline that are correlated with the tidal flow. The excursions may be explained by the hydraulic response of the stratified flow to a lateral constriction.

A simple, two-layer analysis reveals that the tidal pumping is a Stokes transport that results from the vertical distortion of the mean shear flow by vertical fluctuations of the halocline. Thus it is not an independent mechanism of salt flux, and it should not be regarded as a diffusive or dispersive flux, but rather as part of the net, estuarine salt flux.

Introduction

The seaward advection of salt in estuaries due to the river outflow is balanced in the mean by landward fluxes due to spatial and temporal correlation of velocity and salinity. Among the various contributors to the upstream flux, the major terms are generally the estuarine circulation, i.e., the vertical (and sometimes lateral) variations of tidally averaged velocity and salinity, and tidal pumping, i.e., temporal correlation of velocity and salinity [Uncles et al., 1985, Lewis and Lewis, 1983]. Tidal pumping has long been recognized to be an important contributor to horizontal exchange, and a number of physical mechanisms have been described to explain tidal pumping in various flow regimes. Stommel and Farmer [1952] showed that flow separation at the mouths of estuaries results in higher salinities during floods than ebbs, thus producing a net landward salt flux. Schijf and Schonfeld [1953] and Okubo [1973] developed the concept of "tidal trapping", in which fluid parcels are advected into lateral "traps" during one part of the tidal cycle and out in another, resulting in a net upstream salt transport. Oscillatory shear dispersion [Smith, 1977; Fischer et al., 1979] is a mechanism for tidal pumping that is important when the timescale of vertical or transverse mixing is comparable to the tidal timescale.

Buoyancy Effects on Coastal and Estuarine Dynamics
Coastal and Estuarine Studies Volume 53, Pages 213-226
Copyright 1996 by the American Geophysical Union

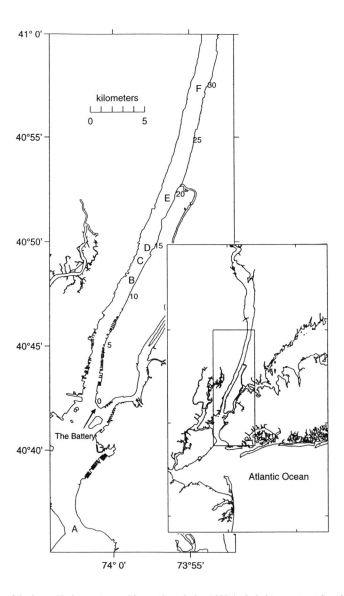

Figure 1. Map of the lower Hudson estuary. Observations during 1992 included transects at locations C and E, and during 1993 at locations B, C and D. The locations of transects performed by Hunkins [1981] are indicated by the letters A and F. The location of the Battery on the southern tip of Manhattan is indicated. Numbers adjacent to the east bank of the river denote kilometers from the Battery.

Dronkers and van de Kreeke [1986] showed that tidal pumping can result from abrupt changes in cross-sectional area in an estuary. Based on their analysis, the tidal pumping can be equated to the difference between the Lagrangian salt flux (defined by a reference frame that moves landward and seaward with the tidal prism) and the Eulerian salt flux, i.e., that it represents a local Stokes transport of salt [see also Jay, 1991]. This type of mechanism differs from other tidal pumping mechanisms in that it is inherently advective; in fact it can be considered as a distortion of the Eulerian circulation by spatial and temporal fluctuations in the structure of the flow.

The observational evidence for tidal pumping in different estuaries shows considerable variability in the magnitude of tidal pumping compared to the other mechanisms of salt flux. Hughes and Rattray [1980] found that tidal pumping was the dominant contributor to landward salt flux in a section across the Columbia River estuary during high discharge observations, and that it was comparable to estuarine circulation during low discharge conditions. Lewis and Lewis [1983] found longitudinal variations in the relative contribution of tidal pumping to the total flux, with the maximum values of tidal pumping occurring near constrictions. Hunkins [1981] found that tidal pumping in the Hudson River estuary was generally small, and in one instance there was significant out–estuary salt flux by tidal pumping, opposing the tendency of the mean salinity gradient.

Some recent observations in the Hudson River estuary indicate a large contribution of tidal pumping during strong stratification conditions, and negligible pumping during moderate stratification conditions. The tidal pumping results from tidal variations in the depth of the halocline that are correlated with the horizontal velocity. The variations in halocline depth appear to result from the internal hydraulic response to a constriction. A two-layer analysis shows that the tidal pumping of salt is simply the result of distortion of the mean, estuarine salt transport by the vertical variations in halocline elevation. The mechanism is consistent with Dronkers and van de Kreeke's [1986] theory for the influence of abrupt changes in topography on the salt flux.

Observational Program

The lower Hudson River estuary extends approximately 30 km northward from New York Harbor to Tappan Zee (Figure 1). It is unusually straight, owing to the north-south trend of a resistant rock formation, the Palisades, which lies along the west bank of the river. The cross-sectional area of the river is also nearly uniform along the estuarine reach. The estuary is moderately stratified, with a vertical salinity difference of 3–10 psu [Abood, 1974]. It maintains a strong longitudinal salinity gradient during both low and high discharge conditions, ranging from 0.3–1 psu km^{-1}. A progressive tidal wave produces tidal currents of approximately 1 m s^{-1}, and non-tidal, estuarine velocities are 10–30 cm s^{-1}. Mean, riverine outflow velocity varies from 1 cm s^{-1} during low discharge conditions to 20 cm s^{-1} during high discharge.

Measurements of water properties and currents were obtained during tidal-cycle surveys in the lower Hudson estuary during low discharge conditions in August, 1992 and high discharge conditions in April–May, 1993. A 1.2 MHz shipboard acoustic Doppler current profiler provided the velocity data, and an Ocean Sensors CTD provided temperature, salinity and light transmission data for estimation of suspended sediment concentrations. Measurements were obtained at several cross-sections (Figure 1). The vessel traversed each section once per hour, with continuous velocity measurements and discrete water properties measurements at five cross-stream locations. Flux calculations were based on the data at these five discrete stations. A two-minute average of the velocity data was obtained for each station.

Estimation of the net salt flux was based on harmonic analysis of the velocity and salinity data, using mean, semi-diurnal, quarter-diurnal, and sex-diurnal components [see also Geyer and Signell, 1990] to yield a periodic approximation of the data. This technique is insensitive to slight deviations from a 12.4 hour tidal cycle, in contrast to simple averaging to obtain the tidal residuals. A depth-proportional (or "sigma") coordinate was used in the vertical to yield a temporally continuous representation of the salinity and velocity data from surface to bottom. This approach separates the barotropic Stokes drift from the tidal pumping, in contrast to estimates based on a fixed vertical coordinate, in which they are combined.

Neglecting small contributions due to triple correlation of sea-level, velocity and salinity, the salt flux can be partitioned as

$$Q_s = u_f S_0 + \overline{\langle u' \rangle \langle S' \rangle} + \left\langle \overline{\tilde{u}\tilde{S}} \right\rangle + \left\langle \tilde{u}'\tilde{S}' \right\rangle \tag{1}$$

where Q_s is the total salt flux, u_f is the cross-sectionally averaged velocity due to the river outflow, S_0 is the temporally and cross-sectionally averaged salinity, the tildes (⁓) represents variations from the time mean, the primes represent deviations from the spatial mean, the overbars indicate cross-sectional averages, and the angle brackets indicate tidal-cycle averages. The first term on the right hand side is the mean, barotropic salt flux (due in general to riverine outflow), the second term is the spatial correlation of salinity and velocity (due principally to the estuarine circulation), the third term is the cross-sectional mean tidal pumping contribution, and the fourth term is the tidal pumping due to deviations from the cross-sectional mean. Note that the influence of barotropic Stokes drift does not appear directly in Equation 1, because u_f represents the mean Lagrangian outflow, i.e., the sum of the Eulerian velocity and the Stokes drift.

Results

During the August 1992 survey, freshwater discharge to the estuary was 200–300 m³ s⁻¹ (typical of low discharge conditions). There was significant salt stratification, with 8 psu difference between surface and bottom waters, averaged over a tidal cycle (Figure 2). Tidally averaged currents showed a pronounced estuarine circulation, with velocity differences of up to 50 cm s⁻¹ between the surface outflow and bottom inflow. The cross-sectionally averaged and tidally averaged flow was upstream at approximately 7 cm s⁻¹, compared to a net downstream flow of 3 cm s⁻¹ caused by the freshwater outflow. This difference is due in part to an unresolved diurnal tidal oscillation, and in part to low-frequency, barotropic motions in the estuary, similar to those observed by Wang and Elliot [1978] in Chesapeake Bay. Tidal currents were at moderate spring conditions, with speeds exceeding 100 cm s⁻¹ (Figure 3). Tidal variations in salinity of up to 5 psu resulted from advection of the longitudinal salinity gradient.

Salt flux was calculated from 12-hour timeseries measurements at sections C and E (Figure 1). In spite of the large tidal variations in salinity, tidal pumping was small at both sections (Table 1; Figure 4), with the salt flux being dominated by the tidal mean velocity and salinity. The weak signal of tidal pumping indicates that the velocity and salinity fluctuations were nearly in quadrature, thus they were weakly correlated. Quadrature between velocity and salinity would result simply from advection of the longitudinal salinity gradient by the tidal flow. Other temporal variations in salinity, due to mixing or vertical motions, would be required to shift the phase of the salinity fluctuations relative to velocity and produce tidal pumping.

River discharge was high during the April 1993 survey, ranging from 1500 to 2500 m³ s⁻¹. Stratification was stronger than the 1992 period (Figure 5), with roughly 12 psu difference between surface and bottom waters for tidally averaged data. The estuarine shears were slightly stronger than the 1992 conditions, but the net outflow due to the river discharge was strongly evident (u_f=20 cm s⁻¹). Tidal currents increased from intermediate to strong spring conditions during the observation period, producing near-surface speeds of up to 140 cm s⁻¹

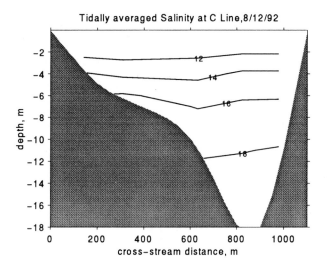

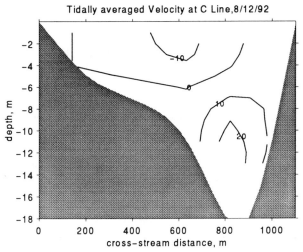

Figure 2. Contours of tidally averaged salinity (psu) and along-channel velocity (cm s⁻¹) at the C cross-section, obtained during low discharge conditions, on 8/12/92. Negative velocities are seaward.

during the ebb (Figure 6). Large temporal salinity variations were evident, with mid-water salinity varying between 12 and 20 psu.

In contrast to the 1992 data, tidal pumping was found to be quite pronounced (Table 1, Figure 7). In fact, salt flux associated with tidal pumping was 2–3 times the magnitude of that due to estuarine circulation. Observations at three different cross-sections during the 1993 observations all showed a dominance of tidal pumping over estuarine circulation (Table 1).

The maximum expression of the tidal pumping was in the middle of the water column (Figure 7). This implies that the maximum covariance of velocity and salinity occurs at this level. The tidal variations of salinity (Figure 6) show that the halocline had large vertical excursions through the tidal cycle, which resulted in maximum variation of salinity in the middle of the water column. Whether or not these variations resulted in tidal pumping depended on the phase of velocity variations relative to salinity variations. The phase shift between velocity and salinity in the 1993 observations was about 60°, which produced a significant in-phase contribution of the salinity and velocity, hence the large magnitude of tidal pumping. In the 1992 observations (Figure 3), there were also significant variations of salinity, but the phase was close to 90°, and so tidal pumping was weak.

Discussion

The Pumping Mechanism

The large magnitude of tidal pumping during the 1993 observations resulted from the correlation of vertical variations of the halocline with the velocity. This suggests that the transport may have been due to a propagating internal wave, which would have maximum salt transport at the mean level of the halocline. Longitudinal salinity sections (Figure 8) indicate that indeed there were large, wave-like variations in the halocline during the ebb, with vertical fluctuations of more than 6-m over several km. However, repeated observations at hourly intervals near sections B, C and D revealed that the "wave" was not propagating; its position remained stationary through the ebb, although its amplitude increased to a maximum in the latter part of the ebb, as indicated in Figure 6.

The variation in halocline elevation near section D during the ebb appears to be the result of internal hydraulic effects of the estuarine geometry. There is a marked constriction in the estuary between km 17 and 15, narrowing from 1.6 to 1.0 km. The outflow velocities of 1.5 m s^{-1} during the ebb were large enough to produce supercritical conditions with respect to the internal hydraulics [Lawrence, 1990]. The influence of a constriction on a supercritical, two-layer flow is to thicken the "active layer" [Lawrence, 1990], which in this case was the upper

Table 1: Forcing Variables and Salt Fluxes
The various contributions to salt flux indicated in Equation 1 are presented in units of cm s^{-1}, normalized by mean salinity for the section. Negative values indicate seaward transport. Section locations are shown in Figure 1. River Discharge is based on U.S. Geological Survey gauge data; tidal range is in meters. The salt flux terms are as follows: mean refers to the salt flux due to the net outflow (or inflow, as in the 1992 observations). The estuarine transport is the spatially varying, tidally averaged salt flux. Pumping is the sum of all of the tidally varying contributors except for the barotropic Stokes drift. The residual is the net flux due to all of the terms. The residual would be zero for steady–state conditions. Data from Hunkins [1981] are shown for comparison.

Date	Section	River Discharge	Tidal Range	Mean	Estua-rine	Pump-ing	Resi-dual
8/12/92	C	250	1.2	7.7	0.9	0.0	8.6
8/13/92	E	250	1.2	8.4	0.5	-0.5	8.4
4/30/93	C	2000	1.2	-16.3	5.5	9.1	-1.5
5/1/93	B	2000	1.4	-18.6	3.5	8.9	-6.2
5/1/93	D	2000	1.4	-18.0	4.6	8.1	-5.3
5/2/93	C	2000	1.5	-14.5	2.1	5.4	-7.0
8/11/77	A*	low		0.9	0.6	-0.2	-0.5
5/18/78	A*	high		-3.0	2.5	-0.5	-1.0
8/31/78	F*	low		-1.3	1.6	-0.9	-0.6

* Data from Hunkins [1981]

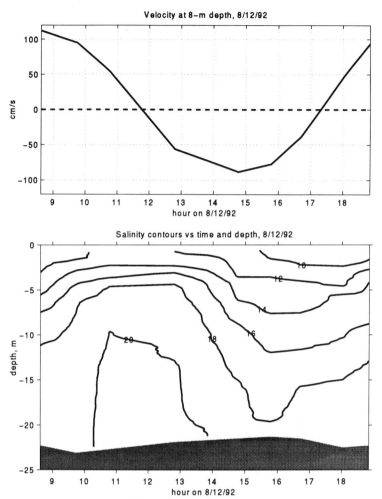

Figure 3. Time-variations of velocity and salinity in the deepest portion of the C cross-section during low discharge conditions, on 8/12/92. The upper panel shows along-channel velocity at 8-m depth, and the lower panel shows contours of salinity (psu) as a function of time and depth.

layer, owing to much higher velocities in the upper water column. Thus the depression of the halocline near section D is consistent with a hydraulic response.

During the flood, the conditions were also supercritical; however the hydraulic response was greatly reduced from the ebb conditions. This is due to the reduced vertical shear during the flood, which rendered both layers nearly equal in their hydraulic response to the constriction, effectively canceling out the vertical response of the halocline.

The downward displacement of the halocline during the ebb due to the hydraulic response to the constriction provided a local mechanism for tidal pumping. It resulted in lower salinities in the middle part of the water column during the ebb than the flood, and thus yielded a net contribution to tidal pumping. It should be noted , however, that this mechanism makes only a local contribution to the salt flux. In wide parts of the estuary (such as km 17),

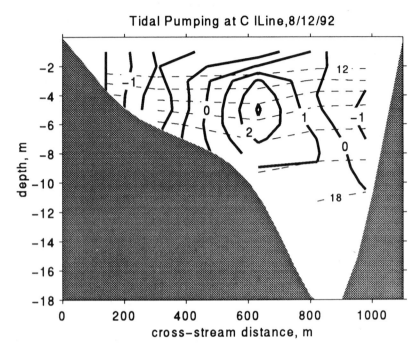

Figure 4. Contours of tidal pumping at the C cross-section (bold contours), obtained during low discharge conditions, on 8/12/92. The salt flux has been normalized by the cross-sectional mean salinity to yield units of cm s^{-1}, to facilitate comparison with the mean advective outflow. The thin contours are tidally averaged salinity contours.

the interface rises during the ebb, causing a reversal of the sign of the tidal pumping by this mechanism. Such large variations in tidal pumping of salt would produce unreasonably large divergences and convergences in salt flux, if they were not compensated by variations in the Eulerian salt flux.

Indeed, a closer consideration of the kinematics of a tidally-varying, two-layer flow indicates that there can be large along-channel variations in the Eulerian flow that compensate for equal and opposite variations in tidal pumping, due simply to the vertical fluctuations of the interface. A simple example is shown in Figure 9 that illustrates the pumping mechanism in the Hudson. This cartoon depicts a two-layer flow in a rectangular channel, with a steady, estuarine circulation and a barotropic tidal flow. No transport occurs across the interface. If the position of the halocline were fixed, there would be a uniform, Eulerian shear flow, and all of the salt flux would be accomplished by the mean shear. However, if there is a zone in which the halocline dips during the ebb, the change in layer thickness will change the velocities in each of the layers, reducing the shears in this zone, even though the transport within each layer remains fixed. The tidally averaged, Eulerian shear at this location can be reduced or even change sign, depending on the amplitude of the interface fluctuations, thus dramatically altering the net, Eulerian salt flux. However, the total salt flux remains the same, since the transport in each of the layers is continuous. The deficit in Eulerian transport is made up by a Stokes drift, or tidal pumping component, in the zone where the interface elevation varies.

This simple analysis demonstrates that temporal variations in interface elevation can produce equal and opposite variations in the Eulerian salt flux and tidal pumping, while not

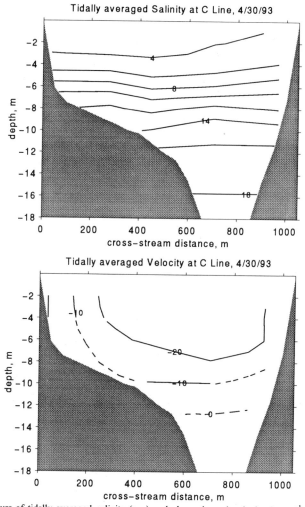

Figure 5. Contours of tidally averaged salinity (psu) and along-channel velocity (cm s^{-1}) at the C cross-section, obtained during high discharge conditions, on 4/30/93.

altering the total, Lagrangian salt flux. The only way that fluctuations in interface elevation could alter the total salt flux would be through a progressive internal wave, which would produce a Stokes drift component of salt transport that was distinct from the Eulerian flow. However, the topographically trapped fluctuations in interface elevation that were observed in the Hudson do not add to the total salt flux; they just redistribute the shear-induced transport by distorting its vertical structure through the tidal cycle.

Variability of the Pumping Mechanism

The magnitude of tidal pumping was large during the high discharge conditions in 1993, but it was negligible during the low discharge observations in 1992 and in Hunkins [1978] observations during both high and low discharge. Sections E and F (Figure 1) were not in regions of constrictions, so the internal hydraulic mechanism would not be expected to operate at those locations. The variation in tidal pumping at the constrictions may be explained in part by variations in the internal Froude number, which would alter the magnitude and structure of interface fluctuations. In addition, non-tidal variations in interface elevation (due, for example, to winds or discharge variations) may also influence the partitioning between the mean, Eulerian salt flux and the tidal pumping component.

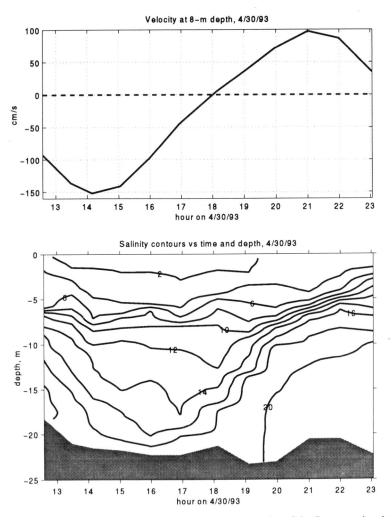

Figure 6. Time-variations of velocity and salinity in the deepest portion of the C cross-section during high discharge conditions, on 4/30/93. The upper panel shows along-channel velocity at 8-m depth, and the lower panel shows contours of salinity (psu) as a function of time and depth.

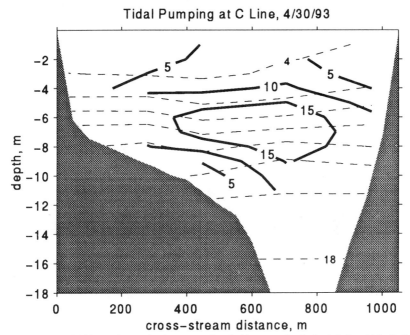

Figure 7. Contours of tidal pumping at the C cross-section (bold contours), obtained during. high discharge conditions, on 4/30/93. The salt flux has again been normalized by the cross-sectional mean salinity to yield units of cm s^{-1}. The thin contours are tidally averaged salinity contours.

Implications

These findings, as well as similar findings in previous observational studies of salt flux [e.g., Hunkins, 1980; Lewis and Lewis, 1983; Dronkers and van de Kreeke, 1986], indicate that tidal pumping is often only of local significance to the overall salt balance in these partially mixed estuaries, and thus that the estuarine salt flux is the dominant contributor to the overall upstream salt transport. The paradigm of Hansen and Rattray [1966], in which a significant fraction of the upstream salt flux results from tidal dispersive processes, is not supported by the observational evidence for these partially mixed, or Type 2, estuaries. Only in reaches where there are significant variations in cross-section will the tidal pumping term become important, and the magnitude and perhaps even the sign of the pumping term will change as a function of along-channel position. A counter-example may be a system such as the Columbia estuary [Hughes and Rattray, 1980], in which the complexity of the geometry may lead to strong tidal dispersion throughout the estuary, due, for example, to chaotic dispersion [Ridderinkof and Zimmerman, 1992; Zimmerman, 1986; Signell and Geyer, 1990]. However, in estuaries of more uniform morphology, the estuarine circulation would be expected to dominate the upstream salt balance at scales larger than the tidal excursion distance.

Summary

This study provides observational evidence for tidal pumping in a lateral constriction in a moderately stratified estuary. The magnitude of tidal pumping exceeded the estuarine circulation in this reach during one observational period, but the tidal pumping was

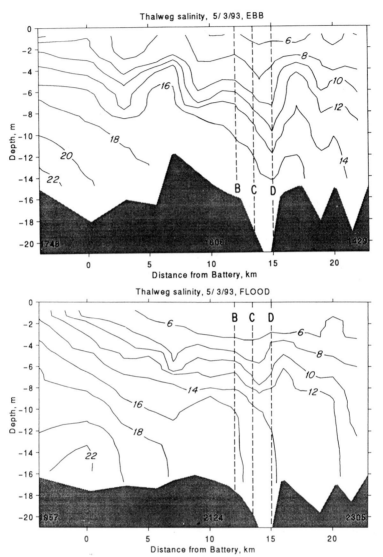

Figure 8. Salinity sections (psu) along the lower Hudson estuary during high run-off conditions in May, 1993 during ebb (upper panel) and flood (lower panel). The locations of the flux sections are indicated. The drop in the interface around km 15 during the ebb appears to be the result of hydraulic response to the constriction.

negligible during another period with weaker stratification. The tidal pumping results from vertical variations in interface elevation that appear to result from the hydraulic response to constrictions during ebb flows. Although the tidal pumping may amount to a large fraction of the total salt flux, a simple, two-layer description reveals that it results from the spatial distortion of the mean, estuarine shear flow. Thus it is not an independent mechanism of salt flux, and it should not be regarded as a diffusive or dispersive flux, in context with the Hansen and Rattray [1966] classification scheme.

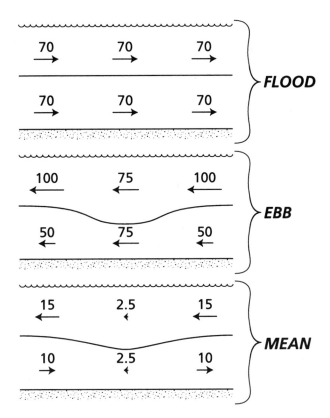

Figure 9. Cartoon illustrating the influence of variations of layer thickness on the mean, Eulerian velocity distribution. During flood, the transports and velocities are uniform in both layers, but during ebb there is more outward transport in the upper layer. In the zone where the halocline gets deeper, the velocity decreases in the upper layer and increases in the lower layer. The tidally averaged velocities (lower panel) indicate that the Eulerian shear vanishes in the zone where the halocline fluctuation occurs. The total salt flux is unaltered by the halocline motion; it is just redistributed between the Eulerian and Stokes transport, or tidal pumping component.

Acknowledgments: The authors would like to acknowledge the insights of Drs. van de Kreeke and Dronkers on the Lagrangian approach to the salt flux, and the discussions with Drs. Chant and Wilson on the dynamics of the Hudson River estuary. This research was supported by the Hudson River Foundation, Grant Nos. 002/92A and 002/94P. This is Woods Hole Oceanographic Institution contribution 9151.

References

Abood, K.A., Circulation in the Hudson estuary, in *Hudson River Colloquium*, edited by O.A. Roels, pp. 38-111, Annals NY Acad. Sci., 1974.

Chant, R., Tidal dynamics of the Hudson River estuary. Ph.D. thesis, State University of New York, Stony Brook, 1995.

Dronkers, J., and J. van de Kreeke, Experimental determination of salt intrusion mechanisms in the Volkerak estuary, *Netherlands Journal of Sea Research, 20,* 1–19, 1986.

Fischer, H. B., E. J. List, R. C. Y. Kho, J. Imberger, and N. H. Brooks, *Mixing in Inland and Coastal Waters,* Academic Press, New York, 483 pp., 1979.

Geyer, W. R., and R. P. Signell, A reassessment of the role of tidal dispersion in estuaries and bays, *Estuaries, 15,* 97-108, 1992.

Hughes, F. W., and M. Rattray, Jr., Salt flux and mixing in the Columbia River estuary, *Estuarine and Coastal Mar. Sci., 10,* 479-493, 1980.

Hunkins, K., Salt dispersion in the Hudson estuary, *J. Phys. Oceanogr., 11,* 729-738, 1981.

Jay, D. A., Estuarine salt conservation: A Lagrangian approach, *Estuarine, Coastal and Shelf Sci.,32,* 547-565, 1991.

Lewis, R. E., and J. O. Lewis, The principal factors contributing to the flux of salt in a narrow, partially stratified estuary, *Estuarine, Coastal and Mar. Sci., 16,* 599-626, 1983.

Okubo, A., Effects of shoreline irregularities on streamwise dispersion in estuaries and other embayments, *Netherlands J. of Sea Res. 8,* 213–224, 1973.

Ridderinkhof, H., and J. T. F. Zimmerman, Chaotic stirring in a tidal system, *Science, 258,* 1107-1111, 1992.

Schijf, J. B., and J. C. Schonfeld, Theoretical considerations on the motion of salt and fresh water, *Proc. Minn. Intl. Hydr. Con., 5th Cong. I.A.H.R.,* 321-333, 1953.

Signell, R. P., and W. R. Geyer, Numerical simulation of tidal dispersion around a coastal headland, in *Residual Currents and Long-Term Transport, Coastal and Estuarine Series,* edited by R. T. Cheng, pp. 210-222, Springer-Verlag: New York, 1990.

Smith, R., Long-term dispersion of contaminants in small estuaries, *J. Fluid Mech., 82,* 129–146, 1977.

Stommel, H., and H. G. Farmer, On the nature of estuarine circulation, Part I. Ref. No. 52-88, Woods Hole Oceanographic Institution, Woods Hole, MA, 1952.

Uncles, R. J., R. C. A. Elliott, and S. A. Weston, Dispersion of salt and suspended sediment in a partly mixed estuary, *Estuaries, 8,* 256-269, 1985.

Wang, D. P., and A. J. Elliott, Non-tidal variability in the Chesapeake Bay and Potomac River: Evidence for non-local forcing, *J. Phys. Oceanogr., 8,* 225-232, 1978.

Zimmerman, J. T. F., The tidal whirlpool: A review of horizontal dispersion by tidal and residual currents, *Netherlands J. Sea Res., 20,* 133-154, 1986.

16

Density Structures in "Low Inflow Estuaries"

John L. Largier, Clifford J. Hearn, and D. Bart Chadwick

Abstract

The study of estuarine hydrodynamics has been dominated by a focus on the defining feature of estuaries - freshwater inflow. Low-inflow "estuaries" (LIE's), or semi-enclosed bays, have received very little attention in spite of their common occurrence. In this paper, we discuss the occurrence of LIE's as a seasonal feature in mediterranean-climate regions (e.g., California, USA). Buoyancy fluxes are dominated by air-water exchange. There is competition between a positive buoyancy flux due to heating and a negative buoyancy flux due to evaporation. In spite of large thermohaline signals, the resultant buoyancy flux is often too weak to bring about vertical stratification and buoyancy-driven exchange in shallow estuarine basins. This results in very long residence times being observed in tidally sheltered waters, such as at the landward end of these LIE basins. Through a simple model of surface fluxes and longitudinal dispersion in LIE's, we explain observed temperature, salinity and density structures in Tomales, San Diego, and Mission Bays (California, USA). In summary, we identify four regions: a marine region in the outer bay (dominated by tidal exchange with the ocean), a thermal region (where heating dominates the buoyancy flux), a hypersaline region (where evaporation dominates the buoyancy flux), and an estuarine regime (where freshwater inflow dominates the buoyancy flux). The clear spatial separation of thermal and hypersaline regimes is owing to the different temporal structure of heating and evaporation in mediterranean-climate estuaries.

Introduction

In much the same way that people who study rivers in arid or semi-arid areas refer to them as water courses, recognizing that run-off is not persistent, so we too need to find a new term or re-define the old one (e.g., Pritchard, 1967) to allow for the fact that many "estuaries" do not experience a persistent freshwater inflow. In arid areas, or during the dry season in mediterranean climate regions, river inflow is negligible or absent. Any estuary-ocean density differences are due to differential heating or evaporation, i.e., due to surface buoyancy fluxes. With the anthropogenic extraction of water from rivers, more estuaries are becoming net evaporative during the dry season (i.e., the fresh water loss from the basin is greater than that supplied by river inflow).

Observations from three Californian estuaries represent the essential spatio-temporal density structures during dry mediterranean-climate summers. However, limited published data from estuaries in other mediterranean regions suggest that the structures observed in California are likely to be found in many of the estuaries of southwestern Australia, southwestern South Africa, Portugal, Morocco, Chile, northwestern Mexico, and so on.

In the following, a simplified estuarine basin is considered (Fig. 1). This basin receives freshwater via river inflow R and precipitation P while it loses freshwater through evaporation E'. Heat is exchanged with the atmosphere (heat flux Q). The temperature, salinity and density of this basin depends on the strength of these inputs/outputs and on the rate of hydrodynamic exchange

Buoyancy Effects on Coastal and Estuarine Dynamics
Coastal and Estuarine Studies Volume 53, Pages 227-241
Copyright 1996 by the American Geophysical Union

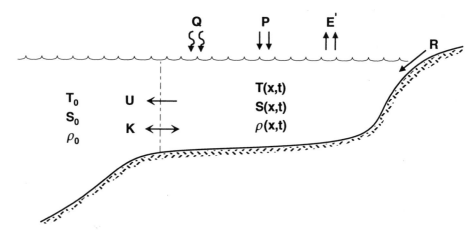

Figure 1. Schematic and definition of significant water and heat exchanges in a low-inflow estuary (variables defined in text).

with the adjacent ocean (the ambient). This hydrodynamic exchange is expressed as the sum of subtidal advection U and subtidal diffusion K. Following Largier et al. (1996), the basin is considered to be a "hypersaline estuary" if its salinity is significantly greater than the ocean salinity (i.e., exceeds ocean salinity by at least one standard deviation of ocean salinity fluctuations). A "thermal estuary" describes a basin in which the density structure is dominated by temperature. An "inverse estuary" is one in which the density of the water exceeds that in the ocean.

Observations

Hypersaline conditions, a clear indicator of low freshwater inflow, have been observed in many estuaries along the Pacific coast of the Californias (Fig. 2). In Tomales Bay, the northernmost of these basins, temperature and salinity data have been collected over a 7-year period (Largier et al., 1996). The seasonal cycle is dramatic, dominating any interannual fluctuations (Fig. 3). Winters are cold and wet, yielding large river inflow and low salinities. Summers are warm and dry, resulting in net evaporation and high salinities. Occupying a rift valley along the San Andreas fault, this basin is long and thin. Temperature, salinity and density data are clearly functions of the distance from the ocean, at one end, and from the major river flowing in at the other end. At station 16, 16 km from the mouth and 4 km from the head, salinities drop into the mid-20's while temperatures drop below 10° C during winter. In summer, salinities exceed ocean values by about 2 and temperatures exceed ocean values by several degrees centigrade. While estuarine temperatures rise during early summer, ocean temperatures remain low as a result of wind-driven upwelling along the ocean coast (Lentz, 1991; Largier et al., 1993). The summer gradient in temperature maintains a positive density gradient (a "thermal estuary"). It is only in the fall, when estuary temperatures drop, that the density gradient approaches zero and may become inverse owing to the hypersalinity of the basin, which persists until the onset of winter rains.

The coarse temporal resolution of Tomales Bay data does not allow one to ascertain whether a steady state salt balance occurs during late summer. In Mission Bay, where data were collected more frequently, a quasi-steady salinity and temperature distribution is observable from June to September (Fig. 4). In this basin, hypersaline conditions are only observed in the one-dimensional side branches and not in the more complex morphology of the middle and outer bay (Fig. 2). As in

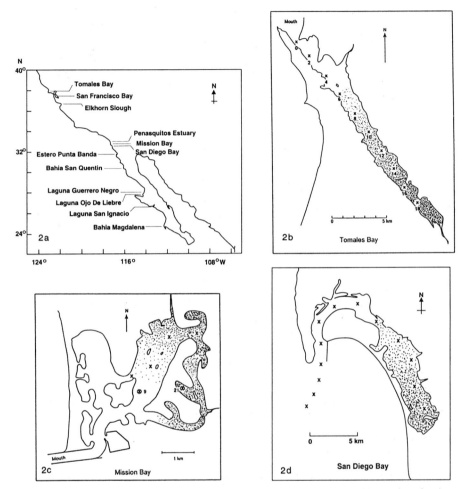

Figure 2. Seasonally hypersaline estuaries along the west coast of North America. (a) Location of various systems in which hypersalinity has been reported. (b) Position of CTD stations in Tomales Bay and typical region of hypersalinity in late summer (shaded). (c) Mission Bay stations and hypersaline region. (d) San Diego Bay stations and hypersaline regions.

Tomales Bay, inverse conditions are only observed following the decrease in water temperatures in the fall and early winter, prior to rainfall. In Mission Bay, the hypersalinity is stronger and when the temperature gradient reverses in early winter a significant inverse structure is observed (Fig. 4). This density head persists, however, indicating the absence of a singular density-driven flushing event in 1992. Rather, a steady salinity indicates that freshwater loss via evaporation is balanced by a seaward salt flux due to tidal diffusion mechanisms.

A similar seasonal cycle and longitudinal structure is observed in San Diego Bay. In this longer system (Figs. 2 and 5), one notices the separation between the region where increasing temperature dominates the density change ("thermal estuary") and the region where the increasing salinity dominates density change ("hypersaline estuary"). The middle and outer bay is thus

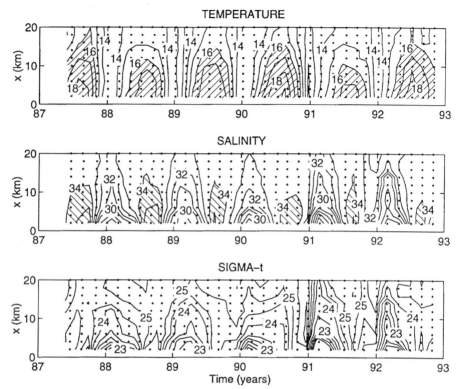

Figure 3. Multi-year seasonal variation of vertically averaged temperature, salinity, and sigma-t in Tomales Bay, as a function of distance from the head ($x=0$). Regions of water that are warmer and saltier than the ocean are hatched (summer). Regions of freshwater influence are seen in winter.

characterized by a positive density gradient while the innermost part of the bay is characterized by a negative density gradient ("inverse estuary"). A density minimum is found in between these outer and inner regions. Reviewing Tomales Bay and Mission Bay data a similar offset of temperature and salinity structures can be seen in the density structure (this is not clearly shown in this paper).

The Longitudinal Salt Balance

Noting the occurrence of hypersalinity in systems with long, narrow basins or in long, narrow branches of systems, Largier et al. (1996) use the following one-dimensional model of vertically averaged salinity to illustrate the essence of the subtidal salt balance of these low-inflow estuaries:

$$\partial_t S = \partial_x \left[\left[\frac{R}{wh} + \frac{Px}{h} - \frac{E'x}{h} \right] S - K \partial_x S \right]$$

(1)

where h is the basin depth, w is the basin width, x is the longitudinal distance from the head of the estuary, t is time, K is the subtidal eddy diffusivity in the longitudinal direction and E', R and P are

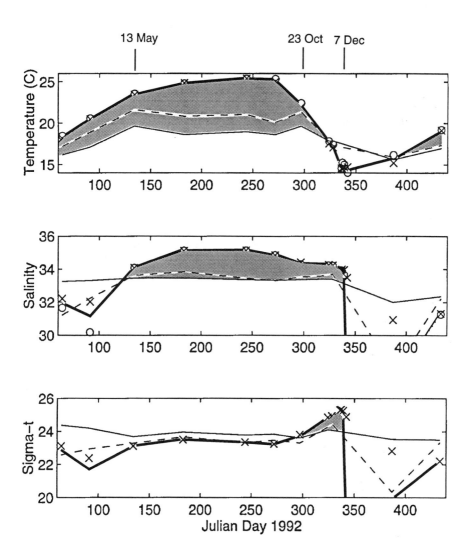

Figure 4. The seasonal variation in temperature, salinity, and sigma-t in Mission Bay. At station 21 (farthest from the ocean, Figure 2c), surface values (open circles), bottom values (crosses), and vertical average values (bold line) are plotted. For comparison, vertical average values are also plotted for station 9, at the landward extent of the well-flushed outer estuary (dashed line), and for Scripps Pier, an ocean station several kilometers north of the mouth of the Bay (solid line). The period during which station 21 is warmer, saltier or denser than the ocean is indicated by shading. Stratification is not observed in the hypersaline inner estuary. The date marks indicate the start and end of hypersaline and inverse conditions.

as earlier (Fig. 1). Using $E = E'-P-R/wx$ as net evaporation, and expecting a quasi-steady salinity distribution in mid-summer ($\partial_t S = 0$) one can solve for $S(x)$. In all three systems studied, longitudinal dispersion appears to be a continuous subtidal process. Using a mixing length expression for the longitudinal diffusivity $K = kx^2$, one obtains the solution

$$\frac{S}{S_0} = \left[\frac{x}{L}\right]^{-E/kH} \tag{2}$$

which fits well to observed data (e.g., San Diego Bay, Fig. 6). From the longitudinal salinity distribution one can estimate the average residence time of water at a particular distance from the ocean (assuming constant salinity S_0), using the bulk expression

$$\tau_{res} = \frac{(S - S_0)}{E\,S}\,h \tag{3}$$

Average residence times increase markedly towards the head of the basin with negligible inflow. This is consistent with the marked decrease in the value of the tidal diffusivity, $K = kx^2$, obtained from a mixing length model. This long residence, relative to the time scales of surface fluxes (e.g., evaporation), is the defining dynamical character of these hypersaline basins.

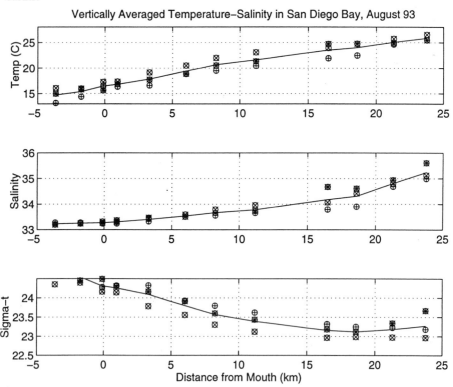

Figure 5. Vertically averaged temperature, salinity, and sigma-t for San Diego Bay on 3 days in August 1993 (symbols). Solid line is average of the three data sets, which exhibit tidal differences. Data are plotted as distance from the mouth.

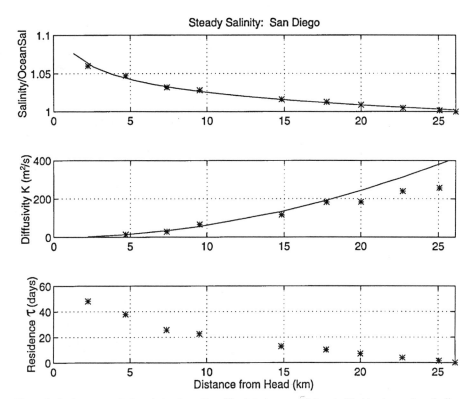

Figure 6. In the top panel, the solution from Eqn. (2), plotted as a solid line, is fitted to observed vertically averaged salinity (from line in Fig. 5.), plotted as asterisks. The middle panel presents diffusivity $K = kx^2$ as a function of distance from the head x, with the value of k obtained from the fit in the top panel. Values plotted as asterisks are obtained from balancing advection and diffusion at each station (assuming steady salinities, see Largier et al., 1996). In the bottom panel, residence time is plotted - values are calculated from Eq. (3).

Largier et al. (1996) find that this simple one-dimensional approach adequately represents the salinity distributions of 5 illustrative examples: San Diego Bay, Mission Bay and Tomales Bay, which are discussed in this paper, and Elkhorn Slough (California) and Langebaan (South Africa). Residence times in the hypersaline inner basin vary from 10 to 100 days.

Surface Buoyancy Fluxes

While salinity is a useful conservative tracer of water residence, to understand density structures and potential buoyancy-driven exchange, one has to consider the simultaneous changes in salinity and temperature. This is particularly true for mid-latitude west coast regions which are characterized by coastal upwelling, cool ocean waters and significant temperature gradients. In tropical systems, where the ocean is warm, temperature gradients are less important (e.g., Wolanski, 1986). In freshwater "estuaries" connected to large lakes, temperature is all important as there is no active salinity effect (e.g., Hamblin and Lawrence, 1990).

As noted, outer and middle San Diego Bay exhibits a significant temperature gradient whereas the maximum salinity gradient occurs farther into the bay. The resultant density minimum is clearly illustrated on a temperature-salinity plot (Fig. 7). This key structure of low-inflow west coast estuaries can be illustrated with a very simple model of heating and evaporation. Bulk formulae for surface heating (e.g., Gill, 1982) can be reduced, through approximation, to a simple dependence on the time that a water parcel has been within the estuary:

$$Q = Q_{other} + \rho_a c_p D_h u (T_a - T)$$

$$= q(T_{eff} + T_a - T)$$

where T is the (surface) water temperature and T_a is atmospheric temperature; Q_{other} represents the heat flux terms other than sensible heat flux. These other heat flux terms are expected to have a much weaker dependence on the increasing surface water temperature, and are treated as constant over the time scale of interest in this simplification. They are represented by an effective constant temperature addition T_{eff} to atmospheric temperature T_a. The coefficient q is a product of air density ρ_a, heat capacity of the air c_p, wind speed u, and a dimensionless heat transfer coefficient (Stanton number) D_h. The change in temperature is then

$$\partial_t T = \frac{Q}{c_w h \rho} = b - cT$$

where

$$b = \frac{q}{c_w \rho h}\left(T_{eff} + T_a\right)$$

and

$$c = \frac{q}{c_w \rho h}$$

ρ is the water density, c_w the heat capacity of the water and h the water depth. Thus, the temperature of a parcel of water that has been resident in the basin for time t is given as

$$T(t) = \frac{b}{c} - de^{-ct} \tag{4}$$

where

$$d = \frac{b}{c} - T(t = 0)$$

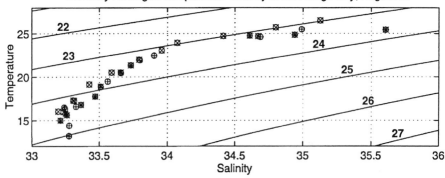

Vertically Averaged Temperature–Salinity in San Diego Bay, August 93

Figure 7. Vertically averaged temperature-salinity data from San Diego Bay (August 1993, see Fig. 5) plotted over a sigma-t field.

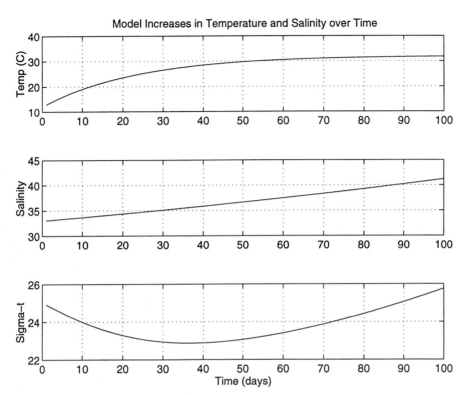

Figure 8. Model results for the increase of temperature from Eq. (4) and the increase of salinity from Eq. (5) as a function of time that a water parcel has been resident in a shallow, zero-inflow basin. Parameter values are those typical of Californian systems. The temporal decrease and subsequent increase in calculated density is shown in the bottom panel.

This gives the rate of increase in water temperature as a simple dependence on the temperature itself (assuming constant wind, air temperature, humidity, insulation and vapor pressure). The water heat content is saturated (under these constant conditions) when T increases to balance $T_a + T_{eff}$.

Under these conditions, evaporation occurs at an approximately constant rate, being only weakly dependent on surface temperature and salinity. Thus, the salinity of a parcel of water that has been resident in the basin for time t is given as

$$S(t) = \frac{S_0}{1 - Et/h} \tag{5}$$

Starting with ocean water of temperature 12° C and salinity 33 at time $t=0$ (as the parcel of water enters the shallow-water basin), and assuming conditions representative of a low-inflow California estuary, the asymptotic increase in temperature and the constant increase in salinity are plotted for a 100-day period (Fig. 8). Initially, density decreases due to warming (net negative buoyancy flux). A density minimum is noted after a time period of about a month. If the San Diego Bay data from August 1993 are plotted against residence time, given by Eq. (3), the same structure is observed (Fig. 9).

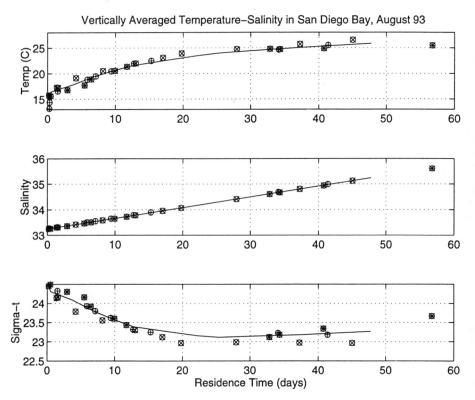

Figure 9. Observed temperature, salinity, and sigma-t values in San Diego Bay (August 1993, Fig. 5) plotted as a function of residence time, estimated from Eqn. (3).

Longitudinal Density Structure

The time history of water in a quasi-steady system is reflected in the distance of water parcels from the ocean. Water further from the ocean is older on average, as described by the one-dimensional diffusion model of these long, narrow basins. Using the solution for S(x) under the assumption of $K = kx^2$, as in Eq. (2), and estimating residence time by the degree of hypersalinity, one obtains a simplified longitudinal variation in residence time:

$$\tau_{res}(x) = \frac{h}{E}\left[1 - \left(\frac{x}{L}\right)^{E/kh}\right]$$

which can be written as

$$x(\tau_{res}) = L\left[1 - \frac{E\tau_{res}}{h}\right]^{kh/E}$$

(6)

where x is the distance from the head at which a given residence time τ_{res} would occur in this simplest one-dimensional model (Fig. 10a). Using this time-to-space conversion, one obtains the expected spatial increase in temperature and salinity due to the temporally integrated model heat input and freshwater removal (Fig. 10b, c). Further, from these $T(x)$ and $S(x)$ distributions, one obtains the longitudinal density distribution $\rho(x)$ (Fig. 10d) and the longitudinal gradient in density (Fig. 10e).

Wherever the longitudinal density gradient is strong enough, one may expect it to drive a stratified exchange flow - cold, dense seawater intruding as a basal flow near the mouth and/or salty, dense inner basin water draining seaward as a basal flow near the head. Following Linden and Simpson (1988), the buoyancy-driven diffusivity is proportional to $|\sigma_x \rho|.n$. With typical model output (Fig. 10), then, one can expect maximum buoyancy effects in the deeper thermal outer estuary and in the high-gradient hypersaline region towards the head of the estuary (Fig. 10h, Fig. 11). In contrast, simple tidal diffusivity, which is proportional to x^2, decreases strongly

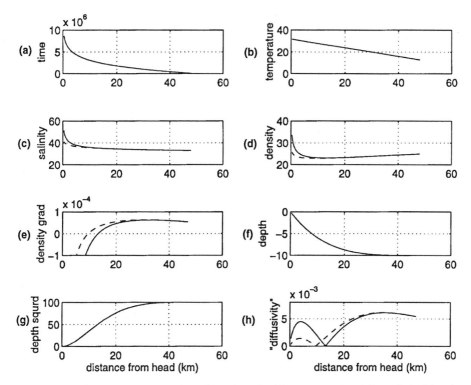

Figure 10. (a) Model residence times obtained from Eq. (6), with parameter values being typical of those in Californian LIE's. (b) Model temperature, given residence time from first panel and heating modeled by Eq. (4). (c) Model salinity, given residence time from first panel and salinity increase over time modeled by Eq. (5). The dashed line is for a constant depth basin, whereas the solid line indicates how salinity values increase if the basin shoals markedly towards the head as illustrated in panel (f). (d) Density calculated from model temperature and salinity. Solid line is for adjusted salinity (as in third panel). (e) Longitudinal density gradient. (f) Illustrative longitudinal depth profile, allowing for enhanced hypersalinity in the shallow regions at the head of the estuary. (g) The square of depth as a function of longitudinal position. (h) The typical longitudinal dependence of buoyancy diffusivity values in a low-inflow estuary, based on the dependence given by Linden and Simpson (1988). Diffusivity units are arbitrary.

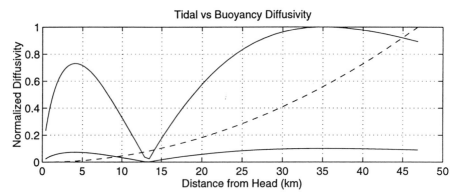

Figure 11. The typical buoyancy diffusivity profile $K \propto |\partial_x \rho| \, h^4$ for a low-inflow estuary (see solid line in Fig. 10h) compared with a typical tidal diffusivity profile $K \propto x^2$. Both diffusivities are normalized; tidal values are represented by a dashed line. An additional solid line represents the shape of buoyancy diffusivities that are an order of magnitude smaller that the tidal values. In this case, buoyancy diffusivities exceed tidal diffusivities only in the strongly hypersaline region close to the head (over the shallow region within 5-10 km from the head of this model basin).

towards the head of the basin where $x \rightarrow 0$ (Fig. 11). Thus, it is in the hypersaline region near the head of the estuary that density-driven exchange is likely to be a significant component of the longitudinal diffusion. Transient hypersaline-related inverse circulation has been observed at the landward ends of both San Diego and Mission Bays. Although weak, the observed structures are clearly due to the relaxation of the inverse density distribution. In Fig. 12, the density minimum as well as the two regions of stratified exchange flow are evident. These coherent vertical structures are transient, disappearing tidally. The detailed hydrodynamic structures of the "thermal estuary" part of San Diego Bay are discussed by Chadwick et al (1996). Aspects of the role of buoyancy exchange in Tomales Bay are addressed through a model study of historical depth changes in Tomales Bay (Hearn et al., 1996).

Strongest evaporation-associated inverse circulation is expected in deep systems, with relatively weak vertical mixing, e.g., the Mediterranean Sea (Lacombe and Richez, 1982) and Spencer Gulf (Nunes Vaz et al., 1990). However, significant inverse circulation, albeit transient, can be expected in evaporative estuaries with extensive shallows, to enhance the net evaporative loss of freshwater, and deep channels, in which density exchange flow can occur. An example is Laguna Ojo de Liebre (Fig. 12a in Postma, 1965). It appears that significant inverse circulation is also found in tidal creeks through mangrove swamps, due to the large evapotranspiration through the mangrove plants (e.g., Maputo Bay, Mocambique, A. Hoguane, pers. comm.).

Conclusion

In many mediterranean-climate, west coast estuaries one can identify up to 4 distinct hydrodynamic regimes (Fig. 13). The presence and extent of these regimes varies seasonally. Closest to the mouth, a "marine" regime is typically found. Residence times are comparable to the tidal period, temperature and salinity are similar to oceanic values and diffusivity is dominated by tidal effects, particularly tidal pumping. With increased distance from the sea (at least one tidal excursion), residence times increase to several days and significant temperature gradients are observed. This "thermal" regime is distinguished by large Q/h and it is absent in regions where the oceanic source water is warm. At longer residence times, found deeper into the basin, the heat content is saturated and an evaporative increase in salinity dominates the surface buoyancy flux (E/h is large). This is the "hypersaline" regime where inverse circulation may be observed. In

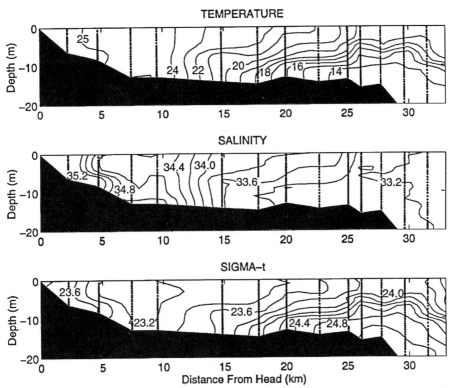

Figure 12. Vertical-longitudinal sections of temperature, salinity and sigma-t during slack highwater on 10 August 1993. Actual data points are indicated by dots (CTD profiles).

some systems, a small freshwater inflow $R \ll EwL$ (where L is the basin length) persists during the dry season and an "estuarine" regime may be observed at $x < R/Ew$. There is some evidence that density-driven circulation occurs in the thermal regime - e.g., San Diego Bay (this paper, and Chadwick et al., 1996), Hamilton Harbor (Hamblin and Lawrence, 1990) and Hilary Harbor (Schwartz and Imberger, 1988) - in the hypersaline regime - e.g., San Diego and Mission Bays (this paper) and Laguna Ojo de Liebre (Postma, 1965) - and in the estuarine regime - e.g., Van Diemens Gulf (Wolanski, 1986). But, it is not clear how important this transient stratified exchange is to the net longitudinal exchange in tidal basins. It is clear, however, that regions of near-zero gradient $\partial_x \rho$ exist and that these regions must be dominated by tidal action. Any closed circulation cells, as suggested by Wolanski (1986), are likely to occur in non-tidal estuaries only.

On mid-latitude west coasts, such as those of the Californias, the seasonal temperature cycle tends to lead the seasonal salinity cycle by a month or two (e.g., Tomales Bay, Fig. 3), with low estuarine salinities in spring. This is due to residual river inflow in spring, the more rapid response of the basin to heating/cooling and the delay in winter rains relative to the day length (and heat cycle). Owing to this temporal lag and the spring-summer presence of cold upwelled water along the open coast, much of a Californian estuary is characterized by a "thermal" regime in early summer and by a "hypersaline" regime in the fall. It is the interaction between these longitudinal density gradients and vertical mixing that will determine the importance of surface buoyancy fluxes in enhancing estuary-ocean exchange in low-inflow systems. A study of vertical mixing processes in shallow LIE's is desirable, given the presence of both tidal and wind stirring, as well as double diffusive salinity-temperature gradients and the potential for nocturnal overturn.

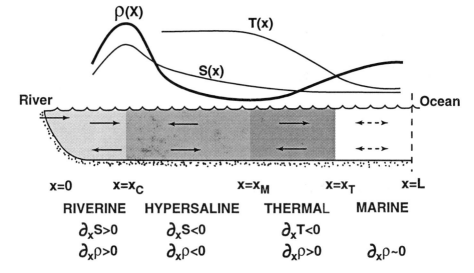

Figure 13. Schematic of the longitudinal temperature, salinity, and density profiles that may result in up to four distinct regimes in a California LIE. The marine outer basin is dominated by tidal diffusion. The thermal regime is characterized by a positive density gradient. The hypersaline regime is characterized by a negative density gradient, and the riverine regime is characterized by a positive density gradient due to river inflow. A density minimum may be observed in mid-estuary and a density maximum at or near the head of the basin (e.g., Fig. 7).

Acknowledgments. We are grateful to our colleagues, Steve Smith, Tim Hollibaugh, and Kimball Millikan, amongst others, for their part in data collection and scientific discussion. In this work, support was received from the National Science Foundation (Land Margin Ecosystem Research), Contracts 89-14833 and 89-14921; the California Department of Boating and Waterways, Interagency Agreements 91-100-080-19, 92-100-060-21, and 93-100-030-16; and the California Regional Water Quality Control Board (San Diego Region), Interagency Agreement 1-188-190-0.

References

Chadwick, D. B., J. L. Largier, and R. T. Cheng, The role of thermal stratification in tidal exchange at the mouth of San Diego Bay, Proceedings of 7th International Conference on the Physics of Estuaries and Coastal Seas, edited by D.G. Aubrey and C.T. Friedrichs, pp. (this volume), 1996.

Gill, A. E., *Atmosphere-Ocean Dynamics*. Academic Press, Oxford, 500 pp., 1982.

Hamblin, P. F., and G. A. Lawrence, Exchange flows between Hamilton Harbor and Lake Ontario, in *Proceedings of 1990 Annual Conference Canadian Society Civil Engineering*, Hamilton, May 1990, vol. 5, 140-8, 1990.

Hearn, C. J., J. L. Largier, S. V. Smith, , J. Plant, and J. Rooney, Effects of changing bathymetry on the summer buoyancy dynamics of a shallow mediterranean estuary; Tomales Bay, California, *Proceedings of 7th International Conference on the Physics of Estuaries and Coastal Seas,* edited by D.G. Aubrey and C.T. Friedrichs, pp. (this volume), 1996.

Lacombe, H., and C. Richez, The regime of the Straits of Gibraltar, in *Hydrodynamics of Semi-enlosed Seas*, edited by J. C. J. Nihoul, pp. 13-73, Elsevier, Amsterdam, 1982.

Largier, J. L., B. A. Magnell, and C. D. Winant, Subtidal circulation over the northern California shelf, *J. Geophys. Res., 98*(C10), 18147-18179, 1993.

Largier, J. L., J. T. Hollibaugh, and S. V. Smith, Seasonally hypersaline estuaries in mediterranean climate regions, *Est. Coast. Shelf Sci.*, in press, 1996.

Lentz, S. J. (ed.), *CODE (Coastal Ocean Dynamics Experiment). A collection of reprints*, 817 pp., Woods Hole Oceanographic Institution, Woods Hole, MA, 1991.

Linden, P. F., and J. E. Simpson, Modulated mixing and frontogenesis in shallow seas and estuaries, *Cont. Shelf Res., 8,* 1107-1127, 1988.

Nunes Vaz, R. A., G. W. Lennon, and D. G. Bowers, Physical behaviour of a large, negative or inverse estuary, *Cont. Shelf Res., 10*(3), 277-304, 1990.

Postma, H., Water circulation and suspended matter in Baja California lagoons, *Neth. J. Sea Res., 2*(4), 566-604, 1965.

Pritchard, D. W., Observations of circulation in coastal plain estuaries, in *Estuaries,* edited by G. H. Lauff, pp. 37-44, AAAS Publ. #83, Washington, 1967.

Schwartz, R. A., and J. Imberger, Flushing behaviour of a coastal marina, *Environmental Dynamics Ref.*, ED-88-259, University of Western Australia, Nedlands, Australia, 1988.

Wolanski, E., An evaporation-driven salinity maximum zone in Australian tropical estuaries, *Est. Coast. Shelf Sci., 22*, 415-424, 1986.

17

Effects of Changing Bathymetry on the Summer Buoyancy Dynamics of a Shallow Mediterranean Estuary; Tomales Bay, California.

Clifford J Hearn, John L Largier, Stephen V Smith, Joshua Plant, and John Rooney

Abstract

Tomales Bay, lying along the line of the San Andraess Fault on the coast of California, is a highly unidirectional, shallow, mediterranean estuary. A numerical model is used to predict the temperature and salinity of Tomales Bay from 1987 to 1993 using detailed hydrological and meteorological data. The model results compare favorably with the seasonal variation of average values observed in the Bay during that seven-year period. The present estuarine basin is shorter, and shallower, than the basin which existed in the 19th century prior to major usage of land and water resources. The numerical model is used to simulate the behaviour of an estuary with the historical bathymetry (from a survey made in 1861) over the same seven year period 1987 to 1993. The model results show that the historical estuary became more inverse in summer with longer net flushing times but faster baroclinic mixing near the head of the basin due to greater longitudinal density gradients.

Introduction

Tomales Bay in California [Smith et al., 1991a, b] was the site of detailed temperature and salinity observations from 1987 to 1993 [LMER Coordinating Committee, 1992]. The Bay (figure 1), which is some 20 km long, lies along the San Andraess Fault, and as a result, is unidirectional with a large aspect ratio of length to width. Topographically, it consists of an Inner Basin (mean depth about 3 m) joined to the ocean via a narrow, and much deeper, tidal channel (depths to 20 m). This structure is typical of estuaries of the present class [Kjerfve, 1994]. This study has developed a numerical model of the salinity, and temperature, behaviour of the Inner Basin. This uses detailed meteorological and riverflow data. The Inner Basin becomes hypersaline in summer (relative to the ocean) but develops only a very limited density difference due to the compensating effect of estuarine heating. Smith et al. [1991a] have shown that the summer estuary has a very long ocean exchange time (of order 100 days) and slow summer flushing is evident in many mediterranean estuaries [McComb et al., 1981] so that they are environmentally very fragile [e.g. Hearn et al., 1994; McComb and Lukatelich, 1995]. It is therefore of some importance to understand the summer dynamics, and in particular, the possible influence of anthropogenic changes. The present paper pursues this aim and uses the model that has been developed to investigate the influence of human changes to the bathymetry of the estuary. This involves a second simulation (over the same seven year

Buoyancy Effects on Coastal and Estuarine Dynamics
Coastal and Estuarine Studies Volume 53, Pages 243-253
Copyright 1996 by the American Geophysical Union

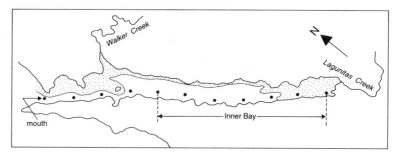

Figure 1. Map of Tomales Bay and inflowing Creeks with shaded areas corresponding to depths of less than 2m. The circles represent the sampling stations used in this study.

period) based on the historical bathymetry of the estuary from a survey made in 1861. The paper concludes with a general discussion of the summer dynamics.

Model of Tomales Bay

The essentially unidirectional form of Tomales Bay, and its large aspect ratio, allows the model to be simplified to two dimensions in the vertical plane but terms are included which describe the longitudinal variation of the lateral geometry (in the manner of Perrels and Karelse, 1981]. The model, which is described by Hearn [1996], is baroclinic and uses a vertical eddy viscosity due to tidal stress [Fischer, 1973; Heaps and Jones, 1987] and a Richardson Number dependence [Munk and Anderson, 1948] is added. The Bay appears to be driven primarily by a combination of this tidal mixing and gravitational processes due to the surface buoyancy flux from riverflow, surface heating/cooling and evaporation [e.g. Bowden, 1983]. In view of the explicit nature of the model (so that the time step is restrained by the CFL Criterion), and the rather long runs of seven years, the restriction to two dimensions in the vertical gives a valuable reduction in computation time. This simplification seems to be justified (in this first modeling study) since very little variations in temperature and salinity were found in some exploratory measurements made laterally across the basin. The data used here are of low resolution in both space and time. They were collected during field excursions made approximately every three weeks at ten sampling stations (Figure 1) separated by distances of 2 km along the axis of the estuary. Generally, measurements were made only at the top and bottom of the water column, but since the estuary is well-mixed in summer, the vertical resolution of the data is adequate for the present study.

At the mouth of the estuary, the model uses predicted tidal elevation [Peak, 1994], and radiation boundary conditions, together with observed temperatures and salinities which are assumed (in accordance with the data) to be constant through the water column. These latter are influenced by coastal fluxes, and are not representative of open shelf conditions, so that salinities at the mouth are severely depressed by winter storms. The model uses 134 cells along the length of the estuary and ten levels in the vertical. Riverflow is derived from detailed stream data using an algorithm due to Smith et al. [1991a]. Two major creeks enter Tomales Bay (Figure 1); Lagunitas Creek at the head and Walker Creek close to the tidal channel. The flow of both creeks is controlled by reservoirs. The model assumes that both streams empty into the surface layer of the nearest non-dry model cell (which changes with the tidal elevation for Lagunitas Creek).

The surface fluxes of heat and mass are given by formulae taken from Fischer et al. [1979]. These are based on the model surface temperature together with the air temperature, humidity, and wind speed (from data collected with Sierra Misco ALERT stations) and the solar short wave flux. The latter is simply calculated as a function of time knowing the latitude of the

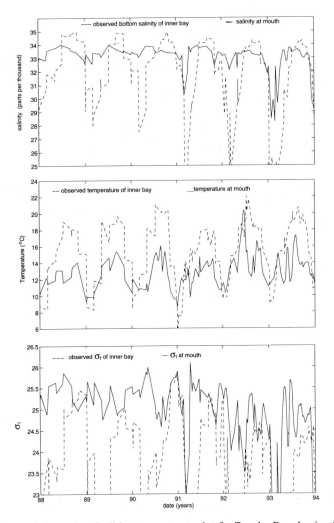

Figure 2. Observed time series of salinity, temperature, and st for Tomales Bay showing the average of observations over the Inner Bay and values at the mouth of the estuary.

estuary and ignoring any cloud cover; this is an acceptable approximation for the summer mediterranean conditions of interest to this paper. The model therefore calculates the evaporative velocity and heat flux from which can be derived the average surface buoyancy flux B (per unit mass) defined as

$$B \equiv -\left\langle \frac{g}{\rho}\frac{\partial \rho}{\partial t} \right\rangle_{Inner\ Bay} \tag{1}$$

$$\frac{1}{\rho}\frac{\partial \rho}{\partial t} = \frac{1}{h}\left(\beta S v - \frac{\alpha Q}{\rho C_p}\right) \qquad (2)$$

in which h is depth, v the net velocity of evaporation allowing for riverflow (distributed over the area of the Inner Bay); α, and β, are the linear coefficients describing thermal expansion and the fractional rate of change of density ρ with salinity S, respectively,

$$\alpha \equiv -\frac{1}{\rho}\frac{\partial \rho}{\partial T} \qquad \beta \equiv \frac{1}{\rho}\frac{\partial \rho}{\partial S} \qquad (3)$$

Q is the net rate of surface heating and C_p the specific heat. B is not explicitly used in the model but is a prime ingredient of the buoyancy dynamics since it controls the density difference between the estuary and ocean. It consists of two terms (of opposite sign) which individually drive the temperature and salinity differences but which may be of similar magnitude in summer producing small net (positive or negative) buoyancy flux B.

Seasonal Variation of Salinity, Temperature and Density

Figure 2 shows the seasonal variation of salinity for the mouth and Inner Bay. There are salinity depressions during winter rains and hypersalinity occurs during the summer mediterranean droughts; 'hypersalinity' is used here to describe a state in which the estuarine salinity exceeds that at the mouth. The accompanying time series in Figure 3 compares the model and observed salinity of the Inner Bay which can be seen to follow the same seasonal trends. The model results are plotted as weekly averages whilst the broken line, which represents observed salinities, simply joins the three-weekly data points. The model does not reproduce individual dips in salinity caused by winter storms probably because of the inadequate time resolution of both the riverflow and observational data. It adequately reconstructs the spring rise of salinity, reaching marine values in early summer, and reproduces the hypersalinity of late summer. The hypersalinity is indicative of long residence times since the evaporative velocity is only a few mm per day; about 100 days are needed for 1 mm per day of evaporation to raise the salinity of a 3.5 m deep basin of marine water by 1 part per thousand. This type of salt budget calculation was performed in detail by Smith et al. [1991a]. The evaporative velocity was calculated from observed temperature, wind speed, and humidity. The budget included the small, but important, freshwater inflow due to the summer release of reservoir water. It gives, in a quite unequivocal way, the ocean exchange time averaged over the Inner Basin and shows that the exchange time varies seasonally and reaches values of order 100 days in late summer.

Figures 2 and 3 show the corresponding time series for σ_t in Tomales Bay. The summer hypersalinity might be expected to produce an inverse estuary (in the sense of Pritchard, 1967, that the density of the Inner Basin exceeds that at the mouth). However, Figure 2 shows that the summer estuary becomes at most only weakly inverse and tends to remain almost neutral due to the compensatory effect of heating. Note that these σ_t values are averages for the Inner Bay; σ_t values at the head of the estuary do sometimes show slightly more inversion (compare Figure 5, which shows the longitudinal distribution of σ_t along the estuary). Notice that the hypersalinity is 1 to 2 ‰ which would produce a σ_t difference of order 0.7 to 1.4 whilst actual σ_t differences are rarely positive, and even then, of much lower magnitude. Figure 2 displays the temperature variation of the mouth and Inner Bay and shows that seasonal heating of the Inner Bay is of order 5 ˚C. This heating is responsible for the nearly complete compensation of the σ_t increase due to hypersalinity. The model has a tendency to underestimate peak summer temperatures by up to 1 °C. The agreement between the predicted and observed σ_t is good with the notable exception of the latter part of 1990, ie the early winter, when the model σ_t is much lower due mainly to the under estimation of the salinity of the Inner Bay. Similar, but smaller, early winter errors can be identified in most years. They coincide with the start of winter rains when the late-summer estuary becomes stratified but the resolution of the data is insufficient to clearly determine the cause of these discrepancies.

Influence of Historical Bathymetry

The model run was repeated using bathymetry for the Inner Bay obtained from a survey conducted in 1861. The historical basin was both longer and deeper than the modern estuary

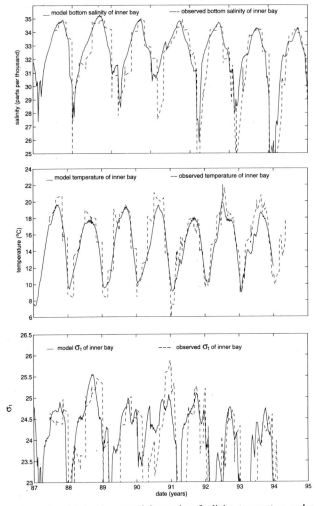

Figure 3. Comparison of modeled and measured time series of salinity, temperature, and σ_t averaged over the Inner Bay. The model uses the present bathymetry of the estuary.

due to sedimentation from the inflowing Creeks and land reclamation. The model parameters were identical in all other respects and the model used the same meteorological record and temperature/salinity (and tidal) data at the mouth of the estuary. The historical form of the outer part of Tomales Bay, which contains the tidal channel, also shows some very complex shifts in the shape of its sediment banks [Peak, 1994]. These are deliberately excluded from the present study since changes in channel topography are dealt with elsewhere [Hearn, 1995] and produce effects which are dominantly tidal and distinct from the summer buoyancy effects of interest here.

Figure 4a plots the salinity difference (between the Inner Bay and mouth) predicted using historical bathymetry against corresponding model values (averaged over the same week) for the present estuary. Figures 4b and 4c give similar plots for temperature and σ_t differences. The results indicate that the historical estuary was more hypersaline, experienced greater heating and became more inverse. Table 1 shows average values of the salinity, temperature, and σ_t differences for the present and historical (entries in parenthesis) estuaries; note that the σ_t difference between the basin and mouth is defined as $- \Delta\sigma_t$. The table displays averages over the periods for which the estuary is either hypersaline, or inverse, and gives the average number of weeks per year for which it is in either of those two states. These results show that the baroclinic exchange, responsible for forcing the estuary towards neutrality, is weaker in the historical estuary. This is due to the increase in length of the Inner Basin and the upper reaches of the historical estuary being very shallow; both of these effects reduce the strength of the baroclinic processes.

Such reductions would be partially offset by the increased depth of the main part of the Inner Basin. The result is a net increase in the ocean exchange time for the entire Inner Basin as is demonstrated by the increased hypersalinity. The concurrent increase in temperature does also increase the rate of evaporation but the major reason for the increase in hypersalinity is the weakened ocean exchange.

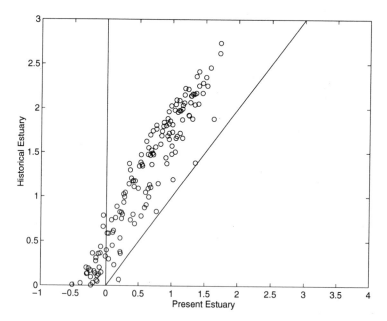

Figure 4a. Scatter diagram of modeled weekly salinity differences (between Inner Bay and mouth of Tomales Bay) based on the historical and present bathymetry.

Table 1. Average salinity, temperature, and σ_t difference (Δs, ΔT, and $-\Delta\sigma_t$) between the Inner Basin and mouth of Tomales Bay for the present and historical (entries in parenthesis) estuaries. Averages are calculated for conditions in which the estuary is hypersaline, or inverse, and the table shows the average number of weeks per year for which these conditions occur.

estuarine state	weeks per year	ΔS (‰)	ΔT ($^\circ C$)	$\Delta\sigma_t$
hypersaline	18.0 (21.9)	0.65 (1.26)	1.75 (2.48)	0.28 (0.08)
inverse	3.9 (9.9)	0.22 (0.82)	0.27 (1.02)	-0.28 (-0.41)

It is of interest to enquire as to the effect of the modified bathymetry on the efficiency of the temperature compensation of the density change due to hypersalinity. Define the ratio of compensation of salinity by temperature as

$$c \equiv \frac{\alpha\Delta T}{\beta\Delta S} \tag{4}$$

where ΔT and ΔS are the temperature and salinity differences; α, β are defined in (3). The compensation ratio can be expected to decrease as the estuary becomes more inverse. This is confirmed by the mean salinity and temperature differences in Table 1 which show that the present hypersaline estuary has a compensation ratio c of 0.90 whilst in the historical hypersaline estuary c is lower at 0.66.

Figure 5 shows the longitudinal distribution of salinity, temperature and σt on a typical (but quite arbitrary) day in late summer (4 September 1990). Observations are presented together with model predictions (averaged over that day) based on present and historical bathymetries. The model agrees with salinity observations to within 0.1‰. Comparing

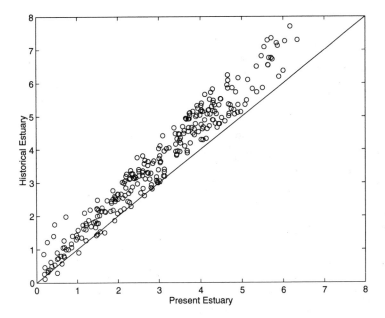

Figure 4b. Scatter diagram of modeled weekly temperature differences (between Inner Bay and mouth of Tomales Bay) based on the historical and present bathymetry.

Figures 8 and 1, it is evident that the observed temperature distribution shows some structure through the channel, and at its junction with the Inner Basin, which may be partly tidal. This is not resolved by the model and there are similar discrepancies in the distribution of σ_t. The model apparently smooths out this thermal structure. The effect may be associated with the lack of the lateral dimension in the model. Furthermore, the assumption of spatially uniform meteorological conditions over the model estuary may not be realistic because of the terrestrial topography. However, the model does evidently reproduce salinity well, and mean values of temperature and σ_t. Changing to the historical bathymetry increases the salinity gradient uniformly along the estuary. Temperature is increased by up to 0.7 C which is smaller than the unresolved spatial structures. The overall effect is that σ_t is increased everywhere in the estuary and in particular there is an increase in the density gradient in the upper part of the Inner Basin. This increased gradient causes increased local horizontal baroclinic mixing near the head.

Discussion and Conclusion

The seven year monitoring program, and model simulation of the summer buoyancy dynamics, of Tomales Bay have shown:

1. The shallow estuary is well mixed in summer and is held close to neutral buoyancy difference relative to the ocean by baroclinic ocean exchange. This is achieved by limiting the surface buoyancy flux through compensation of the heat and mass flux terms.

2. The quasi-neutrality of the estuary produces very long ocean exchange times creating hyper-salinity (the density effect of which is compensated by heating).

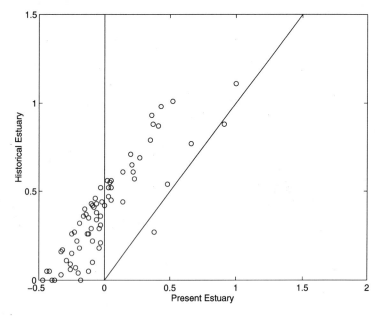

Figure 4c. Scatter diagram of modeled weekly σ_t differences (between Inner Bay and mouth of Tomales Bay) based on the historical and present bathymetry.

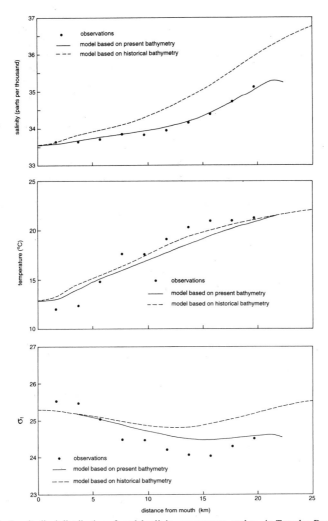

Figure 5. The longitudinal distribution of model salinity, temperature, and σ_t in Tomales Bay averaged over the day of 4 September 1990 and data collected at the sampling stations (shown in Figure 1) on that day. The two lines represent model simulations based on present and historical (broken lines) bathymetry.

3. The historical estuary is more hypersaline and inverse. This is due to weakening of baroclinic processes by the greater length of the historical basin and also by the existence of very shallow regions near its head (which have since been reclaimed). However, the increased density gradients near the head increase local baroclinic mixing in that region. Detailed tracer experiments performed within the model runs show an increase in the mixing time at the head of a least a factor of 2. Such an increase is environmentally important to the mixing into the estuarine basin of material entering the head of the estuary. It occurs only locally and is in contrast to (and a consequence of) the net reduction of exchange to the ocean of the estuary as a whole. Compensation of salinity by heating is less effective in the model historical estuary.

4. The average buoyancy fluxes per unit mass B are shown in Table 2 for the hypersaline and inverse states of the present and historical (entries in parenthesis) estuaries. The table also shows the separate averages of positive and negative B; the peak negative values of B are shown. The present estuary is so weakly inverse that the net flux is positive. The historical estuary, with its weaker ocean exchange, has greater negative buoyancy flux.

In a steady state, the estuary is either classical ($\Delta\sigma_t > 0$), or inverse ($\Delta\sigma_t < 0$), according to the sign of B but the basin does have a finite response time to changes in this buoyancy flux. The sign of B controls, and is controlled by, the baroclinic exchange in these shallow mediterranean estuaries and appears to be, along with $\Delta\sigma_t$, a basic parameter of the dynamical system. The term 'negative' estuary is often used interchangeably with 'inverse' but it is suggested here that it could usefully be reserved to describe basins in which B < 0. With the present bathymetry the model shows Tomales Bay to be 'negative' for 10.7 weeks per year whilst with the historical bathymetry it is 'negative' for 14.1 weeks per year. These values exceed the corresponding periods for the estuary to be inverse given in Table 1. The number of weeks in which the estuary is 'mixed' in the sense of being either inverse and 'positive', or classical and 'negative', are virtually identical at 9.6 for the present bathymetry and 9.7 for the historical estuary.

Table 2. Average buoyancy fluxes per unit mass B x 10^{10} (m s-3) from model simulations with present and historical (entries in parenthesis) bathymetries.

estuarine state	average B	average B for B > 0	average B for B < 0	peak B for B < 0
hypersaline	4.72 (-6.35)	10.4 (10.6)	-5.7 (-17.0)	-27.1 (-59.6)
inverse	4.23 (-6.96)	11.6 (10.5)	-7.4 (-17.6)	-27.1 (-65.0)

Acknowledgments. The model used here was developed under a grant from the Australian Research Council. The Tomales Bay data collection and analysis were funded by National Science Foundation Grants OCE-8613647, OCE-8816709, OCE-8914833 (to SVS), and OCE-8616469, OCE-8914921 to J T Hollibaugh, who with other members of the LMER/BRIE projects, made major contributions to this study. The US. Geological Survey (Ken Markham) and Marin Municipal Water District (Dana Roxon, Randy Arena) provided hydrological data. One of us (CJH) is grateful to the Department of Oceanography, University of Hawaii, for its hospitality during the initial phase of this research and to Dr Hua Wang for assistance with model development.

References

Bowden, K. F., *Physical Oceanography of Coastal Waters*, 302 pp., John Wiley and Sons New York, 1983.

Fischer, H. B., Longitudinal dispersion and turbulent mixing in open-channel flow, *An. Rev. Fluid Mech., 5,* 59-78, 1973.

Fischer, H. B., E. J. List, R. C. Y. Koh, J. Imberger, and N. H. Brooks, *Mixing in Inland and Coastal Waters,* 483 pp., Academic Press, New York, 1979.

Heaps, N. S. and J. E. Jones, Estimation of storm generated currents, in *Three Dimensional Models of Marine and Estuarine Dynamics*, edited by J. C. J. Nihoul and B. M. Jamart, pp 505-538, Elsevier, Amsterdam, 1987.

Hearn, C. J., Water exchange between shallow estuaries and the ocean, in *Eutrophic Shallow Estuaries and Lagoons*, edited by A. J. McComb, pp. 151-172, CRC Press, Boca Raton, 1995.

Hearn, C. J., Application of the model SPECIES to Kaneohe Bay, Oahu, Hawaii, in *Proc 4th Int, Conf. on Estuarine and Coastal*, edited by M.L. Spaulding and R.T. Cheng, pp. 355-366, ASEC, New York, 1996.

Hearn, C. J., A. J. McComb and R. J. Lukatelich, Coastal lagoon ecosystem modelling, in *Coastal Lagoon Processes* edited by B. Kjerfve, pp. 449-484, Elsevier, Amsterdam, 1994.

LMER Coordinating Committee, Understanding Changes in Coastal Environments: The LMER Program, *Eos Trans. AGU, 73,* 481-485, 1992.

Kjerfve, B., Coastal lagoon ecosystem modelling, in *Coastal Lagoon Processes* edited by B. Kjerfve, pp 1-8, Elsevier, Amsterdam, 1994.

McComb A. J. and R. J. Lukatelich, The Peel-Harvey estuarine system, Western Australia, in *Eutrophic Shallow Estuaries and Lagoons,* edited by A. J. McComb, pp. 5-17, CRC Press, Boca Raton, 1995.

McComb, A. J., R. P. Atkins, P. B. Birch, D. M. Gordon, and R. J. Lukatelich, Eutrophication in the Peel-Harvey estuarine system, Western Australia, in *Nutrient Enrichment in Estuaries* edited by B. J. Neilsen and L. E. Cronin, Humana Press, New Jersey, pp. 332-342, 1981.

Munk, W. H. and E. R. Anderson, Notes on the formation of the thermocline, *J. Marine Research, 7,* 276-295, 1948.

Peak, S., The effect of changing bathymetry on salinities and flushing times in Tomales Bay, California: 1861-1994, honours thesis, University of New South Wales, Australian Defence Force Academy, ACT, Australia, pp. 174, 1994.

Perrels, P. A. J. and M. Karelse, A two dimensional laterally averaged model for salt intrusion in estuaries, in *Transport models for inland and coastal waters,* edited by H. B. Fischer, pp 483 -535, 1981.

Pritchard, D. W., Observations of circulation in coastal plain estuaries, in *Estuaries* edited by G. H. Lauff, AAAS Publ. #83, Washington pp. 37-44, 1967.

Smith, S. V., J. T. Hollibaugh, S. J. Dollar and S. Vink, Tomales Bay Metabolism: C-N-P stoichiometry and ecosystem Heterotrophy at the land-sea interface, *Estuarine, Coastal and Shelf Science, 33,* 223-257, 1991a.

Smith, S. V., J. T. Hollibaugh, S. J. Dollar and S. Vink, Tomales Bay, California: A case for carbon controlled nitrogen cycling, *Limnology and Oceanography, 34,* 37-52, 1991b.

18

A Note on Very Low-Frequency Salinity Variability in a Broad, Shallow Estuary

William W. Schroeder, William J. Wiseman, Jr., Jonathan R. Pennock and Marlene Noble

Abstract

Wind stress and river discharge data from the Mobile Bay region indicate significant variance at frequencies below the weather band, but above the semi-annual signal. However, the two records are only very weakly coherent. Two years of salinity data were analyzed in an effort to determine the role that forcing in this band may play in the dynamics of the estuary. These analyses suggest that the salinity response at 50 to 80-day periods is largely due to river discharge. This is also consistent with the estimated 50-day flushing time of Mobile Bay during periods of low discharge. Some wind influence within this band is also apparent.

Introduction

Mobile Bay (Figure 1), on the northern coast of the Gulf of Mexico, is a broad, shallow estuary with depths on the order of 3 m, except for 13 to 15 m depths within a narrow, longitudinal ship channel. The bay is the estuary of the Mobile River system which on average discharges approximately $1850 \, m^3 s^{-1}$ of fresh water. River input is highly seasonal, flooding normally occurring in late winter and spring while low flow conditions can prevail from mid summer to early winter and the annual discharge varies considerably from year to year [Figure 2; Schroeder and Wiseman, 1986]. Winds are also highly seasonal with predominantly gentle breezes from the south during the summer in contrast to strong prefrontal flow from the south and post-frontal flow from the north associated with cold-air outbreaks during the winter [Schroeder and Wiseman, 1985]. Tides in this region of the northern Gulf of Mexico are principally diurnal, with a mean range of 0.4 m, a maximum tropical range of 0.8 m, and a minimum equatorial range of 0.0 m [Marmer, 1954; Seim et al., 1987]. On average this results in very weak tidal currents except within inlets and passes [Schroeder and Lysinger, 1979]. Previous work [Schroeder et al., 1990] on the salinity stratification-destratification cycle within the bay indicates that river flow appears to be the dominant control, the winds being important only in the absence of large freshwater discharge.

Two years (April, 1990 to April, 1992) of near-surface and near-bottom salinity measurements were collected at upper-bay and lower-bay mooring sites (Figure 1). Extensive breaks occur in the data records as a result of the retrieval/re-deployment cycle (e.g. see Figure 3). Otherwise, though, the records are of high quality. The data sets indicate the presence of significant longitudinal and vertical salinity gradients (Figure 3). Based on a comparison of these mooring data with data collected on sixteen extensive bay-wide fair-weather hydrographic surveys [Pennock et al., 1994], performed during the mooring deployments, these salinity gradients appear to be characteristic of the salinity regime throughout the bay and not just representative of specific sites.

Buoyancy Effects on Coastal and Estuarine Dynamics
Coastal and Estuarine Studies Volume 53, Pages 255-263
Copyright 1996 by the American Geophysical Union

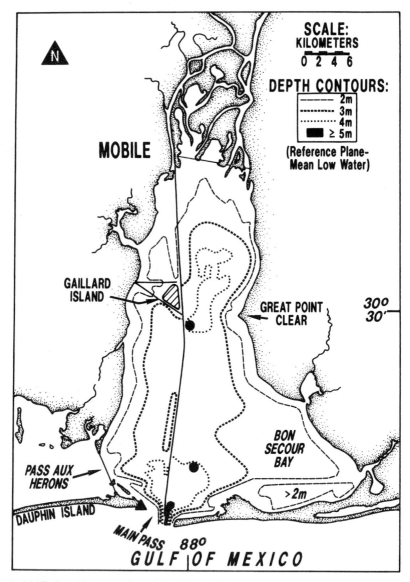

Figure 1. Mobile Bay, Alabama northern Gulf of Mexico, USA. Locations of the instrument mooring sites and the Dauphin Island Sea Lab meteorological station are represented by black circles and a black triangle respectively.

Methodology

The basic data sets we utilized for our analysis are 38-hour low-passed time series subsampled at 6-hour intervals. Records of river discharge and wind stress covered a four year period including the two year period of the salinity records. Spectra of these records (not shown) were red with significant energy below the semi-annual period. Each record also exhibited either an energy peak or plateau between 0.0008 and 0.0005 cph (50 and 80-day periods, respectively). Wind stress spectra also exhibited a rise in energy within the weather band (0.02 to 0.004 cph or 2 to 10-day periods). The response of the estuary to these weather band forcings has been discussed elsewhere [Schroeder et al., 1993; Noble et al., 1994].

In the present note, our interest lies with the lower frequency signals. The gappy nature of the salinity records (Figure 3) precludes estimation of the spectra using normal time series analysis techniques. Instead, we follow the general methodology suggested by Thompson [1971] and Sturges [1992]. We first interpolate linearly between the gaps in the 38-hour low-passed records. We then low-pass filter with a 10-day cut-off, cosine-bell the first and last 10 percent of the interpolated records and subsample at 2.5-day intervals, retaining the sampled data only when it falls at times corresponding to the original data. Least squares amplitudes of the first 30 harmonics of the resulting data set were estimated using MATLAB [MATLAB, 1992]. The MATLAB routine for estimating the Moore-Penrose inverse utilizes a singular value decomposition of the model matrix. Since the result is dependent upon reliable estimates of the singular values, the inverse was calculated twice. During the first pass, the singular values were examined to determine the level at which the singular values dropped below a noise threshold. The second pass retained only those values above a subjective estimate of this threshold. In the present application, this technique retained approximately half the estimated singular values. The resultant harmonic amplitudes were used to estimate the spectra of salinity at the different stations.

The harmonic amplitudes determined by least-squares as described in the preceding paragraph were also used to estimate the coherence squared between the salinity records and the external forcing. The interpolated salinity data was, also, band-pass filtered to retain the signal components between 0.002 and 0.0004 cph (20 and 100-day periods). The resultant band-pass filtered time-series were subsampled at 10-day intervals retaining only the data points occurring where the original records contained data.

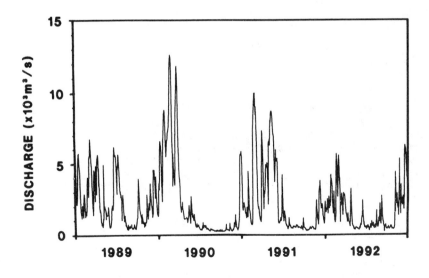

Figure 2. River discharge from the Mobile River system, 1989 to 1992.

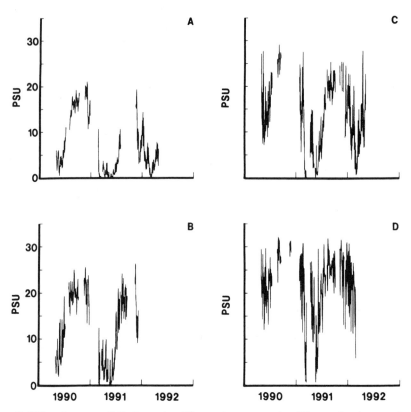

Figure 3. 38 hour low-pass salinity time series: (A) upper-bay upper meter; (B) upper-bay lower meter; (C) lower-bay upper meter; and (D) lower-bay lower meter.

Results

The spectrum estimates resulting from the least-squares estimation technique are presented in Figure 4. Note that the validity of the spectra estimated with this novel technique is suggested when the low frequency, least squares spectrum estimates are compared to spectrum estimates from a standard analysis performed on a long continuous record from the same site (Figure 5); the spectrum levels and slopes are similar in the frequency bands of overlap. Each spectrum in figure 4 is red and has significant energy at the annual period, representing the seasonal cycle of salinity within the estuarine system. Energy at the semi-annual period reflects the non-sinusoidal character of this seasonal cycle, while energy at other sub-annual frequencies reflects the observed interannual variability. Each estimated salinity spectrum shows a peak within the 0.0008 to 0.0005 cph (50 to 80-day) band.

The coherence squared between river discharge and salinity at the four measurement locations exhibited similar variability (Figure 6). Major coherence peaks occurred at the very lowest frequencies resolved, slightly above .0005 cph, near .001 cph, and slightly below .0015 cph. The low number of degrees of freedom allowed by the data suggests that the peak near 0.0005 cph is statistically significant at the 80% level only at the upper bay near-surface meter while the peak near 0.0015 cph is significant at the same level only at the lower bay

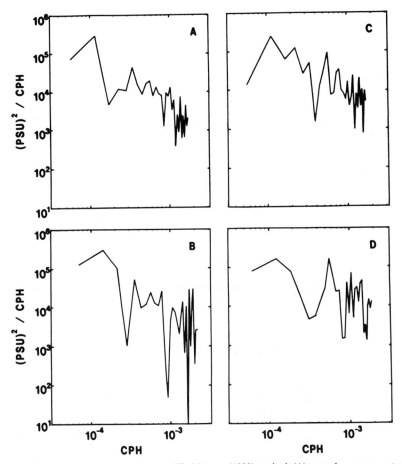

Figure 4. Salinity spectra estimated using a modified Sturges (1992) method: (A) upper-bay upper meter; (B) upper-bay lower meter; (C) lower-bay upper meter; and (D) lower-bay lower meter.

near-surface meter. The coherence squared at the lowest frequencies resolved is significant at all meters except the lower bay near-surface meter. At the upper bay near-surface site, the salinity and discharge phase relationship is linear, suggesting a constant time delay, near the frequencies of significant coherence squared. At the lower bay near-surface meter, the salinity and discharge signals are π radians out of phase near 0.0015 cph. This suggests that the upper bay salinity responds to the river discharge through advection while other processes are important in the lower bay.

We also estimated the coherence squared between the salinity and both the low-pass filtered eastward and northward wind stress components (Figure 6). At the upper bay near-surface meter, the very low-frequency salinity fluctuations are coherent with the northward wind stress. At the same site, there is significant coherence with the wind stress between 0.001 and 0.0015 cph, but the variance in the salinity signal is reduced at these frequencies. The salinity at the upper bay near-bottom meter is coherent with the eastward wind stress in the 0.0005 to 0.0008 cph band as well as exhibiting a large, but statistically non-significant

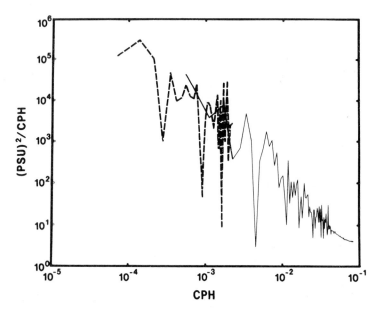

Figure 5. Spectrum of the least-squares estimated spectrum of salinity at the near-bottom meter at the upper bay site (dashed) and the spectrum estimated from the periodogram of the longest continuous record available at the same site (solid).

coherence with river discharge. Both meters at the upper bay mooring show significant coherence with the northward wind stress at the very lowest frequencies resolved.

At the lower bay mooring, the only statistically significant coherence at the near-surface site was with the northward wind stress at high frequencies (.001-.00125 cph) where the salinity variance levels were low. At the near-bottom meter, the salinity was coherent with the wind stress in this same band. At the near-bottom meter, there is also a statistically non-significant peak in coherence with the northward wind stress in the 0.0005-0.0008 cph band.

Discussion

The observed weak coherence between river discharge and salinity in this river-dominated system is due to a variety of causes. Variability in the salinity of the offshore coastal ocean which serves as the salt reservoir for the estuary will clearly influence the salinity structure of the estuary, particularly at stations in the lower bay. Recent work [Kelly, 1991] indicates that the mean salinity of the coastal ocean waters which exchange with Mobile Bay may vary by as much as 6 psu during the course of a two year period. Lysinger [1982] documents similar salinity variations within Main Pass. River discharge represents a strong forcing for the Mobile Bay estuary. It clearly can drive the system to zero salinity during flooding events. Beyond this point, increased freshwater discharge has no effect on the salinity. The system functions as a hard limiter, a highly non-linear system, under these conditions. Finally, high frequency cycles of stratification and destratification [Schroeder et al., 1990], also influence the salinity variability in a highly non-linear fashion.

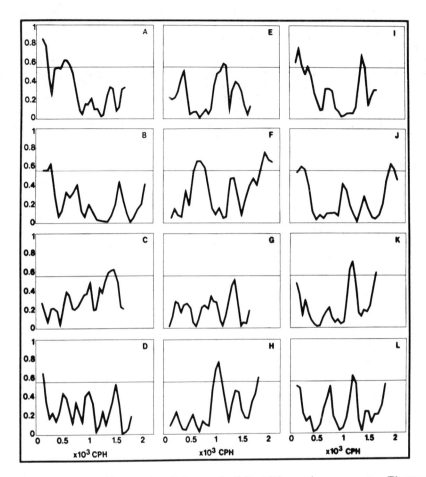

Figure 6. Coherence squared between river discharge and salinity at (A) upper bay upper meter; (B) upper-bay lower meter; (C) lower-bay upper meter; and (D) lower-bay lower meter; between eastward wind stress and salinity at (E) upper bay upper meter; (F) upper-bay lower meter; (G) lower-bay upper meter; and (H) lower-bay lower meter; and between northward wind stress and salinity at (I) upper bay upper meter; (J) upper-bay lower meter; (K) lower-bay upper meter; and (L) lower-bay lower meter. The horizontal line is drawn at the 80% significance level.

During low river discharge the middle and lower bay regions often switch from the more persistent two-layered, vertically stratified system to one that is nearly vertically homogeneous. Previous work [Schroeder, 1978; 1979] has shown that during periods of low to moderate river discharge, consistent with northern hemisphere conditions, river waters from the Mobile River system generally flow along the western half (right-hand side) of Mobile Bay as they move to the south through the bay while Gulf of Mexico waters flow along the eastern half (right-hand side) of the bay as they move towards the north. The locations of the two moorings (Figure 1) place them more often under the influence of conditions prevalent along the eastern side of the bay. During periods when the system undergoes this switching, the change in governing dynamics [Pritchard, 1967] would preclude the success of a linear predictor and reduce the observed coherence between the signals.

Perhaps more important in the failure to identify a linear model based solely on discharge is the apparent influence of low-frequency wind stress fluctuations in driving the system. One normally considers wind forcing in estuaries to be important in the 'weather band', periods of a few days to a few weeks. The present analyses suggest that wind forcing is an important control on the estuarine salinity variability at scales up to a few months, at least. Similar effects have been noted elsewhere [Wm. Boicourt, personal communication]. Whether this apparent response to the wind is due to local forcing or to a larger scale far-field response, e.g. the low frequency winds driving a shelf circulation which alters the salinity field at the mouth of the estuaries and, thus, the salt reservoir for the estuary, is not clear.

Summary

Nearly four years of wind stress and river discharge data from the Mobile Bay region indicate significant variance at frequencies below the weather band, but above the semi-annual signal. These are very weakly coherent between the records. In an effort to determine the role that forcing in this band may play in the dynamics of the estuary, two years of salinity data were analyzed. These analyses suggest that the salinity response at 50 to 80-day periods is largely due to river discharge. This is consistent with the estimated 50-day flushing time of Mobile Bay, during periods of low discharge, derived by Austin [1954] using a modified tidal prism technique. Longer records, though, will be required to definitively separate the influence of wind-driven and riverine signals in the salinity field.

Acknowledgments. This work is a result of research sponsored by the U. S. Geological Survey/Geological Survey of Alabama [14-08-0001-A0075], National Science Foundation [OSR-9108761], The University of Alabama Marine Science Program and the Marine Environmental Sciences Consortium of Alabama. Partial support was also received from U. S. Geological Survey contract 14-08-0001-23411 with Louisiana State University. Contribution No. 229 of the Aquatic Biology Program, The University of Alabama: Contribution No. 275 of the Marine Environmental Sciences Consortium, Dauphin Island, Alabama. We wish to thank Ms. Celia Harrod for preparing the figures and Ms. Carolyn Wood for keyboarding the manuscript.

References

Austin, G. B., On the circulation and tidal flushing of Mobile Bay, Alabama, Part 1., Texas A&M College Research Foundation Project 24, Technical Report 12, College Station, TX, 1954.

Kelly, F. J., Physical Oceanography/Water Mass Characterization, in Mississippi-Alabama Continental Shelf Ecosystem Study: Data Summary and Synthesis, Vol. II, edited by J. M. Brooks, pp. 10-1 to 10-151, Technical Narrative. OCS Study MMS 91-0063. U.S. Dept. of the Interior, Minerals Mgmt. Service, Gulf of Mexico OCS Regional Office, New Orleans, LA. 1991

Lysinger, W.R., An analysis of the hydrographic conditions found in the Main Pass of Mobile Bay, Alabama. Master's thesis. The University of Alabama, Tuscaloosa, AL, 1982.

Marmer, H. A., Tides and sea level in the Gulf of Mexico, in Gulf of Mexico, Its Origin, Waters, and Marine Life, edited by P. S. Galtsoff,. pp. 101-118, U.S. Fish and Wildlife Service, Fishery Bulletin 89, 1954.

MATLAB, MATLAB Reference Guide, 548 pp., The Math Works Inc., Natick, MA, 1992.

Noble, M., W.W. Schroeder, and Wm. J. Wiseman, Jr., Sheared subtidal circulation patterns in a shallow highly-stratified estuary, EOS Trans., AGU, 75(3), 81-82, 1994.

Pennock, J. R., F. Fernandez, and W. W. Schroeder, Mobile Bay Data Report, MB-01 to MB-34 Cruises (May 1989 - January 1993), Dauphin Island Sea Lab Technical Report 94-001, Dauphin Island, AL, 271 pp., 1994.

Pritchard, D.W., Observations of circulation in coastal plain estuaries, in Estuaries, edited by G.H. Lauff, pp. 37-44, AAAS, 1967.

Schroeder, W. W., Riverine influence on estuaries: A case study, in Estuarine Interactions, edited by M. L. Wiley, pp. 347-364, Academic Press, Inc., New York, 1978.

Schroeder, W. W., The Dispersion and impact of Mobile River system waters in Mobile Bay, Alabama, WRRI Bull. 37, Water Resources Research Institute, Auburn, Alabama, 48 pp., 1979

Schroeder, W. W. and W. R. Lysinger, Hydrography and circulation of Mobile Bay, in Symposium on the Natural Resources of the Mobile Bay Estuary, edited by H. A. Loyacino and J. P. Smith, pp. 75-94, U.S. Army Corps of Engineers, Mobile District, Mobile, AL, 1979.

Schroeder, W. W., and Wm. J. Wiseman, Jr., An analysis of the winds (1974-1984) and sea level elevations (1973-1983) in coastal Alabama, Mississippi-Alabama Sea Grant Consortium Publ. No. MASGP-84-024, 106 pp., 1985

Schroeder, W. W., and Wm. J. Wiseman, Jr., Low-frequency shelf estuarine exchange processes in Mobile Bay and other estuarine systems on the northern Gulf of Mexico, in Estuarine Variability, edited by D. A. Wolfe, pp. 355-367, Academic Press, New York, 1986.

Schroeder, W. W., S. P. Dinnel, and Wm. J. Wiseman, Jr., Salinity stratification in a river-dominated estuary, Estuaries 13, 145-154, 1990.

Schroeder, W. W., M. A. Noble, G. A. Gelfenbaum, and Wm. J. Wiseman, Jr., The role stratification in controlling circulation in a broad, shallow estuary, in 12th Biennial International Estuarine Research Federation Conference, Hilton Head Island, SC, 14-18 November, Abstracts: 113, 1993.

Seim, H. E., B. Kjerfve, and J. E. Sneed, Tides of Mississippi Sound and the adjacent continental shelf, Estuaries, Coastal and Shelf Sci. 25, 143-156, 1987.

Sturges, W., The spectrum of Loop Current variability from gappy data, J. Phys. Oceanogr. 22 (11), 1245-1256, 1992.

Thompson, R. O. R. Y., Spectral estimation from irregularly spaced data, IEEE Trans. Geosci. Electron, GE-9, 107-110, 1971.

19

Tidal Interaction with Buoyancy-Driven Flow in a Coastal Plain Estuary

Arnoldo Valle-Levinson and James O'Donnell

Abstract

A series of numerical experiments were carried out to assess the effects of the interaction of tidal forcing with density-induced flow in a coastal plain estuary with a channel bathymetry. In particular, the focus was to describe the mechanisms responsible for transverse variability in density and flow fields in those systems. The experiments were motivated by observations in the lower Chesapeake Bay that indicate the influence of bathymetry on mean flow and density transverse structure. The numerical results suggest the importance of the interaction among tidal flow, density-induced flow and bathymetry in shaping the longitudinal and transverse circulation in a coastal plain estuary. They also point to the relevance of the transverse circulation in determining the position and development of regions of flow convergence and aggregation of material. The results presented here are consistent with observations of the transverse structure of the longitudinal flow and the transverse circulation in different coastal plain estuaries. The results also suggest that the position, strength, and pattern of transverse flows and front-generating convergences in estuaries are a function of the estuary width, the tidal forcing strength, the water column stratification and the bathymetry.

Introduction

The typical bathymetry of a coastal plain estuary consists of a system of shoals and channels that influence the flow and density fields. Hydrographic observations in coastal plain estuaries with prominent channels show a tendency for front formation along the channels. These fronts have been characterized as shear fronts associated with the differential advection of the estuarine longitudinal density gradient [O'Donnell, 1993]. The differential advection is caused by the transverse shear in the barotropic tidal current, which results from differences in the relative importance of bottom friction over the channels and shoals. The amplitude of the current velocity over the channels is typically greater than over the shoals. Examples of shear fronts have been described in the York River, Virginia [Huzzey and Brubaker, 1988] and in Delaware Bay [Sarabun, 1980]. The transverse structure of the longitudinal flow has also been related to the formation of near-surface convergence zones related to the deepest parts of channels as observed in the Conway River, North Wales [Nunes and Simpson, 1985]. These studies have pointed out the transitional character of these fronts and convergence zones, which occur preferentially during flood stages.

Similar along-channel fronts have been observed with aerial photography in the lower Chesapeake Bay [Nichols et al., 1972]. In the same area, strong transverse gradients in instantaneous and mean current velocity, associated with abrupt bathymetry changes, have been observed with high resolution acoustic Doppler current profiler measurements [Valle-Levinson et al, 1994]. These measurements also show that the greatest instantaneous flow

Buoyancy Effects on Coastal and Estuarine Dynamics
Coastal and Estuarine Studies Volume 53, Pages 265-281

magnitudes, as well as mean inflows, appear over the channels, whereas net outflows develop over the shoals. This flow transverse structure, which agrees with the findings of Wong [1994] in Delaware Bay, might be responsible for the formation of the fronts frequently observed in this area. The purpose of this study is to describe the mechanisms that generate transverse variability both in density and flow fields in systems of abrupt bathymetry changes. This is pursued through a series of simple numerical experiments that illuminate the effects of rotation, bathymetric slope, and tidal forcing on gravitational circulation in an estuary with a channel.

Approach

The numerical model used in this study is modified from that developed by Simons [1974; 1980], as described in Koutitonsky et al. [1987]. It uses a C-grid for spatial discretization of the finite difference form of the three-dimensional momentum, continuity, and salt conservation equations, for which the equation of state is a linear function of salinity [Valle-Levinson, 1992]. Time integration is performed with a leapfrog scheme. Calculations are carried out in a split mode: external mode for transports, and internal mode for shears. Salt advection is formulated with an antidiffusive scheme [Smolarkiewicz, 1983]. Non-linear terms are expressed as in Blumberg and Mellor [1987].

A total of 8 numerical experiments were performed to meet the objectives (Table 1). The first two examine the bathymetric effects on the homogeneous mean tidal flow, without and with rotation influences. The third through sixth experiments consider the effects of channel position and slope (depth) on the density-induced flow. The last two experiments look at the interaction between density-induced and tidally induced flows under the influence of a channel.

Table 1. Summary of numerical experiments performed. The Coriolis parameter f denotes the presence or absence of rotational effects. The amplitude of the tidal forcing is represented by U_0. The density contrast between the head and mouth of the estuary is $\Delta\rho$. The distance of the deepest part of the channel from the south boundary is y_p. Maximum channel depth is h_0.

Exp. No.	$f(s^{-1})$	U_0 (m/s)	$\Delta\rho$ (kg/m³)	y_p (km)	h_0 (m)
1	0	0.5	0	7.5	25
2	8.8×10^{-5}	0.5	0	7.5	25
3	0	0	8	7.5	25
4	8.8×10^{-5}	0	8	7.5	25
5	8.8×10^{-5}	0	8	7.5	17
6	8.8×10^{-5}	0	8	11.5	25
7	8.8×10^{-5}	0.3	8	7.5	25
8	8.8×10^{-5}	0.5	8	7.5	25

Solutions are computed for several cases for an interval of 3 days (72 h at which variables are close to steady-state) in a domain consisting of an estuary 15 km wide, 55 km long and 10 m deep with a longitudinally uniform gaussian channel that extends to a depth of 25 m in the

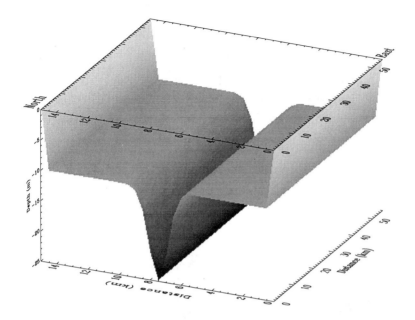

Figure 1. Domain of the numerical experiments. The estuary is 15 km wide, 55 km long and 10 m deep with a longitudinally uniform gaussian channel that extends to a depth of 25 m in the middle of the estuary. The estuary is open at the seaward and landward boundaries. The horizontal grid size is 500 m and the thickness of each of the 10 levels is 2.5 m.

middle of the estuary (7.5 km from lateral boundaries). The estuary is open at the seaward and landward boundaries. The horizontal grid size is 500 m and the thickness of each of the 10 levels is 2.5 m (Figure 1). Bottom friction is parameterized with a linear formulation $t_b = ru_b$, where r is a friction parameter that equals 8×10^{-4} m/s [e.g. Csanady, 1982] and u_b is the near-bottom (1.25 m from bottom) velocity vector. Higher values of r produce slightly weaker flows and vice versa. The qualitative character of the results remains unaffected by the choice of r.

Surface and bottom salinity fluxes, as well as surface (wind stress) momentum fluxes are set equal to zero. At the open boundaries the momentum transports are normal to the boundaries and are radiated out of the domain using an Orlanski-type condition as described by Miller and Thorpe [1981]. Salinity is extrapolated to the boundary for both inflows and outflows through an advective condition similar to the aforementioned condition for momentum (Orlanski-type). Vertical fluxes of momentum and salt are parameterized through time- and space-dependent eddy viscosities and diffusivities. The parameterization is carried out with the Munk-Anderson (1948) scheme:

$$A_V = A_{vo} [1 + 10.0 \ Ri]^{-0.5}$$

$$C_V = C_{vo} [1 + 3.33 \ Ri]^{-1.5}$$

where A_v, C_v, are the eddy viscosity and eddy diffusivity of salt, respectively, that depend on the Richardson number Ri. A_{vo} and C_{vo} are neutral values of A_v, C_v and equal 5×10^{-5} m^2/s.

The first experiment consists of a uniform density fluid forced with a tidal flow of amplitude 0.50 m/s and period of 12 h. The second experiment adds Earth's rotation effects (Coriolis parameter = 8.8×10^{-5} s^{-1}) to the first one. For the rest of the experiments the initial

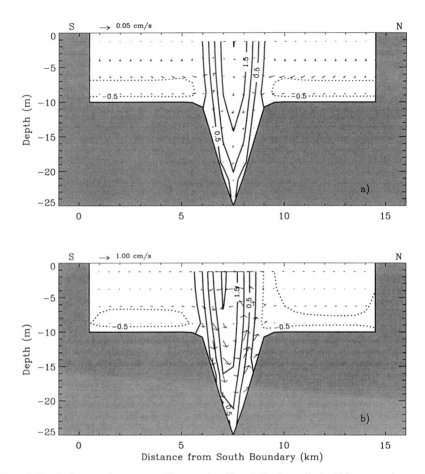

Figure 2. Results for mean, homogeneous flows produced by a 0.50 m/s amplitude tidal current. a) numerical experiment 1 (without Coriolis effect), and b) numerical experiment 2 (with Coriolis effect). Looking seaward in the northern hemisphere. Contour lines represent longitudinal flows at intervals of 0.005 m/s in (a) and 0.01 m/s in (b). Positive values denote net inflows. The vector field represents the transverse flow structure. The vertical component is exaggerated 100 times.

salinity field consists of a vertically uniform longitudinal salinity gradient that varies linearly in the longitudinal direction only. The initial salinity at the landward open boundary is 17 and at the seaward open boundary is 25. The adjustment of this salinity gradient is assessed with and without earth's rotation influences. Then, the effects of the position of the channel across the estuary and the depth of the channel are examined. Finally the effects of tidal forcing of the order of the gravity current speed ($c_g = 0.5[g'H]^{0.5}$, where g' is the reduced gravity), and greater than c_g are explored. For the present estuary and salinity configurations $c_g \sim 0.4$ m/s. The speed c_g is one half of the internal wave speed [Turner 1973].

For the first two experiments, the mean flows of the sixth tidal cycle are examined at one cross section in the center of the estuary (28 km from mouth). For the other experiments, solutions for the flow and salinity fields are examined at three sections across the estuary at distances of 18 km (10 km seaward of the center of the estuary), 28 km (center of the estuary) and 38 km (10 km landward of center of estuary) from the mouth. For experiments 7 and 8, tidal forcing starts after 6 h of adjustment of the salinity gradient and results are presented at a section 28 km from the mouth for the sixth tidal cycle. This is a procedure similar to that used by Valle-Levinson and Wilson, 1994. Areas of transverse flow near-surface convergence and of inflows/outflows are compared among experiments.

Results

The first two experiments examine homogeneous fluids forced by tidal flows for which the ratio of surface elevation (η) to water depth is 0.05, i.e., $\eta = 0.5$ m. The subsequent four experiments present cases that have not been forced externally. The last two depict the influence of tidal forcing on density-induced flow. Tidal excursions for the experiments forced by tidal currents are given by U_oT/π., where U_o is the amplitude of the tidal current (0.3 and 0.5 m/s), and T is its period (12 h). Hence, tidal excursions are 4.1 km for $U_o = 0.3$ m/s and 6.9 km for $U_o = 0.5$ m/s.

The first experiment illustrates an homogeneous fluid forced by semidiurnal tidal forcing. Results are presented as the mean flow of the sixth tidal cycle simulated (Figure 2a). The mean longitudinal flow (without rotation influences) consists of inflow over the channel and outflow over the shoals. The inflow is concentrated in a symmetrical jet over the middle of the channel with strongest speeds at the surface. The net outflows are relatively weak as they are distributed over a greater area than the net inflow. The flood-dominated asymmetry of the vertical shears over the channel with respect to the ebb-dominated asymmetry over the shoals produce the mean flow structure to appear as it does. Over the channel, the strongest vertical shears occur after maximum flood (tide going down) during flood periods and after maximum ebb (tide going up) during ebb periods. The magnitude of the shears that develop when the tide is going down is greater than that of the shears that develop when the tide is going up and therefore, the flood-dominated asymmetry responsible for net inflows appears over the channel. A similar behavior occurs over the shoals but in favor of ebb-dominated asymmetry in the vertical shears. There, maximum vertical shears occur before maximum flood and ebb. The weak non-linearities in the tidal forcing are responsible for the weak transverse flow that shows convergence towards the middle of the channel at the surface and divergence at mid-depth.

The effects of rotation on the homogeneous fluid (experiment 2) are to lean net inflows toward the south and to strengthen the transverse circulation (Figure 2b). The core of maximum inflow and the outflow along the north wall are slightly stronger than in the non-rotating case, and the mean transverse circulation depicts a vertical-plane clockwise gyre over both shoals of the estuary.

The third experiment describes the fluid adjustment to the initial salinity gradient without Earth's rotation influences. After 3 days the flow has reached a quasi-steady state with the baroclinic pressure gradient balancing the bottom stress as in Wong [1994]. The intrusion of high salinity fluid is confined to depths below 7 m and becomes more confined to the deep channel as distance from the mouth increases (Figure 3). As a consequence, the inflow associated with this high salinity is of slightly greater magnitude (0.25 m/s) at a distance of 38 km from the mouth than that at 18 km from the mouth of the estuary. Similarly, the magnitude of the outflow velocity slightly increases seaward. This behavior is due to the thinning of the low salinity layers with seaward distance and of the high salinity layers with landward distance. A cross-estuary structure is apparent only in the longitudinal flow as the strongest inflow is found in the channel. There is also evidence for transverse flow near-surface convergence in the apex of the channel.

The fourth experiment illustrates the effects of the Earth's rotation on the gravitational flow over an estuary with a channel. After 3 days of adjustment, a well-defined transverse structure in the flow and salinity fields becomes evident. Close to the mouth the salinity and flow fields are not greatly affected by the channel in the estuary (Figure 4). High-salinity

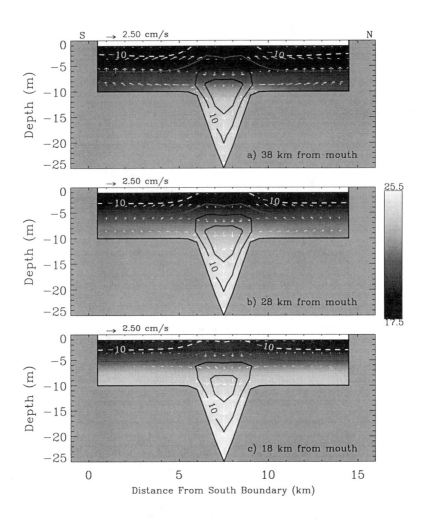

Figure 3. Flow and salinity structure at three different cross-sections along the estuary for experiment 3, baroclinic adjustment without rotation. Flow representation is the same as in Figure 2 except for the vertical component, which is exaggerated 400 times. Contour interval is 0.10 m/s. Salinity field is represented by different shades.

inflow is weak and occupies most of the cross-section, whereas low-salinity outflow is strong and restricted to a near-surface distance of approximately one internal radius of deformation ($R_d = (g'h)^{0.5}/f \sim 8$ km) from the coast. The distance between the zero isotach and the north boundary decreases with depth and becomes zero before reaching bottom so that the outflow is detached from the bottom. The transverse flow shows strong downwelling and near-surface convergence associated with the large shear of the longitudinal flow. Further landward, the outflow weakens and appears over both the south and north boundaries of the channel. Inflow is concentrated over the deep channel and extends to the surface. Its magnitude increases landward as the high salinity fluid becomes restricted to the channel. The transverse flow

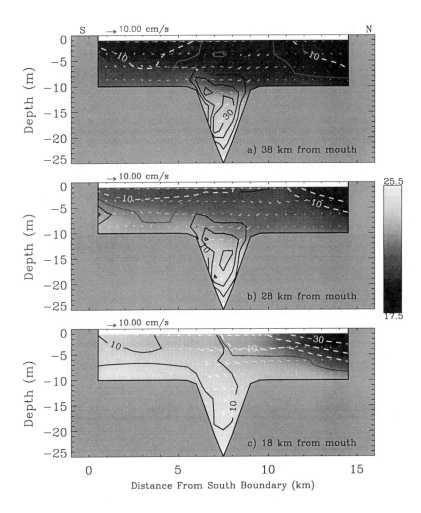

Figure 4. Flow and salinity structure at three different cross-sections along the estuary for experiment 4, baroclinic adjustment with rotation. Flow and salinity representations are the same as in Figure 3. Flow vertical component is exaggerated 200 times.

constitutes a clockwise gyre in the vertical plane of the upper 10 m that is limited to the south by the boundary and to the north by the zero isotach. Another clockwise gyre occurs inside the channel associated with the core of maximum-magnitude inflow. The area of strongest near-surface convergence of transverse flow occurs over the northern slope of the channel.

In a separate calculation (fifth experiment), the effects of different transverse slope of the channel are depicted. For this purpose, a channel with a maximum depth of 17 m was used under the same experimental settings as in the previous experiment (4), i.e., fluid adjustment of an initial salinity gradient under rotation influences. The shallower depth of the channel induces weaker flows with lower inflow salinities than in the case with the deeper (25 m) channel (Figure 5).

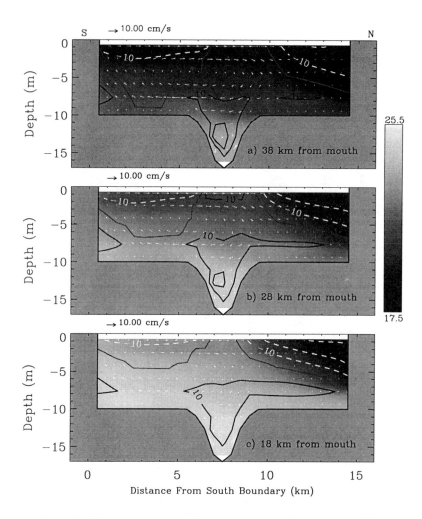

Figure 5. Flow and salinity structure at three different cross-sections along the estuary for experiment 5, baroclinic adjustment with rotation and reduced channel depth. Flow and salinity representations are the same as in Figure 4.

The sixth experiment demonstrates the influence of the position of the channel on the gravitational flow and density fields. When the channel is near the north boundary the inflow region appears mainly over the channel (Figure 6). As in the other experiments with rotation, inflow reaches the surface over the area of the channel and favors strong transverse shears in the areas of maximum bathymetry gradient. Weak inflow also develops near the bottom in the vicinity of the south wall as a consequence of rotational effects. This weak inflow disappears as the distance from the mouth increases. Outflow is restricted to an area less than 2 km (much smaller than R_d) near the north wall and roughly preserves its width along the estuary. This is due to the presence of the channel, which guides the inflow. Outflow also develops to the south of the estuary and, far from the mouth (38 km), can reach to the bottom in the region

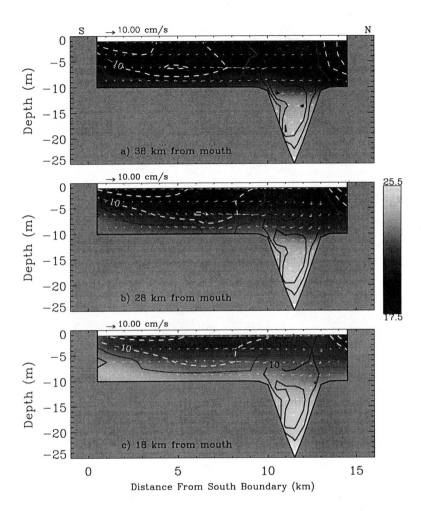

Figure 6. Flow and salinity structure at three different cross-sections along the estuary for experiment 6, baroclinic adjustment with rotation and channel near the north boundary. Flow and salinity representations are the same as in Figure 4.

between the south wall and the channel. The transverse circulation is similar to previous experiments with the development of two clockwise gyres (in the vertical plane) and greatest near-surface convergence over the northern slope of the channel than elsewhere across the estuary.

The seventh experiment examines the interaction of tidal flows of the order of c_g with the buoyancy flows produced by the initial salinity gradient. The ratio of surface elevation to water depth, which establishes the magnitude of the nonlinearly induced residual circulation, is 0.03, i.e., small relative to the tidal current amplitude. The tidal excursion (L_t) for this experiment is 4.1 km. Instantaneous behavior during the sixth tidal cycle shows that the transverse structure of salinity is similar to that of previous experiments. This structure

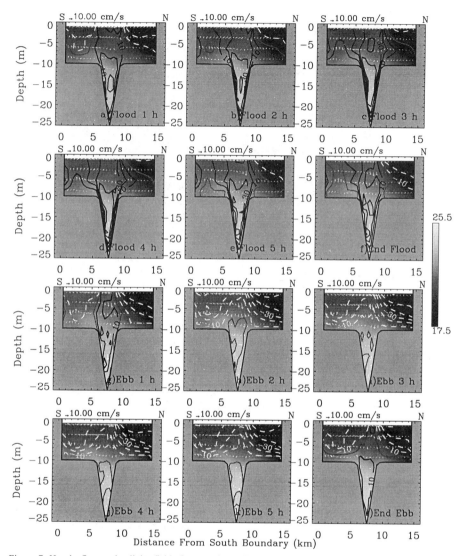

Figure 7. Hourly flow and salinity fields for experiment 7. Interaction of tidal flow with amplitude of the order of the characteristic speed c_g and density-induced flow. Flow and salinity representations are the same as in Figure 4.

changes little throughout the tidal cycle at a cross-section in the middle between the open boundaries (Figure 7). High salinity fluid is confined to the deep channel and the lowest salinity is restricted to the north wall as described in the unforced case. Local variations of salinity depict fluctuations of less than 2 with maxima on flood and minima on ebb. Isohalines migrate vertically with upward motions during flood and downward during ebb periods. The flow field displays well-established transverse structures. During early flood, outflow occurs over the shoals and inflow over the channel from the bottom to the surface. The core of

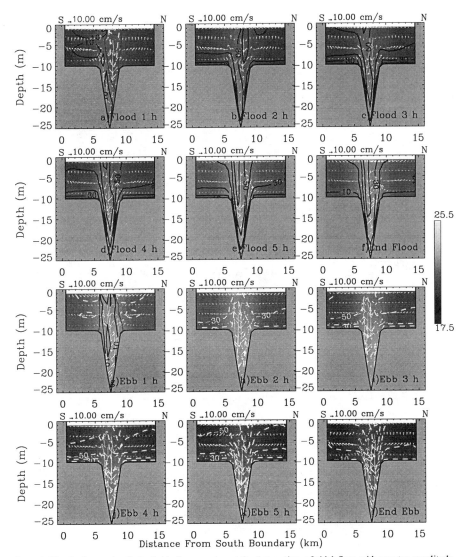

Figure 8. Hourly flow and salinity fields for experiment 8. Interaction of tidal flow with greater amplitude than the characteristic speed c_g and density-induced flow. Flow and salinity representations are the same as in Figure 4.

maximum inflow migrates upward from the channel to the surface as the tidal forcing increases in strength. The greatest transverse gradients in the longitudinal flow develop over the regions of strongest bathymetry gradient. The transverse gradients to the north are greater than those to the south due to the persistent outflow on the north shoal. Even during maximum flood, near-surface outflow is present near the north wall owing to Earth's rotation effects. The weak barotropic forcing of this experiment is unable to counteract the effects of the longitudinal baroclinic pressure gradient in that region of the estuary. During ebb, the

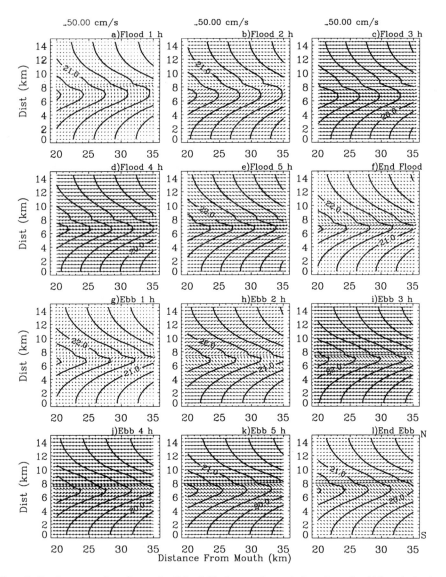

Figure 9. Hourly near-surface flow and salinity fields for experiment 8 throughout the tidal cycle in the central portion of the estuary. Flow fields are represented by vectors and salinity by contours (interval of 0.5). The north wall is at the top of each frame.

magnitude of the outflows is always greatest near the surface and decreases with depth. In the channel it is close to zero except during slack periods when inflow is present. The transverse flow shows that the zone of near-surface flow convergence migrates from the abrupt bathymetry change at the north (during early flood) to the middle of the channel (end of flood, early ebb) and back to the north (end of ebb). The strongest near-surface convergence occurs over the

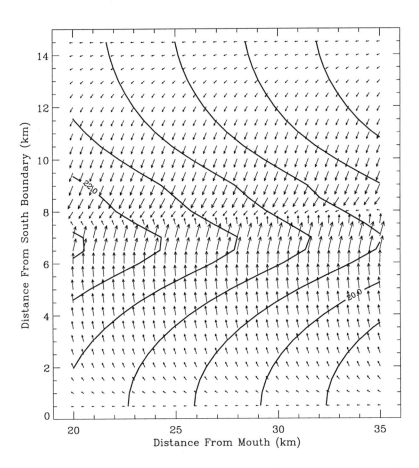

→ 20.00 cm/s

Figure 10. Mean surface flow and salinity fields of those presented in Figure 9. Same representation as in Figure 9.

channel at the end of flood-beginning of ebb. This is the time of the tidal cycle when the longitudinal flow is the weakest and the transverse baroclinic pressure gradient the strongest. This pressure gradient arises from high salinity over the channel and low salinity over the north shoal at this tidal stage. This is also the time of the tidal cycle when aggregation of material should develop, which is consistent with the findings of Nunes and Simpson [1985] and Guymer and West [1991]. Mean fields after one tidal cycle (not shown) reveal that weak forcing does not appreciably alter the unforced distributions of salinity, longitudinal flow and transverse flow obtained from experiment 4 (Figure 4).

The eighth experiment shows the interaction of tidal forcing greater than c_g with the gravitational flow. The ratio of surface elevation to water depth is 0.05, i.e., $\eta = 0.5$ m. The tidal excursion (L_t) of this experiment is 6.9 km. Instantaneous fields during the tidal cycle depict the effects of tidally induced vertical mixing on the salinity fields (Figure 8). The vertical structure of salinity is relatively uniform compared to the weak forcing case. Relatively low salinity appears over the shoals and high salinity over the channel throughout the tidal cycle. This transverse structure agrees with that described by Huzzey [1988] in the narrower York river during flood tides. It does not agree with her observations during ebb periods thus suggesting that the width of the channel and the magnitude of the tidal forcing are also important in determining the transverse density field. In this case, strongest flood flows occur over the apex of the channel, but in contrast to Huzzey [1988] and Huzzey and Brubaker [1988] strongest ebb flows do not develop over the apex of the channel. They appear 1 km to the north (for 2 hours) and are not high enough to reverse the transverse salinity gradient established during flood periods.

Tidal forcing is strong enough, however, to counteract the effects of the baroclinic pressure gradient near the north wall during flood and over the channel during ebb. The transverse flow shows persistent near-surface convergence in the middle of the channel at every tidal stage but strongest at end of flood-early ebb (Figure 9). Maximum intrusion of high salinity fluid occurs over the channel throughout the estuary. During flood, isohalines (e.g. 21) are advected landward approximately 8 km over the channel. This distance is greater than L_t due to reinforcement of tidal flows by the baroclinic pressure gradient force that acts landward over the channel. In contrast, isohaline advection over the south shoal (between 0 and 2 km from south boundary) is close to L_t (7 km) due also to the baroclinic pressure gradient force, which acts seaward over the shoals. Over the north shoal (between 12 and 14 km on Figure 9) isohaline advection during flood is roughly 6 km, lower than L_t to additional seaward acceleration from Coriolis effects. During ebb, L_t is longer over the shoals than over the channel. This results, as expected, in net transport of salt into the estuary over the channel and net seaward transport of buoyant fluid over the shoals.

Mean fields after the tidal cycle discussed above (sixth cycle of the simulation), display a coherent near-surface convergence pattern throughout the estuary (Figure 10). The near-surface salinity field is a reflection of the differential tidal advection of the longitudinal salinity gradient. Mean transverse flows are of greater magnitude than mean longitudinal flows due to stronger transverse pressure gradients. These fields also show greater near-surface flow convergence and weaker longitudinal flows than the weak forcing and unforced cases (Figure 11). This is due to the increased transverse salinity gradients produced by enhanced tidally induced vertical mixing, which is reflected by the presence of vertically uniform low salinity fluid over the shoals and high salinity over the channel. This transverse flow exhibits a pair of counter-rotating gyres in the vertical plane, which resemble the axial convergence pattern described by Nunes and Simpson [1985] in a well-mixed estuary. The distributions of mean longitudinal and transverse flows and salinity obtained in this experiment are similar to those observed in the lower Chesapeake Bay in early October 1993 [Valle-Levinson and Lwiza, 1995]. Thus, it is proposed here that the position, strength, and pattern of transverse flows and front-generating convergences in estuaries are a function of the estuary width, the tidal forcing, the water column stratification and the bathymetry.

Discussion and Summary

A series of numerical experiments have been performed to elucidate the mechanisms that generate transverse variability in density and flow fields in coastal plain estuaries with channel bathymetry. The experiments illustrate the effects of bathymetry and rotation on both homogeneous and stratified fluids, of the position and slope of a channel on density-induced flows, and of the tidal forcing on density-induced flows over channel bathymetry.

The presence of a channel under homogeneous conditions and without Coriolis effects causes the tidal residual flow (mean after the sixth tidal cycle) to develop net inflow over the channel and net outflow over the adjacent shoals. This is due to ebb-flood asymmetric vertical shears in the longitudinal flow, which favor strongest vertical shears in flood over the channel

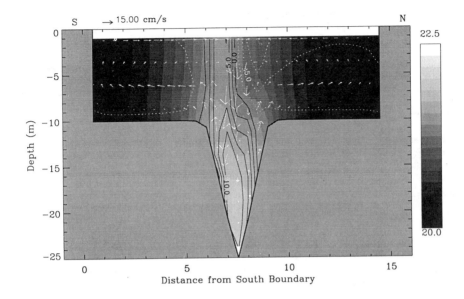

Figure 11. Mean flow and salinity fields for experiment 8. Contour interval for longitudinal flow is 0.025 m/s. Vertical component of the flow is exaggerated 200 times.

and in ebb over the shoals. Coriolis accelerations tend to lean the mean inflow in the direction of the south wall and to strengthen the transverse circulation. This results in two clockwise gyres in the vertical plane over both shoals of the estuary. Maximum inflows develop near the surface, and for the rotating case, maximum outflows develop along the north wall and also near the surface.

The density-induced flow in an estuary is also influenced by the presence of a channel. Inflow tends to develop in the channel even under Coriolis influences and even if the channel is located within one internal radius of deformation from the north boundary, where outflow is present in a flat channel. The channel depth is important in determining the strength and salinity of the inflow. A shallow channel shows weaker inflows and lower inflow salinities than a comparatively deep channel. Maximum inflows occur in the interior of the channel and the region of inflow may extend to the surface depending on the distance from the mouth of the estuary.

The interaction of tidal forcing of the order of the characteristic speed c_g with density-induced flow produces a lateral migration of the zone of transverse flow convergence near the surface. This zone of near-surface convergence moves from the northern slope of the channel during early flood to the middle of the channel by the end of flood/early ebb. The strongest near-surface convergence occurs over the middle of the channel at the end of flood-beginning of ebb, when the longitudinal flow is at its weakest and the transverse baroclinic pressure gradient at its strongest, in agreement with Nunes and Simpson [1985] and Guymer and West [1991]. The lateral migration of the convergence zone is not apparent when the tidal forcing is greater than c_g but the strongest near-surface convergence also appears at the end of flood/beginning of ebb.

Tidal forcing of the order of c_g does not produce marked modifications to the mean flows or salinity fields as compared to the unforced scenario. However, tidal forcing greater than c_g produces enhanced vertical mixing, increased near-surface flow convergence, and weaker longitudinal mean flows than the scenarios with weak forcing and without tidal forcing.

These results are consistent with hydrographic and high resolution flow measurements in the lower Chesapeake Bay [Valle-Levinson and Lwiza, 1995]. They are also consistent with the observations of Nunes and Simpson [1985], Simpson and Turrell [1986], Huzzey [1988], and Guymer and West [1991], with the laboratory experiments of Sumer and Fischer [1977], and with the theoretical results of Smith [1994]. These studies show freshening of shoals and near-surface convergence over channels during flood flows. It is concluded that, for a given salinity longitudinal gradient in estuaries, the position, strength, and pattern of transverse flows and front-generating convergences are a function of the estuary width, the tidal forcing strength, the water column stratification and the bathymetry.

Acknowledgments. This work was financed by the Center for Coastal Physical Oceanography, Oceanography Department, Old Dominion University. The comments of R. Signell and an anonymous reviewer helped to improve this manuscript.

References

Blumberg, A. F., and G. L. Mellor, A description of a three-dimensional coastal ocean circulation model, in *Three Dimensional Coastal Ocean Models*, Coastal Estuarine Sci. Ser., vol. 4, edited by N.S. Heaps, pp. 1-16, AGU, Washington, D.C., 1987.

Csanady, G. T., *Circulation in the Coastal Ocean*, 279 pp., D. Reidel Publishing, Boston, Mass., 1982.

Huzzey, L. M., The lateral density distribution in a partially mixed estuary, *Estuar., Coast. Shelf Sci., 9,* 351-358, 1988.

Huzzey, L. M., and J. M. Brubaker, The formation of longitudinal fronts in a coastal plain estuary, *J. Geophys. Res., 93(C2),* 1329-1334, 1988.

Koutitonsky, V. G., R. E. Wilson, D. Ulmann, and C. Toro, "SWK3D": An enhanced version of the VANERN three-dimensional hydrodynamical model for applications to stratified semi-enclosed seas, Contract Rep. FP 707-6-5361, 110 pp., Inst. Natl. de la Rech. Sci. Oceanol., Rimouski, Que., 1987.

Miller, M. J., and A. J. Thorpe, Radiation conditions for the lateral boundaries of limited-area numerical models, *Q. J. R. Meteorol. Soc., 107,* 615-628, 1981.

Munk, W. H., and E. R. Anderson, Notes on a theory of the thermocline, *J. Mar. Res., 7,* 276-295, 1948.

Nichols, M., M. Kelly, G. Thompson, and L. Castiglione, Sequential photography for coastal oceanography, SRAMSOE Rep. 95, VIMS, Gloucester Point, 1972.

Nunes, R. A., and J. H. Simpson, Axial convergence in a well-mixed estuary. *Estuar., Coast. Shelf Sci., 20,* 637-649, 1985.

O'Donnell, J., Surface fronts in estuaries: a review. *Estuaries, 16(1),* 12-39, 1993.

Sarabun, C. C. Structure and formation of Delaware bay fronts. Ph.D. Thesis, 229 p., The Univ. of Del., Newark, Del., 1980.

Simons, T. J., Verification of numerical models of Lake Ontario, 1, Circulation in spring and early summer, *J. Phys. Oceanogr., 4,* 507-523, 1974.

Simons, T. J., Circulation models of lakes and inland seas, *Can. Bull. Fish. Aquat. Sci., 203,* 146 pp., 1980.

Simpson, J. H., and W. R. Turrell, Convergent fronts in the circulation of tidal estuaries, in *Estuarine Variability*, edited by D.A. Wolfe, pp. 139-152, Academic, Orlando, Fla., 1986.

Smith, R., Combined effects of buoyancy and tides upon longitudinal dispersion, 7th Int. Biennial Conf. Phys. Estuar. Coast. Seas, Buoyancy effects on coastal dynamics, Meeting Abstracts, Woods Hole, MA., 1994.

Smolarkiewicz, P. K., A simple positive definite advection scheme with small implicit diffusion, *Mon. Wea. Rev., 111,* 479-486, 1983.

Sumer, S. M., and H. B. Fischer, Transverse mixing in partially stratified flow, *J. Hydraul. Div. ASCE, 103,* 587-600, 1977.

Turner, J. S., *Buoyancy Effects in Fluids,* Cambridge University Press, 368 pp., 1973.

Valle-Levinson, A., and R. E. Wilson, Effects of sill bathymetry, oscillating barotropic forcing and vertical mixing on estuary/ocean exchange. *J.Geophys. Res. 99(C3),* 5149-5169, 1994.

Valle-Levinson, A., and K. M. M. Lwiza, Effects of channels and shoals on exchange in the lower Chesapeake Bay, *J. Geophys. Res., 100,* 18551-18563, 1995.

Valle-Levinson, A., K. M. M. Lwiza, and B. D. Connolly, Flow lateral structure in the lower Chesapeake Bay (abstract), *EOS, Trans. AGU,* 75(16), Spring Meeting Suppl., 198, 1994.

Valle-Levinson, A., Sill processes and barotropic forcing effects on estuary/ocean exchange, Ph.D. Thesis, 237 pp., Mar. Sci. Res. Center, State University of NY at Stony Brook, 1992.

Wong, K.-C., On the nature of transverse variability in a coastal plain estuary. *J. Geophys. Res. 99(C7),* 14,209-14,222, 1994.

20

Effects of Channel Geometry on Cross Sectional Variations in Along Channel Velocity in Partially Stratified Estuaries

Carl T. Friedrichs and John M. Hamrick

Abstract

Analytic solutions for along-channel velocity through an estuarine cross-section with laterally varying depth are compared to observations from an array of current meters deployed over a nearly triangular cross-section of the James River estuary. Analytic results suggest that the transverse structure of along-channel velocity at this cross-section is primarily due to simple density-driven circulation modified by bathymetry. Comparisons of analytic solutions for the amplitude and phase of tidal velocity to observations suggest that linear models which include realistic lateral depth variation should also incorporate across-channel variation in eddy viscosity. Solutions for various contributions to mean velocity are then derived which incorporate a power-law dependence of eddy viscosity on local depth. Comparison to observations from the James River suggests that density-induced circulation is the dominant contribution to along-channel mean velocity and that riverine discharge also provides a measurable contribution. Nonlinear tides may account for much of the remaining discrepancy between observations and the linear analytic solution. Finally, applications of an existing three-dimensional numerical model of the James River suggest (i) that inclusion of Coriolis acceleration does not greatly effect the cross-sectional distribution of along-channel mean velocity, and (ii) that the form of across-channel variation in eddy viscosity in the analytic model is consistent with the behavior of the numerical model's more sophisticated turbulence closure scheme.

Introduction

The most widely quoted analytic solution for mean circulation in partially-mixed estuaries [Hansen and Rattray, 1965], and other more recent efforts [e.g., Chatwin, 1976; Officer, 1976; Oey, 1984; Prandle, 1985; Jay and Smith, 1990; Scott, 1993] have made the assumption that the cross-sectional form of the estuary channel is rectangular. To motivate their discussion, for example, Hansen and Rattray [1965] cite observations taken along and lower James and Delaware Rivers. Yet these two estuaries exhibit strong across-channel variation in depth which clearly affects the cross-sectional distribution of along-channel velocity [Kuo et al., 1990; Wong, 1994]. In past decades analytic solutions have been derived which indicate the potential importance of lateral depth variation on gravitational circulation [Fischer, 1972; Imberger, 1977; Hamrick, 1979]. However the classical view that Coriolis acceleration is the primary cause for transverse variation in along-channel velocity in coastal plain estuaries [Pritchard, 1967] has persisted in the review literature [e.g., Open University, 1989; Pritchard, 1989]. The purpose of the present discussion is to further investigate the impact of such depth variation on the classical description of estuarine circulation.

Buoyancy Effects on Coastal and Estuarine Dynamics
Coastal and Estuarine Studies Volume 53, Pages 283-300
Copyright 1996 by the American Geophysical Union

Improved observation of lateral variability in coastal plain estuaries has motivated renewed interest in the effect of bathymetry on along-channel velocity [Wong, 1994; Valle-Levinson and Lwiza, 1995]. From the trajectories of Lagrangian drifters, Wong [1994] observed net surface flow to be seaward over shallows along both shores of the Delaware Bay estuary, but observed little net surface flow over the central deep channel. Wong [1994] showed these results to be qualitatively consistent with a linear analytic solution (previously derived by Hamrick [1979]) for circulation through a triangular cross-section. The numerical experiments of Valle-Levinson and O'Donnell [this volume] were motivated by acoustic Doppler profiler measurements across the lower Chesapeake Bay [Valle-Levinson and Lwiza, 1995] which showed a similar lateral segregation of flow, with seaward-directed mean flow concentrated over the shoals and landward-directed mean flow concentrated over the deep channels.

This paper utilizes an existing data set from the James River estuary that is well-suited to the study of velocity patterns resulting from depth variation over an estuarine cross-section. This data set is unique in that it spans a nearly triangular cross-section in a partially-mixed, coastal plain estuary for several weeks. The observations are sufficiently detailed to allow conclusive insights into the ability of simple analytical models to adequately represent first-order processes over complex estuarine topographies. In this paper, simple analytic solutions are found which incorporate lateral variation in eddy viscosity and which suggest that the two-dimensional (y,z) structure of along-channel velocity at this James River cross-section is primarily due to simple density-driven circulation modified by bathymetry.

In the following section, the James River data set is described, followed by derivations of linearized models for the observed patterns of tidal and mean along-channel velocity. Finally, analytic and numerical results are used to gain insight into the potential roles of nonlinear tides and Coriolis effects in determining lateral variability in velocity.

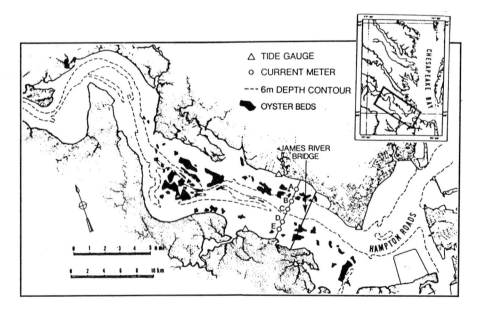

Figure 1. The James River estuary, with locations of current meter and tide gauge stations. Modified from Kuo et al. [1990].

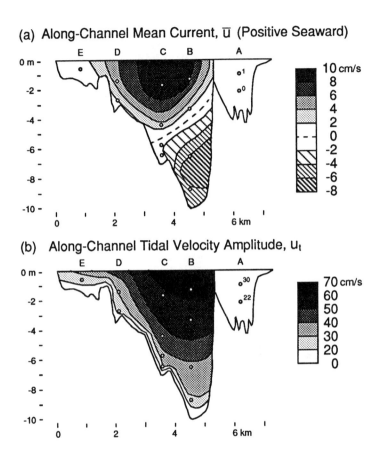

Figure 2. Contour plots of (a) mean and (b) M_2 tidal components of along-channel velocity observed at the James River cross-section during July 1985. Circles indicate current meter locations.

James River Data Set

During July 1985, an array of fourteen current meters were maintained on five moorings spanning a section of the James River estuary just upstream of the James River Bridge (Figure 1). The current meter array was part of a multi-disciplinary study by the Commonwealth of Virginia to better understand the processes which determine the transport pathways of oyster larvae from their spawning grounds in the James estuary to their eventual settling points within the James estuary and elsewhere [Kuo et al., 1990]. The current meters over the deepest portion of the channel (line B) were maintained from June to November 1985, whereas the others where installed for the month of July only. The single current meter at station E provided just 10 days of data, and one of the three meters along line A failed entirely. A detailed account of the field experiment is provided by Hepworth and Kuo [1989].

Figure 2 displays the resulting mean and M_2 tidal components of the along-channel velocity, where positive along-channel is defined by the course of the river to be 130° east of true north. In Figure 2, line A has not been included in the contouring because the focus of this paper is the effect of a nearly triangular-shaped cross-section on along-channel velocity. (Values for line A are plotted directly, and it is evident that the net residual component at station A was almost zero.) From the contour plots in Figure 2, it is striking how closely the tidal velocity amplitude follows the local channel depth. In contrast, the mean current is clearly skewed toward the shallower portion of the "triangular" cross-section, and the lines of constant mean velocity clearly tilt up towards the northeast.

Linear Model

Consider the linearized along-channel momentum balance in an idealized estuary where across-channel velocity and transverse shear stress are presumed to be negligible:

$$\frac{\partial u}{\partial t} = -\frac{1}{\rho_0}\frac{\partial p}{\partial x} + \frac{\partial}{\partial z}\left(A_z\frac{\partial u}{\partial z}\right).$$

(1)

In (1), u is along-channel velocity, t is time, ρ_0 is mean density, p is pressure, A_z is eddy viscosity, and x and z are positive seaward and upward, respectively (Figure 3). The along-channel pressure gradient may reasonably be represented as

$$\frac{1}{\rho_0}\frac{\partial p}{\partial x} = g\frac{\partial \eta_t}{\partial x} + g\frac{\partial \overline{\eta}}{\partial x} + \frac{g(\overline{\eta} - z)}{\rho_0}\frac{\partial \rho}{\partial x},$$

(2)

where g is the acceleration of gravity, ρ is the perturbation density, and the elevation of the free surface, η, is composed of tidal and mean components, η_t and $\overline{\eta}$. In (2) we have assumed that the longitudinal density gradient is independent of vertical position and that density is independent of time, assumptions which are consistent with the weak stratification approximation [e.g., Hamrick, 1990].

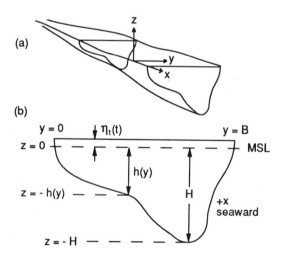

Figure 3. Schematic sketch of idealized estuary: (a) perspective, (b) cross-section.

Substituting (2) into (1) then gives the following governing equations for the tidal and mean components of velocity, u_t and $\overline{u}$:

$$\frac{\partial u_t}{\partial t} = -g\frac{\partial \eta_t}{\partial x} + \frac{\partial}{\partial z}\left(A_z\frac{\partial u_t}{\partial z}\right), \quad \text{and} \quad 0 = \frac{g}{\rho_0}\frac{\partial \rho}{\partial x} + \frac{\partial^2}{\partial z^2}\left(A_z\frac{\partial \overline{u}}{\partial z}\right). \tag{3}$$

If the pressure gradient and boundary conditions are prescribed, then the only additional quantity needed in solving (3) is A_z. The same magnitude and form of A_z will apply in both the tidal and mean solutions because the problem is linear. Since our main interest is the transverse structure of $\overline{u}$, we can use the tidal solution to constrain the magnitude and structure of A_z. Then no free parameters will remain in the mean solution, and we can better evaluate the adequacy of a linear approach in explaining the observed transverse variation in $\overline{u}$.

Tidal Solution and Constraints on A_z

In solving for u_t, it is further assumed that $\partial \eta_t/\partial x$ is independent of y and A_z is independent of z. The first assumption is reasonable and is also required given the limitations of available observations. The latter assumption is not necessary to maintain linearity, but it greatly simplifies the form of the resulting analytic solutions. Furthermore, the finite stratification present in the James River estuary inhibits application of other simple forms for A_z which may be more appropriate in shallow well-mixed estuaries, such as $A_z \sim \kappa u_*(h + z)$ where h is local channel depth. With A_z independent of z, the boundary conditions $u_t = 0$ at z $= -h$, and $\partial u_t/\partial z = 0$ at z $= 0$ then give

$$u_t = \frac{g}{\omega}\left|\frac{\partial \eta_t}{\partial x}\right|\, \text{Real}\left\{i\left(1 - \frac{\cosh \alpha z/h}{\cosh \alpha}\right)\exp i(\omega t + \phi)\right\}, \tag{4}$$

where ω is the tidal radian frequency, ϕ is the phase of $\partial \eta_t/\partial x$ relative to η_t (with positive ϕ indicating $\partial \eta_t/\partial x$ leads η_t), and $\alpha = (i\omega h^2/A_z)^{1/2}$. Equation (4) makes no assumptions concerning transverse variation in h or A_z and holds for arbitrary lateral variations in either.

Previous investigators have suggested a variety of z-independent values for A_z in tidal estuaries. For example, Hansen and Rattray [1965] applied $A_z = 2.5$ cm^2/s to the James in their classic paper. In an often cited paper on tidally-induced residual circulation, Ianniello suggested $A_z \approx 60$ cm^2/s for a tidal channel with u_t and h similar to the James. To constrain the magnitude of the eddy viscosity, we compare (4) to observations of along-channel tidal velocity collected at line B in the deepest portion of the cross-section. Hourly measurements of η_t collected simultaneously for one month at the north and south tide gauges in December 1985 (the south gauge was not deployed in July) give $|\partial \eta_t/\partial x| = 8.0 \times 10^{-6}$, and $\phi = 94°$ for the M$_2$ component. A value of $A_z \approx 12$ cm^2/s then does a reasonable job of reproducing the amplitude and phase of u_t observed at line B in July 1985 (Figure 4).

Authors who have analytically examined the effect of transverse depth variation on gravitational circulation have generally let A_z be independent of both z and h(y) [Imberger, 1977; Hamrick, 1979; Wong, 1994], although Fischer [1972] assumed A_z to be proportional to h. Various relationships between A_z and h(y) may be inferred if one assumes $A_z \sim u'L$, where u' is the turbulent velocity scale and L is the length scale of turbulent eddies. If u' is scaled by tidal velocity and L is scaled by local depth, then

$$A_z(y) \sim U_t(y)\, h(y), \tag{5}$$

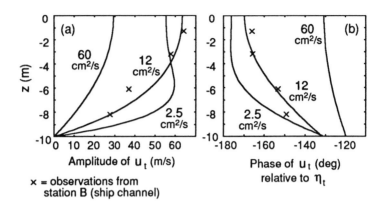

Figure 4. (a) Amplitude and (b) phase of M_2 tidal velocity predicted by Equation (4) with $A_z = 2.5$, 12 and 60 cm²/s, compared with observations from line B.

where U_t is the depth-averaged tidal amplitude. If a tidal channel is frictionally-dominated, the relevant momentum balance gives $g|\partial\eta_t/\partial x| \approx c_d U_t^2/h$, where c_d is the bottom drag coefficient. Then $U_t(y) \sim h^{1/2}$, and from (5), $A_z \sim h^{3/2}$. If a channel is nearly frictionless, the relevant momentum balance gives $\omega U_t \approx g|\partial\eta_t/\partial x|$. Then U_t is independent of h, and from (5), $A_z \sim h$. If the water column is not well-mixed, however, stratification may limit the size of turbulent eddies and, if stratification increases with h, L may increase more slowly than h. In that case, a nearly frictionless, partially-stratified channel may have $A_z \sim h^\beta$ where $\beta < 1$.

To include various dependencies of A_z on h(y), we now let

$$A_z = A_{zH}\left(\frac{h}{H}\right)^\beta \tag{6}$$

in all further derivations, where H is the maximum depth of the cross-section, and $A_{zH} = 12$ cm²/s for the James cross-section. Figure 5 qualitatively compares observations of u_t for the James to $u_t(y,z)$ predicted by (4) for $\beta = 0$, 1 and 3/2. If like most authors, one assumes $\beta = 0$, then predicted velocity amplitude in the shallow part of the channel is too low and predicted phase is too high. Conversely, $\beta = 3/2$ causes amplitude in the shallow margin to be too high and phase to be too low. $\beta \approx 1$ does a reasonably good job of reproducing both velocity amplitude and phase. This result is qualitatively consistent with a weakly stratified, weakly frictional scenario.

Linear Contributions to the Mean Solution

Having used observed tidal velocities to constrain the likely form and magnitude of A_z, we now examine various contributions to the mean along-channel velocity. Contributions to $\overline{U}$ to which can be considered in a straightforward linear manner include density-induced circulation ($\overline{U}_\rho$), riverine velocity ($\overline{U}_r$), flow induced by low-frequency sea-level variations ($\overline{U}_\eta$), and wind-forced velocity ($\overline{U}_w$).

Density-induced circulation

With A_z given by (6), the governing equation for $\overline{U}_\rho$ becomes

$$0 = \frac{g}{\rho_0}\frac{\partial\rho}{\partial x} + A_{zH}\left(\frac{h}{H}\right)^\beta\frac{\partial^3\overline{u}_\rho}{\partial z^3}. \tag{7}$$

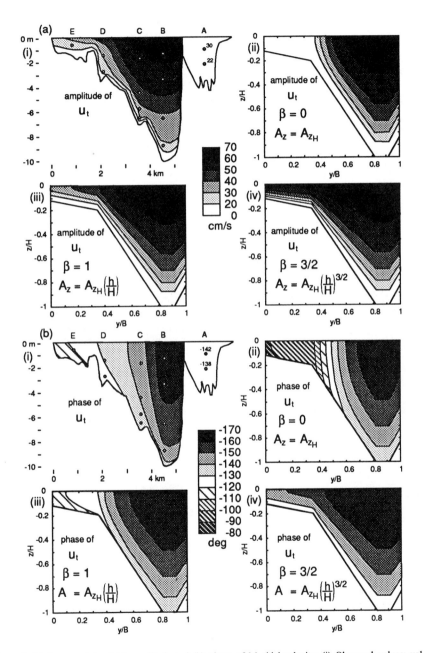

Figure 5. (a) Contour plots of (a) amplitude and (b) phase of M_2 tidal velocity: (i) Observed values; values predicted by Equation (4) with (ii) $\beta = 0$, (iii) $\beta = 1$, and (iv) $\beta = 3/2$.

In solving for $\overline{u}_\rho$ in (7), it is assumed that $\partial\rho/\partial x$ is independent of both y and z. This is probably a reasonable assumption at lowest order for many partially-mixed estuaries, and it is commonly made in analytic solutions for circulation in these systems [e.g., Hansen and Rattray, 1965; Fischer, 1972; Wong, 1994]. However a salinity survey taken during the July 1985 James River experiment (Figure 6) indicates potentially significant z-variation in $\partial\rho/\partial x$ (and y-variation in $\partial\rho/\partial x$ is also likely). Nonetheless, assuming $\partial\rho/\partial x$ to be constant over the cross-section greatly simplifies the integration of (7).

Since (7) is a third-order P.D.E., three external conditions are required for its solution. These are: (i) $\overline{u}_\rho = 0$ at z = -h; (ii) $\partial\overline{u}_\rho/\partial z = 0$ at z = 0; and (iii) $\iint \overline{u}_\rho$ dy dz = 0. In satisfying (iii) (zero net transport over the section), it is convenient to assume a simple analytic shape for the cross-section. For a right-triangle the solution to (7) is

$$\overline{u}_\rho = \frac{gH^3 \widetilde{y}^{-\beta}}{60\,\rho_0\,A_{zH}} \frac{\partial\rho}{\partial x} \left\{ \frac{1-\beta/4}{1-\beta/5} 9\,(\widetilde{y}^2 - \widetilde{z}^2) - 10\,(\widetilde{y}^3 + \widetilde{z}^3) \right\}, \tag{8}$$

where $\mathbf{y} = y/B$ and $z = \widetilde{Z}/H$. For $\beta = 0$, (8) reduces to the solution previously presented by Hamrick [1979] and Wong [1994].

Figure 7 displays mean observed along-channel velocity for the James River cross-section along with $\overline{u}_\rho(y,z)$ as predicted by (8) for $\beta = 0$ and $\beta = 1$. Although tidal velocities suggest $\beta = 1$ to be more appropriate (see Figure 5), results for $\beta = 0$ are displayed in Figure 7 to give some indication of the solution's sensitivity to β. A density gradient of $\partial\rho/\partial x = 3.3 \times 10^{-4}$ kg/m^{-4} was used in (8) based on the average gradient observed between the 7‰ and 22‰ isohalines in Figure 6 and between the same isohalines from an additional salinity survey performed on July 9, 1985. The predicted distributions of $\overline{u}_\rho$ in Figure 7(b-c) are qualitatively similar to the observed distribution of $\overline{u}$, but (8) does not account for the net seaward flow apparent in Figure 7(a). Specifically, $\overline{u}_\rho$ with $\beta = 1$ accounts for more of the observed near surface flow than $\beta = 0$, but $\beta = 1$ over predicts the observed landward flow in the deepest part of the channel to a greater extent.

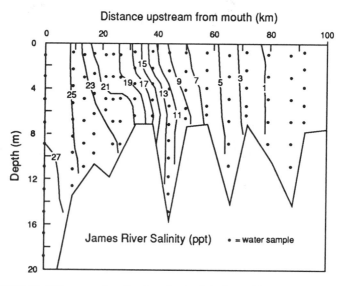

Figure 6. Results of salinity survey along the channel axis of the James River estuary at slack water before flood on July 17, 1985.

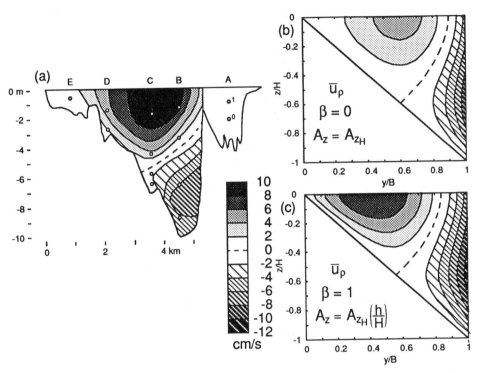

Figure 7. Contour plots of (a) mean velocity observed at the James River cross-section along with density-driven circulation predicted by Equation (8) with (b) $\beta = 0$ and (c) $\beta = 1$.

River Flow and Flow Induced by Low-Frequency Sea Level

The governing equations for $\overline{u}_r$ and $\overline{u}_\eta$ are simply:

$$0 = A_{zH}\left(\frac{h}{H}\right)^\beta \frac{\partial^3 \overline{u}_r}{\partial z^3}, \quad \text{and} \quad A_{zH}\left(\frac{h}{H}\right)^\beta \frac{\partial^3 \overline{u}_\eta}{\partial z^3}. \qquad (9)$$

The three conditions on $\overline{u}_r$ and $\overline{u}_\eta$ are also nearly identical: (i) no flow at $z = -h$; (ii) no stress at $z = 0$; and (iii) $\iint \overline{u}_r \, dy \, dz = Q_r$ and $\iint \overline{u}_\eta \, dy \, dz = Q_\eta$ for $\overline{u}_r$ and $\overline{u}_\eta$, respectively. Q_r is freshwater discharge, and $Q_n = - A_s \langle \partial \eta / \partial t \rangle$, where A_s is the surface area of the estuary upstream of the cross-section, and $\langle \partial \eta / \partial t \rangle$ is the rate of change of η averaged over the (long) time period of interest. Specifically, the averaging period must be much longer than the time required for a gravity wave to travel the length of the estuary. The solutions to (9) are then

$$\overline{u}_r = \frac{6Q_r}{BH} \tilde{y}^{-\beta} (1 - \beta/4) (\tilde{y}^2 - \tilde{z}^2), \quad \text{and} \quad \overline{u}_\eta = -\frac{6A_s}{BH} \left\langle \frac{\partial \eta}{\partial t} \right\rangle (1 - \beta/4) (\tilde{y}^2 - \tilde{z}^2), \qquad (10)$$

where B is the width of the channel. For $\beta = 0$, (10a) and (10b) reduce to solutions previously presented by Hamrick [1979] and by Wong [1994], respectively.

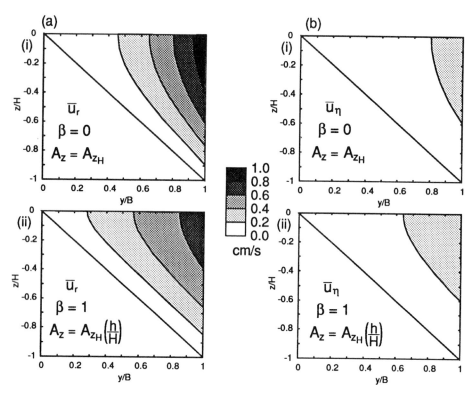

Figure 8 suggests Figure 8. Contour plots of (a) river velocity and (b) velocity forced by low-frequency sea level predicted by Equation (10) with (i) $\beta = 0$ and (ii) $\beta = 1$.

Figure 8 displays $\overline{U}_r(y,z)$ and $\overline{U}_\eta(y,z)$ for the James River as predicted by (10) for $\beta = 0$ and $\beta = 1$. The average river discharge and sea level variation for the period July 5 to 26 at the James River cross-section were $Q_r = 84$ m^3/s and $\langle \partial\eta/\partial t \rangle = -5.3$ mm/day. Figure 8 suggests that $\overline{U}_r$ contributed measurably to the overall observed mean velocity while $\overline{U}_\eta$ did not. Over periods of a few days to a week, which are more characteristic of meteorological events, $\overline{U}_\eta$ can be expected to play a more important role. Figure 8 also indicates that for $\beta > 0$ (i.e., smaller A_z in shallower water) a given discharge is distributed more evenly over the channel cross-section, whereas $\beta = 0$ tends to concentrate flow over the deepest portion of the channel. Finally, Figure 8 implies that river discharge is insufficient to entirely explain the net seaward flow apparent in Figure 7(a).

Wind-Forced Velocity

The governing equation for $\overline{U}_w$ is identical to that for $\overline{U}_r$ and $\overline{U}_\eta$, but with different boundary conditions, namely: (i) $\overline{U}_w = 0$ at $z = -h$; (ii) $\rho_0 A_z \partial \overline{U}_w/\partial z = \tau_{wx}$ at $z = 0$; and (iii) $\iint \overline{U}_w \, dy \, dz = 0$. The solution for $\overline{U}_w$ is then

$$\overline{u}_w = \frac{\tau_{wx} H^3}{\rho_0 A_{zH}} \tilde{y}^{-\beta} \left\{ \tilde{y} + \tilde{z} - \frac{1 - \beta/4}{1 - \beta/3} (\tilde{y}^2 - \tilde{z}^2) \right\}. \tag{11}$$

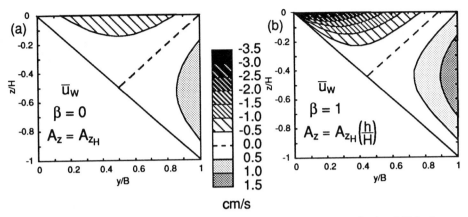

Figure 9. Contour plots of wind-driven circulation predicted by Equation (11) with (a) $\beta = 0$ and (b) $\beta = 1$.

Figure 9 displays $\overline{u}_w(y,z)$ for the James River as predicted by (11) for $\beta = 0$ and $\beta = 1$. Wind speed and direction during July 1985 were recorded at the Norfolk Airport, 30 km southeast of the James River cross-section. The mean observed along-channel wind stress for the period July 5 to 26 was $\tau_{wx} = -0.005$ Pa, where wind stress is calculated from wind speed according to the formulation of Hicks [1972].

During July 1985, the wind stress component along the axis of the James River was much weaker than the across-channel wind-stress, with instantaneous values for $|\tau_{wy}|$ typically about eight times greater than instantaneous $|\tau_{wx}|$. Thus the above value for mean τ_{wx} is highly sensitive to the definition of along- versus across-channel. This potential source of error along with the relatively large distance between the cross-section and location of the wind measurements casts doubt on the significance of the resulting predictions of $\overline{u}_w(y,z)$. Figure 9 is included primarily to illustrate the general form of (11) and its sensitivity to β. Figure 9 indicates that for a given along-channel wind stress, $\beta > 0$ tends to enhance surface velocities in the shallower portion of the cross-section. This is sensible since a smaller A_z in shallower water requires greater shear to balance a given level of applied stress.

Non-Linear Tidal Contributions

Tidal Pumping

If a tidal wave is partially progressive (as is the case along the James River), then non-zero correlations between tidal oscillations in water depth and velocity will lead to a net landward transport of water which is unresolved by current observations alone [e.g., Uncles and Jordon, 1979]. By continuity, the landward transfer of mass associated with this tidal pumping requires a compensating time-averaged seaward return flow. This seaward Eulerian mean current is resolved by the current meter array and appears as an apparent net discharge through the cross-section.

The net landward transport of water associated with tidal pumping, Q_{tp}, can be estimated by integrating the correlation between $\overline{u}_t$ and η_t over the channel cross-section:

$$Q_{tp} = \int_0^B \int_{-h}^0 \frac{\overline{u_t \eta_t}}{h}\, dz\, dy, \quad \text{where} \quad \overline{u_t \eta_t} = \frac{1}{T}\int_0^T u_t \eta_t\, dt. \tag{12}$$

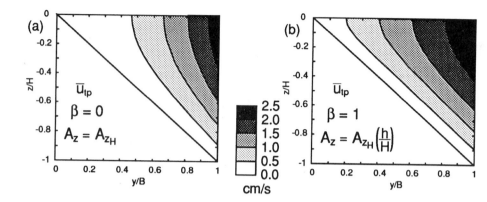

Figure 10. Contour plots of the compensating current brought about by tidal pumping as predicted by Equation (14) with (a) $\beta = 0$ and (b) $\beta = 1$.

Substituting (4) into (12) then gives:

$$Q_{tp} = \frac{a\,g\,B}{2\,\omega}\left|\frac{\partial \eta_t}{\partial x}\right| Real\left\{i\,e^{i\phi}\int_0^1 \left(1 - \frac{\tanh \alpha}{\alpha}\right)d\tilde{y}\right\}.$$ (13)

The compensating mean current, $\overline{u}_{tp}$, is estimated by treating the compensating transport ($-Q_{tp}$) as if it were a riverine discharge:

$$\overline{u}_{tp} = -\frac{6\,Q_{tp}}{BH}\,\tilde{y}^{-\beta}\,(1 - \beta/4)\,(\tilde{y}^2 - \tilde{z}^2).$$ (14)

Figure 10 displays $\overline{u}_{tp}(y,z)$ for the James River as predicted by (14) for $\beta = 0$ and $\beta = 1$. In Figure 10, $|Q_{tp}|$ and $\overline{u}_{tp}$ are larger for $\beta = 1$ because the phase relation between $\overline{u}_t$ and η_t for $\beta = 1$ is more nearly progressive. Inclusion of $\overline{u}_{tp}$ accounts for much of the net seaward flow observed at the James River cross-section, significantly more than was accounted for by the riverine component. Figure 11 qualitatively compares the observed mean flow to the combined contributions of $\overline{u}_\rho$, $\overline{u}_r$ and $\overline{u}_{tp}$. With $\beta = 1$, these three components reproduce the observed seaward and landward magnitudes of $\overline{u}$ reasonably well. With $\beta = 0$, however, the landward-directed near surface velocity is significantly under predicted. For both model cases, the structure of $\overline{u}$ is reproduced reasonably well, however the contour of zero mean velocity (dashed line in Figure 11) appears somewhat too steep in the analytic results. As discussed in the following section, additional nonlinear tidal processes may play an important role.

Tidal Rectification

Nonlinear tidal rectification is examined through application of an existing three-dimensional numerical model, formally called the environmental fluid dynamics computer code (EFDC), which has been previously applied to the James River [Hamrick, 1992]. The EFDC is a time-stepping finite-difference model which resolves tides, and solves the fully nonlinear, hydrostatic equations of motion coupled with conservation equations for turbulent kinetic energy, salinity, and temperature. In simulating the James River, a square 370 m Cartesian grid with six sigma layers in the vertical, and a total of approximately 27,000 cells is used. The

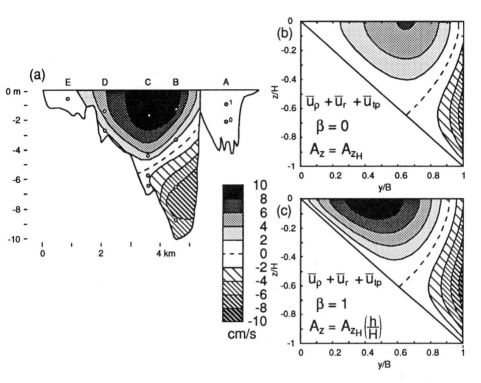

Figure 11. Contour plots of (a) mean velocity observed at the James River cross-section along with mean velocity predicted by $\mathbf{U}_\rho + \mathbf{U}_r + \mathbf{U}_{tp}$ for (b) $\beta = 0$ and (c) $\beta = 1$.

EFDC was not recalibrated for this paper, but run based on a previous James River calibration which, at the time, involved only the adjustment of bottom roughness.

To examine the isolated role of tidal nonlinearities, the EFDC model was forced by the M_2 tide for the case of zero salinity, zero river discharge, and the Coriolis term turned off. Figure 12 displays the velocity output of this numerical experiment, averaged over a tidal cycle, for a cross-section as near as possible to the observed data location. Output from the EFDC model is formulated in terms of mass transport in sigma coordinates rather than velocity in a vertically-fixed frame, and the residual velocities in Figure 12 are calculated by averaging mass transport over the tidal cycle before converting to residual velocity. In other words, mass transport by tidal pumping has already been removed from the problem and the resulting tidally-averaged "Eulerian transport velocity" conserves mass. This approach makes the EFDC output conducive to Lagrangian studies of particle transport.

The tidal rectification displayed in Figure 12 results from the nonlinear terms in the equations of motion, including advection, intratidal variations in eddy viscosity, and quadratic bottom stress. The qualitative pattern of seaward directed flow over the deeper portion of the cross-section and landward directed flow over the shallower portion has been documented previously in other tidal estuaries and channels [e.g., Dyer, 1977; Uncles and Kjerfve, 1986; Friedrichs et al., 1992]. Much of the remaining discrepancy in Figure 11 between the observed and modeled mean flows could be explained by an addition of seaward flow over the deep channel similar to Figure 12.

The Role of Coriolis Acceleration

Geostrophy undoubtedly plays an important role in the transverse momentum balance, but because the across-channel component of velocity is typically much smaller than the along-channel component, the role of the earth's rotation in the along-channel balance is less straightforward. The importance of Coriolis acceleration to the along-channel balance is probably indirect, with its role in the lateral balance affecting across-channel exchange, which in turn affects the salt balance. The distribution of salt then affects vertical eddy viscosity and the longitudinal density gradient, two of the primary factors determining the strength of the along-channel mean current as predicted by (8).

In an attempt to isolate the role of the earth's rotation, EFDC was run using the mean freshwater discharge conditions observed in July 1985, both with and without the Coriolis term. In each case the numerical model was run until the salinity field had reached a steady-state. Figure 13 compares the resulting tidally-averaged along-channel velocity predicted by EFDC, with and without Coriolis acceleration. The numerical model results suggest that the presence of Coriolis acceleration does indeed enhance the strength of the along-channel mean current. However the differences between Figures 13(a) and (b) are primarily in terms of current magnitude and less in terms of current structure. In other words, the cross-sectional distribution of along-channel velocity in the James River estuary does not appear to be overly sensitive to the earth's rotation.

It can be argued that the analytic solutions in Figure 11 implicitly account for the major effects of the earth's rotation on along-channel velocity. The density gradient used in applying (8) and the values chosen for the vertical eddy viscosity are based on field observations which must, by their very nature, include the actual role of Coriolis acceleration. In applying the EFDC model, in contrast, the salinity gradient and vertical eddy viscosity distribution were not tuned to observations, but were determined dynamically. This may also explain why the numerical results over predict the observed values for tidally-averaged along-channel velocity.

Across-Channel Variation of A_z

Finally, the EFDC model provides an opportunity to compare the form of across-channel variation in eddy viscosity assumed in the analytic solutions with that predicted by a more sophisticated turbulence closure scheme. Eddy viscosity in EFDC [Hamrick, 1992] is

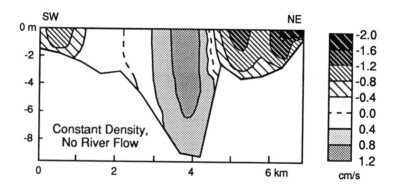

Figure 12. Contour plot of tidally-averaged Eulerian transport velocity output from the three-dimensional EFDC model for the James River under constant density, zero river flow conditions and zero Coriolis acceleration.

determined by the second moment closure scheme of Mellor and Yamada [1982], which involves the use of analytically determined stability functions and the solution of transport equations for turbulent kinetic energy and the turbulent macro-scale. Figure 14 displays depth-averaged values for tidally-averaged eddy viscosity from EFDC as a function of local channel depth across the same section of the James River displayed in Figures 12 and 13, both with and without salinity. Super-imposed on the numerically calculated values for depth-averaged A_z are least-squares fits to the power-law relation $A_z = A_{zH}(h/H)^{\beta}$.

The Mellor-Yamada turbulence closure scheme implemented in EFDC produces $\beta = 1.5 \pm 0.2$ for the uniform density case and $\beta = 0.74 \pm 0.38$ for the more realistic partially-stratified case, where the uncertainty equals two times the standard error of the least-squares fit. Both of these relationships are closer to $\beta = 1$, as favored by the analytic solutions here, than to the $\beta = 0$ case considered previously by Hamrick [1979] and by Wong [1994]. These trends are also consistent with the discussion following Equation (5), which predicts $\beta \geq 1$ for well-mixed channels and $\beta \leq 1$ for stratified channels subject to very weak friction. The values for A_{zH} estimated from the numerical model output are significantly larger than that applied in the analytic solution. A better choice for comparing the overall magnitude of A_z in depth-varying and depth-independent formulations may be a weighted average biased toward values nearer the bottom. Velocity profiles based on depth-varying eddy viscosity are most sensitive to the magnitude of $A_z(z)$ just above the bottom, where shear is highest and $A_z(z)$ is lower than its depth-averaged value.

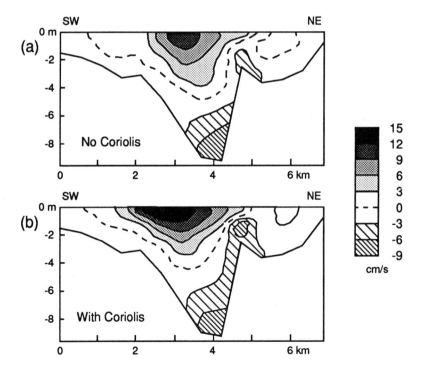

Figure 13. Contour plot of tidally-averaged Eulerian transport velocity output from the three-dimensional EFDC model for the James River (a) without and (b) with Coriolis.

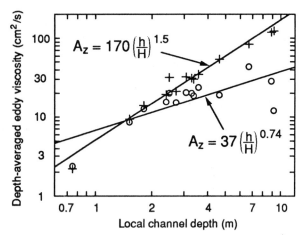

Figure 14. Depth-averaged values for tidally-averaged eddy viscosity as a function of local channel depth, as calculated by the turbulence closure scheme in the three-dimensional EFDC model for the James River: o = with, + = without salinity.

Summary and Conclusions

Most analytic models for mean circulation in partially-mixed estuaries represent the estuary cross-section as rectangular. Although analytic solutions have also been derived which indicate the potential importance of lateral depth variation, the classical view that Coriolis acceleration is the primary cause for transverse variation in along-channel velocity in coastal plain estuaries such as the James River has persisted.

This study utilizes an existing data set from the James River estuary well-suited to the study of velocity variations over a nearly triangular estuary cross-section. During July 1985, an array of fourteen current meters were maintained on five moorings spanning a section of the James as part of a multidisciplinary study to better understand the net transport of oyster larvae. In this paper, linear analytic solutions to the along-channel momentum equation (with the pressure gradient externally imposed) are derived for both tidal and mean components of along-channel velocity.

Comparisons of analytic solutions for the amplitude and phase of tidal velocity to observations from the James River suggest that any linear model which incorporates realistic lateral depth variation should also recognize the likelihood of across-channel variation in eddy viscosity. Solutions for various contributions to mean velocity are then derived which incorporate a power-law dependence of eddy viscosity on local depth. Tidal observations suggest a linear dependence of eddy viscosity on local depth may be adequate for the James.

Linear contributions to mean along-channel velocity are derived resulting from (i) density-induced circulation, (ii) riverine discharge, (iii) flow induced by low-frequency sea-level variations, and (iv) wind-forced circulation. Comparison to observations from the James River suggest that density-induced circulation is the dominant contribution to mean velocity. Riverine discharge provides a measurable contribution, even under relatively low flow conditions; however contributions due to sea level change over time scales of several weeks appear negligible. Significant contributions by winds are possible, although winds were oriented primarily across-channel during July 1985.

The contribution to mean velocity along the James River estuary by nonlinear tides may account for much of the remaining discrepancy between the observations and the linear analytic solution. Analytic estimates of tidal pumping suggest significant Eulerian seaward return flows are likely. An existing three-dimensional numerical model of the James River is used to isolate the potential role of additional tidal nonlinearities, and numerical results suggest tidal rectification results in seaward flow over the deep portion of the channel and landward flow over the shallows.

Numerical model results are also used to investigate (i) the role of the earth's rotation in determining the distribution of along-channel mean velocity, and (ii) the across-channel dependence of viscosity predicted by a more sophisticated turbulence closure scheme. Numerical experiments suggest that the Coriolis term enhances mean current strength, but does not greatly effect its structure. However the earth's rotation is probably included implicitly in the analytic solutions by way of the observationally determined pressure gradient and eddy viscosity. Finally, power-law fits to numerically calculated values for eddy viscosity as a function of local channel depth are found to be consistent with the form of across-channel variation assumed in the analytic solutions.

Acknowledgments. This paper is contribution number 1969 of the Virginia Institute of Marine Science. This work was supported by the Hypoxia Project funded by the Virginia Chesapeake Bay Initiative Programs. The assistance of Mac Sisson in accessing the James River field data is greatly appreciated.

References

Chatwin, P. C., Some remarks on the maintenance of the salinity distribution in estuaries, *Estuarine Coastal Marine Sci.*, *4*, 555-566, 1976.

Dyer, K. R., Lateral circulation effects in estuaries, in *Estuaries, Geophysics, and the Environment*, pp. 22-29, National Academy of Sciences, Washington, D.C., 1977.

Fischer, H. B., Mass transport mechanisms in partially stratified estuaries. *J. Fluid Mech.*, *53*, 671-687, 1972

Friedrichs, C. T., D. R. Lynch, and D. G. Aubrey, Velocity asymmetries in frictionally-dominated tidal embayments: longitudinal and lateral variability, in, *Dynamics and Exchanges in Estuaries and the Coastal Zone, Coastal and Estuarine Studies*, vol. 40, edited by D. Prandle, pp. 277-312, American Geophysical Union, Washington, D.C., 1992.

Hamrick, J. M., Salinity intrusion and gravitational circulation in partially stratified estuaries, PhD Thesis, 451 pp, University of California, Berkeley, 1979.

Hamrick, J. M., The dynamics of long-term mass transport in estuaries, Proceedings of the 2nd International Conference, in *Residual Currents and Long-term Transport, Coastal and Estuarine Studies*, vol. 38, edited by R. T. Cheng, pp. 17-33, Springer-Verlag, New York, 1990.

Hamrick, J. M., Estuarine environmental impact assessment using a three-dimensional circulation and transport model, in *Estuarine and Coastal Modeling*, edited by M. L. Spaulding et al., pp. 292-303, American Society of Civil Engineers, New York, 1992.

Hepworth, D., and A. Y. Kuo, James River seed oyster bed project, physical data report I, *Data Report No. 31*, Virginia Institute of Marine Science, School of Marine Science, College of William and Mary, Gloucester Point, VA, 1989.

Hicks, B. B., Some evaluations of drag and bulk transfer coefficients over water bodies of different sizes, *Boundary-Layer Meteorol.*, *3*, 201-213, 1972.

Ianniello, J. P., Tidally induced residual currents in estuaries of constant breadth and depth. *J. Mar. Res.*, *35*, 755-786, 1977.

Jay, D. A., and J. D. Smith, Residual circulation in shallow estuaries, 2. weakly stratified and partially mixed, narrow estuaries, *J. Geophys. Res.*, *95*, 733-748, 1990.

Hansen, D. V., and M. Rattray, Gravitational circulation in straits and estuaries. *J. Mar. Res.*, *23*, 104-122, 1965.

Kuo, A. Y., J. M. Hamrick and G. M. Sisson, Persistence of residual currents in the James River Estuary and its implications to mass transport, in *Residual Currents and Long-term Transport, Coastal and Estuarine Studies*, vol. 38, edited by R. T. Cheng, pp. 389-401, Springer-Verlag, New York, 1990.

Mellor, G.L., and T. Yamada, Development of a turbulence closure model for geophysical fluid problems, *Rev. Geophys. Space Phys.*, *20*, 851-875, 1982.

Oey, L.-Y., On steady salinity distribution and circulation in partially mixed and well mixed estuaries, *J. Phys. Oceanogr.*, *14*, 629-645, 1984.

Officer, C. B., *Physical Oceanography of Estuaries (and Associated Coastal Waters)*, pp. 116-125, Wiley, New York, 1976.

Open University, Tidal flats and estuaries, in *Waves, Tides and Shallow-Water Processes*, pp. 112-128, Pergamon, New York, 1989.

Prandle, D., On salinity regimes and the vertical structure of residual flows in narrow tidal estuaries, *Estuarine Coastal Shelf Sci.*, *20*, 615-635, 1985.

Pritchard, D. W., Observations of circulation in coastal plain estuaries, in *Estuaries*, edited by G. H. Lauff, pp. 37-44, American Association for the Advancement of Science, Washington, D.C., 1967

Pritchard, D. W., Estuarine classification - a help or a hindrance, in *Estuarine Circulation*, edited by B. J. Neilson, A. Kuo, and J. Brubaker, pp. 1-38, Humana Press, Clifton, N.J., 1989.

Scott, C. F., Canonical parameters for estuary classification, *Estuarine Coastal Shelf Sci.*, *36*, 529-540, 1993.

Uncles, R. J., and M. B. Jordan, Residual fluxes of water and salt at two stations in the Severn estuary, *Estuarine Coastal Marine Sci.*, *9*, 287-302, 1979.

Uncles, R. J., and B. Kjerfve, Transverse structure of residual flow in North Inlet, South Carolina. *Estuaries*, *9*, 39-42, 1986.

Valle-Levinson, A., and J. O'Donnell. Tidal interaction with buoyancy-driven flow in a coastal plain estuary, this volume.

Valle-Levinson, A., and K. M. M. Lwiza, The effects of channels and shoals on exchange between the Chesapeake Bay and the adjacent ocean, *J. Geophys. Res.*, *100*, 18,551-18,563, 1995.

Wong, K.-C., On the nature of transverse variability in a coastal plain estuary, *J. Geophys. Res.*, *99*, 14,209-14,222, 1994.

21

Effect of Variation in Vertical Mixing on Residual Circulation in Narrow, Weakly Nonlinear Estuaries

Kyeong Park, and Albert Y. Kuo

Abstract

This paper addresses the effect of variation in vertical mixing on residual circulation in narrow, weakly nonlinear estuaries such as Chesapeake Bay and its tributaries. As vertical mixing varies, it has two opposing effects on residual circulation. One is a direct effect, in which increased vertical mixing weakens circulation by enhancing vertical momentum exchange. The other is an indirect effect, in which increased vertical mixing modifies the salinity distribution, strengthens the longitudinal salinity gradient ($\partial s/\partial x$), and results in stronger circulation. To examine these two opposing processes, some theoretical models are reviewed and a numerical model in x-z plane is applied to the Rappahannock Estuary, a western shore tributary of Chesapeake Bay. The results indicate that the variation in residual circulation as the vertical mixing varies is determined by the time scale of variation in vertical mixing relative to the response time of longitudinal salinity distribution. In the Rappahannock Estuary, the fortnightly spring-neap cycle is short compared to the response time of the longitudinal salinity distribution, which is on the order of months. Hence, the increase in $\partial s/\partial x$ from neap tide to spring tide is not as important as the increase in vertical mixing, which results in a decrease in residual circulation during spring tide. The numerical model results also show that the boundary conditions on salinity at the river mouth and freshwater discharge at the fall line can modify $\partial s/\partial x$ and overcome the direct effect of vertical mixing on residual circulation.

Introduction

In estuaries, the long-term, large scale transport and distribution of waterborne materials such as salt, sediments, pollutants, phytoplankton and larvae are primarily determined by residual circulation. In narrow estuaries, the flow and salinity distributions are essentially two dimensional in the longitudinal (x) and vertical (z) directions. Two-layer residual circulation in narrow estuaries consists of several modes. Hansen and Rattray [1965] showed the residual currents, when neglecting tidal nonlinearities, consisting of three modes: barotropic flow due to freshwater discharge, baroclinic flow induced by longitudinal density gradient, and wind-induced flow due to wind stress. Ianniello [1977 and 1979] derived the Eulerian and Lagrangian residual currents, respectively forced by the Stokes drift return flow and by the nonlinear convective acceleration. Jay [1990] identified another mode driven by nonlinear stress divergence, i.e., internal tidal asymmetry. In estuaries with small ratios of

Buoyancy Effects on Coastal and Estuarine Dynamics
Coastal and Estuarine Studies Volume 53, Pages 301-317
Copyright 1996 by the American Geophysical Union

tidal amplitude to depth, the density induced residual current dominates the nonlinearly induced ones [Ianniello, 1981]. In estuaries with small freshwater velocities (discharge rates divided by the cross-sectional areas), the internal ebb-flood asymmetry mode is not significant [Jay, 1990]. This paper focuses on the systems with weak tidal forcing and low freshwater velocities such as Chesapeake Bay and its tributaries, in which the residual circulation driven by the vertically integrated longitudinal density (salinity) gradient is dominant over the ones induced by nonlinearities in tide and stress divergence.

The longitudinal salinity gradient ($\partial s/\partial x$) is affected by many processes. Increase in freshwater discharge pushes the seawater downriver increasing $\partial s/\partial x$ [Pritchard, 1989]. Increase in vertical mixing shortens the salt intrusion, increasing $\partial s/\partial x$ [Rigter, 1973; Jay and Smith, 1990; Geyer, 1993]. Increase in salinity in the incoming seawater at the river mouth may increase $\partial s/\partial x$ if the freshwater discharge and thus the salt intrusion remains relatively constant. Characteristic differences in the salinity of the water transported into an estuary in response to wind events have been observed for tributary estuaries in Chesapeake Bay: Rappahannock River [Kuo and Park, 1992], Potomac River [Wang and Elliott, 1978] and Choptank River [Sanford and Boicourt, 1990], and for fjord estuaries: Puget Sound [Cannon et al. 1990]. The residual circulation, which is a function of $\partial s/\partial x$, modifies the salinity distribution and thus $\partial s/\partial x$ by controlling the salt intrusion [Bowden, 1983].

This paper addresses the effect of variation in vertical mixing on residual circulation in narrow, weakly nonlinear estuaries, excluding wind-induced mode. First, the interactions among vertical mixing, vertical stratification ($\partial s/\partial z$), $\partial s/\partial x$ and residual circulation are qualitatively described. Dimensionless ratios, γ and γ_P, are identified from the analysis of the theoretical models of Hansen and Rattray [1965] and Prandle [1985] respectively, which were derived by neglecting tidal nonlinearities. These ratios are measures of the relative importance of two opposing mechanisms on residual circulation as vertical mixing varies, fractional changes in $\partial s/\partial x$ and vertical mixing, and thus dictate variation in residual circulation. The ratios, γ and γ_P, are expressed in terms of three parameters, $\partial s/\partial x$, vertical eddy viscosity (A_z) and tidal current amplitude (U_t). The vertical two-dimensional hydrodynamic model described in Park and Kuo [1994] is applied to the Rappahannock Estuary, a western shore tributary of Chesapeake Bay, to estimate these parameters. The objectives of this model application are to examine the behavior of the ratios, γ and γ_P, and to study the dominant processes in determining the effect of variation in vertical mixing on residual circulation.

Theoretical Consideration

Qualitative Description

As vertical mixing changes, positive and negative feedback loops exist among vertical mixing, $\partial s/\partial z$, $\partial s/\partial x$ and residual circulation. The interactions among the four parameters are illustrated in Figure 1 for weakly nonlinear estuaries. Increase in vertical mixing reduces $\partial s/\partial z$ (arrow A in Figure 1), which in turn increases vertical mixing (arrow B), thus resulting in a positive feedback loop. Increase in vertical mixing weakens residual circulation by enhancing vertical momentum exchange (arrow C). Increase in vertical mixing may increase $\partial s/\partial x$ by shortening the salt intrusion (arrow E). The shortening of the salt intrusion during spring tide has been argued using the steady-state momentum and salt balance equations [Geyer, 1993], and shown using field data from the Columbia River [Jay and Smith, 1990] and data from a tidal flume [Rigter, 1973]. The change in tidal mixing over spring-neap cycle tends to have shorter time scale than the response of mass (salinity) distribution [Fischer, 1980]. Since the response time of mass distribution tends to be long for large estuaries, the process represented by the arrow E in Figure 1 may not be pronounced over spring-neap cycle in large estuaries.

Increase in $\partial s/\partial x$ strengthens residual circulation (arrow F), which in turn decreases $\partial s/\partial x$ (arrow G) by increasing the length of salt intrusion [Bowden, 1983], thus resulting in a

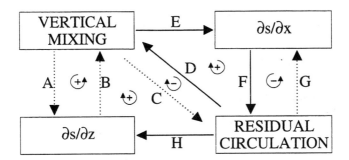

Figure 1. A schematic diagram showing interactions among vertical mixing, vertical stratification ($\partial s/\partial z$), longitudinal salinity gradient ($\partial s/\partial z$), and residual circulation. A straight line with an arrow indicates direct effect of one parameter on the other: solid line indicates positive effect and dashed line indicates negative effect, i.e., increase in one parameter decreases the other. An open circle with an arrow indicates a positive (+) or a negative (-) feedback loop.

negative feedback loop. Enhanced residual circulation increases $\partial s/\partial z$ by straining (arrow H) as well as residual velocity shear ($\partial u/\partial z$). Increased $\partial s/\partial z$ weakens vertical mixing (arrow B) by decreasing Richardson number (R_i). The processes represented by the arrows C, H and B in Figure 1 form another positive feedback loop. Increased $\partial u/\partial z$ by strengthened circulation, however, not only intensifies turbulence but also decreases R_i, resulting in increase in vertical mixing (arrow D). In Figure 1, the processes represented by the arrows E, F and D form a positive feedback loop and the processes represented by the arrows E, F, H and B form a negative feedback loop. The importance of these feedback loops depends on the time scale of variation in vertical mixing relative to the response time of salinity distribution, i.e., arrow E. In summary, Figure 1 illustrates that the complex feedback loops exist among the four parameters (vertical mixing, residual circulation, $\partial s/\partial x$ and $\partial s/\partial z$), which make it difficult to discern the effects of variation in one parameter on the other.

This paper examines the effect of variation in vertical mixing on residual circulation in narrow, weakly nonlinear estuaries. As vertical mixing increases, it has two opposing effects on residual circulation. One is a direct effect, in which increased vertical mixing weakens circulation by enhancing vertical momentum exchange (arrow C). The other is an indirect effect, in which increased vertical mixing changes salinity distribution, strengthens $\partial s/\partial x$, and results in increase in circulation (arrows E and F). The importance of the latter indirect effect, which involves the response of mass (salinity) distribution, depends on the time scale of the adjustment of salinity distribution relative to that of variation in vertical mixing. The theoretical models of Hansen and Rattray [1965] and Prandle [1985] for narrow, weakly nonlinear estuaries are analyzed to quantify the relative importance of these two opposing processes. These theoretical models have been derived by imposing a known longitudinal density (salinity) gradient. However, the processes illustrated in Figure 1 indicate that the residual circulation and the salinity distribution (and thus the longitudinal gradient) are a coupled problem with positive and negative feedback loops [Jay and Smith, 1990; McCarthy, 1993]. The theoretical models are revisited in the next section with this inter-dependency of the circulation and the salinity distribution in mind.

Quantitative Description

Hansen and Rattray [1965] derived, neglecting tidal nonlinearities, an analytical solution of estuarine circulation for narrow estuaries. With no wind stress, the residual velocity, U, may be expressed as:

$$\frac{U}{U_f} = \frac{3}{2}(1 - \zeta^2) + \frac{v \cdot Ra}{48}(1 - 9 \cdot \zeta^2 + 8 \cdot \zeta^3) \tag{1}$$

where U_f = integral mean velocity, i.e., freshwater discharge velocity; ζ = dimensionless depth ($\zeta = 0$ at the surface and $\zeta = 1$ at the bottom). The two dimensionless parameters, Rayleigh number (Ra) and gradient parameter (v), are expressed as:

$$Ra = \frac{g \cdot k \cdot S_o \cdot h^3}{(K_x)_o} \cdot \frac{1}{A_z} \qquad v = \frac{B \cdot h \cdot (K_x)_o}{R \cdot S_o} \cdot \frac{\partial S}{\partial x} \tag{2}$$

where g = gravitational constant; k = constant relating salinity to density; h = total depth; B = river width; R = freshwater discharge rate = $U_f \cdot B \cdot h$; S = tidal mean salinity; K_x = horizontal turbulent diffusivity. The subscript "o" designates the variables at a reference point, i.e., at x = 0. From Equations 1 and 2, the maximum upriver residual velocity at the bottom layer, U_m, may be expressed as:

$$U_m = \frac{3}{2} \cdot U_f(1 - \zeta_m^2) + \frac{g \cdot k \cdot h^3}{48} \cdot \frac{\partial S}{\partial x} \cdot \frac{1}{A_z}(1 - 9 \cdot \zeta_m^2 + 8 \cdot \zeta_m^3) \tag{3}$$

The depth, ζ_m, at which U_m occurs, may be expressed as a function of $v \cdot Ra$, $\zeta_m = 3/4 + 6/(v \cdot Ra)$, which ranges from 0.95 to 0.751 as $v \cdot Ra$ varies from 30 to 6000. Note that two-layer flow exists for $v \cdot Ra > 30$.

Assuming constant U_f and ζ_m, the variation in U_m as vertical mixing increases may be expressed as a function of a ratio of fractional change in $\partial s / \partial x$ to that in A_z:

$$U_S - U_N \propto \left(\frac{\partial S}{\partial x}\right)_S \frac{1}{\left(A_z\right)_S} - \left(\frac{\partial S}{\partial x}\right)_N \frac{1}{\left(A_z\right)_N} \propto \gamma - 1 \tag{4}$$

where

$$\gamma = \frac{\Delta S_x}{\Delta A_z} = \frac{\left(\frac{\partial S}{\partial x}\right)_S}{\left(\frac{\partial S}{\partial x}\right)_N} \left[\frac{\left(A_z\right)_S}{\left(A_z\right)_N}\right]^{-1} \tag{5}$$

where the subscript "S" designates the variables during high vertical mixing (e.g., spring tide) and the subscript "N" designates those during low vertical mixing (e.g., neap tide). The same relationship as Equations 4 and 5 can be obtained using the theoretical development in Officer (1976, Eq. 4-106). Assumption of constant U_f makes Eq. 4 represent change in baroclinic gravitational flow of estuarine circulation. Equations 4 and 5 suggest that as vertical mixing increases, residual circulation increases if the dimensionless ratio γ is larger than unity, and decrease if $\gamma < 1$. This ratio γ is equivalent to the ratio of fractional change in $v \cdot Ra$, i.e., $\gamma = (v \cdot Ra)_S / (v \cdot Ra)_N$: neither Ra nor v is easy to be estimated though. Equation 5 cannot be used to compare two estuaries with different intensity of vertical mixing because of the depth (h) dependency of residual velocity (Eq. 3).

Prandle [1985] derived, neglecting tidal nonlinearities, a residual velocity structure for narrow estuaries. The residual velocity due to freshwater discharge and longitudinal density gradient may be expressed as:

$$\frac{U}{U_f} = 0.89\left(-\frac{\xi^2}{2}+\xi+\frac{\pi}{4}\right)+\frac{S_P}{F}\left(-\frac{\xi^3}{6}+0.269\cdot\xi^2-0.037\cdot\xi-0.029\right) \tag{6}$$

where ξ = dimensionless depth (ξ = 1 at the surface and ξ = 0 at the bottom). The two dimensionless parameters, F and S_P, are defined as [Prandle, 1985]:

$$F = \frac{\kappa\cdot U_f\cdot U_t}{g\cdot h} \qquad S_P = k\cdot h\cdot\frac{\partial S}{\partial x} \tag{7}$$

where U_t = tidal current amplitude; κ = bed stress coefficient. From Equations 6 and 7, the maximum upriver residual velocity at the bottom layer, U_m, may be expressed as:

$$U_m = 0.89\cdot U_f\left(-\frac{\xi_m^2}{2}+\xi_m+\frac{\pi}{4}\right)$$

$$+\frac{g\cdot k\cdot h^2}{\kappa}\cdot\frac{\partial S}{\partial x}\cdot\frac{1}{U_t}\left(\frac{\xi_m^3}{6}+0.269\cdot\xi_m^2-0.037\cdot\xi_m-0.029\right) \tag{8}$$

The depth, ξ_m, at which U_m occurs, may be expressed as $\xi_m = 0.074 - 1.788\cdot(F/S_p)$. For the S_P/F values reported in Prandle [1985, Table 2], the ξ_m ranges from 0 to 0.057 as S_P/F varies from 24 to 100. Note that two-layer flow exists for $S_P/F > 24$.

Assuming constant U_f and ξ_m, the variation in U_m as vertical mixing increases may be expressed as a function of a ratio of fractional change in $\partial s/\partial x$ to that in U_t:

$$U_S - U_N \propto \left(\frac{\partial S}{\partial x}\right)_S\frac{1}{\left(U_t\right)_S}-\left(\frac{\partial S}{\partial z}\right)_N\frac{1}{\left(U_t\right)_N}\propto \gamma_P - 1 \tag{9}$$

where

$$\gamma = \frac{\Delta S_x}{\Delta U_t} = \frac{\left(\frac{\partial S}{\partial x}\right)_S}{\left(\frac{\partial S}{\partial x}\right)_N}\left[\frac{\left(U_t\right)_S}{\left(U_t\right)_N}\right]^{-1} \tag{10}$$

Assumption of constant U_f makes Eq. 9 represent change in baroclinic gravitational flow of estuarine circulation. Equations 9 and 10 indicate that as vertical mixing increases, residual circulation increases if the dimensionless ratio γ_P is larger than unity, and decreases if $\gamma_P < 1$. This ratio γ_P is equivalent to the ratio of fractional change in S_P/F, i.e., $\gamma = (S_P\cdot F^{-1})_S/(S_P\cdot F^{-1})_N$. For a small amplitude wave, a reasonable assumption for tidal wave in estuaries, U_t is proportional to the tidal amplitude of surface elevation [Ippen, 1966]. Since the tidal amplitude is a measure of the intensity of vertical mixing as A_z is, the two ratios, γ and γ_p, should behave in the same manner.

It should be noted that the theoretical models reviewed above were derived by neglecting tidal nonlinearities, and thus that the resulting ratios, γ (Eq. 5) and γ_p (Eq. 10), are applicable only to such systems with weak tidal forcing. The ratios are expressed in terms of three parameters, $\partial s/\partial x$, A_z and U_t, which are either difficult (even impossible) or too expensive to measure over an entire estuary. Hence, the vertical two-dimensional hydrodynamic model described in Park and Kuo [1994] is applied to the Rappahannock Estuary to estimate these parameters.

Model Description

A laterally-integrated, two-dimensional hydrodynamic model described in Park and Kuo [1994] is used to examine the effect of variation in vertical mixing on residual circulation. The model is based on the principles of conservation of volume, momentum and mass. With a right-handed Cartesian coordinate system with the x-axis directed seaward and the z-axis directed upward, the governing equations are:

the laterally and cross-sectionally integrated continuity equations

$$\frac{\partial (uB)}{\partial x}+\frac{\partial (wB)}{\partial z}=0 \qquad \frac{\partial}{\partial t}(B_\eta \eta) + \frac{\partial}{\partial x}\int_{-H}^{\eta} (uB)\,dz = q \qquad (11)$$

the laterally integrated momentum balance equation

$$\frac{\partial (uB)}{\partial t}+\frac{\partial (uBu)}{\partial x}+\frac{\partial (uBw)}{\partial z}=-\frac{B\partial p}{\rho \partial x}+\frac{\partial}{\partial x}\left(A_x B\frac{\partial u}{\partial x}\right)+\frac{\partial}{\partial z}\left(A_z B\frac{\partial u}{\partial z}\right) \qquad (12)$$

the hydrostatic equation

$$\frac{\partial p}{\partial z}=-\rho \cdot g \qquad (13)$$

the laterally integrated mass balance equation for salt

$$\frac{\partial (sB)}{\partial t}+\frac{\partial (sBu)}{\partial x}+\frac{\partial (sBw)}{\partial z}=\frac{\partial}{\partial x}\left(K_x B\frac{\partial s}{\partial x}\right)+\frac{\partial}{\partial z}\left(K_z B\frac{\partial s}{\partial z}\right)+S_{SA} \qquad (14)$$

the equation of state

$$\rho = \rho_o(1 + k\cdot s) \qquad (15)$$

and the temporally and spatially varying vertical mixing coefficients are calculated using

$$A_z = \alpha Z^2\left(1-\frac{Z}{h}\right)^2 |\frac{\partial u}{\partial z}|\left(1+\beta R_i\right)^{-\frac{1}{2}}+\alpha_w \frac{H_w^2}{T}\exp\left(-\frac{2\pi}{L}Z\right) \qquad (16)$$

$$K_z = \alpha Z^2\left(1-\frac{Z}{h}\right)^2 |\frac{\partial u}{\partial z}|\left(1+\beta R_i\right)^{-\frac{3}{2}}+\alpha_w \frac{H_w^2}{T}\exp\left(-\frac{2\pi}{L}Z\right) \qquad (17)$$

$$R_i = -\frac{g}{\rho}\frac{\partial \rho}{\partial z}\left(\frac{\partial u}{\partial z}\right)^{-2} \tag{18}$$

where t = time; η = free surface position above mean sea level; u and w = laterally averaged velocities in x and z directions, respectively; s = laterally averaged salinity; A_x = horizontal turbulent viscosity; K_z = vertical turbulent diffusivity; p = pressure; ρ = water density; ρ_o = freshwater density; q = lateral inflow; B_η = width at the free surface including side storage area; H = depth below mean sea level; Z = distance from the free surface; H_w, T and L = height, period and length of wind-induced waves, respectively; α, β and α_w = constants determined through model calibration [= 0.0115, 0.25 and 0.005 respectively in Park et al., 1993]; S_{SA} = source or sink of salt due to exchange with storage area.

The governing equations were solved by a two-time level, finite difference method with spatially staggered grid. The implicit treatment of the vertical mixing terms resulted in a tri-diagonal matrix in the vertical direction. To ensure stability, the pressure gradient term in Eq. 12 was evaluated using the surface elevation at a new time step. The horizontal advection term in Eq. 14 was solved using the QUICKEST scheme [Leonard, 1979]. A full description of the model can be found in Park and Kuo [1994].

Model Results

The model described in the previous section has been applied to the tidal Rappahannock River (Figures 2 and 3), a western shore tributary of Chesapeake Bay. The lower portion of this river is a partially mixed estuary and this estuarine portion is the study area of this paper. Hydrodynamic characteristics of the tidal Rappahannock River can be found in Kuo et al. [1991] and Park et al. [1993]. The model was calibrated and verified using the field data collected from the tidal Rappahannock River during summer of 1987 and 1990. The calibration and verification of the model are described in detail in Park et al. [1993]. The model gives a good description of the prototype behavior for surface elevation and current velocity, both tidal and subtidal components, and salinity [Park and Kuo, 1994].

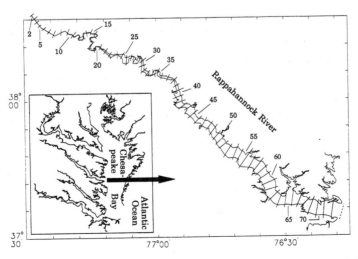

Figure 2. The tidal Rappahannock River in Virginia showing the model segmentation (Δx = 2.5 km). The insert shows lower Chesapeake Bay and its major tributaries.

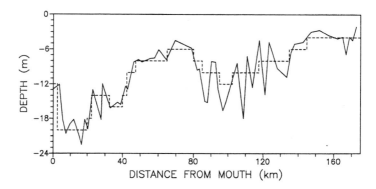

Figure 3. Longitudinal bathymetry of the tidal Rappahannock River ($\Delta z = 2$ m): field survey (solid line) and model input (dashed line).

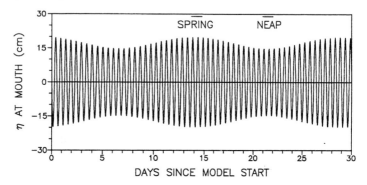

Figure 4. Surface elevation at the river mouth with M_2 and S_2 components only: long-term mean amplitude in the tidal Rappahannock River is 17.22 cm for M_2 and 2.53 cm for S_2 component. The horizontal lines indicate the averaging intervals of two tidal cycle for spring and neap tides.

The calibrated model is used to study the effect of variation in vertical mixing on residual circulation. First, the residual velocities calculated by the model with constant boundary conditions are compared between spring and neap tides. Then, the model calculated residual velocities with the measured time-varying boundary conditions are presented and discussed. The ratios, γ and γ_p, are calculated by the model and their behavior is examined.

With Constant Boundary Conditions

A model simulation using constant boundary conditions is performed. A constant freshwater discharge of 10 m^3 sec^{-1} at the fall line and a constant salinity profile at the mouth (17.0, 17.5, 18.0, 19.0, 19.5 and 20.0 ppt from the surface to the bottom every 2 m) are used for the boundary conditions. A harmonic tide with long-term mean amplitude of M_2 (17.22 cm) and S_2 (2.53 cm) components at the mouth (Figure 4) is used to force the model. The model results are presented in Figure 5. In Figure 5, the vertical mixing is stronger (Figure 5a) and $\partial s/\partial z$ is slightly weaker (Figure 5b) during spring tide than neap tide. The spring tide is ca. 7 days apart from the neap tide (Figure 4). This spring-neap cycle is short compared to the response time of salinity distribution, resulting in no appreciable change in $\partial s/\partial x$ (Figure 5c).

Then, the increased vertical mixing during spring tide results in weaker residual circulation (Figure 5d). The ratios, γ and γ_P, are less than unity throughout the estuarine portion of the river (Figure 5e).

If the increased vertical mixing exists long enough to change the salinity distribution, then $\partial s/\partial x$ will increase as the vertical mixing increases, which may increase the residual circulation if the effect of increased $\partial s/\partial x$ overcomes that of increased vertical mixing, i.e., if γ (or γ_P) > 1. Two model simulations are performed for 4 months: one with spring tide amplitude (19.75 cm) and the other with neap tide amplitude (14.69 cm) at the river mouth. All other conditions are the same as the ones used for Figure 5. The results from two model simulations are compared in Figure 6. Compared to Figure 5, the vertical mixing increases (Figure 6a) and $\partial s/\partial z$ decreases (Figure 6b) more with spring tide amplitude owing to the longer duration of spring tide mixing. The spring tide mixing is long enough to increase $\partial s/\partial x$ in the lower part of the river between km 0-40 (Figure 6c). The residual circulation with spring tide mixing is approximately the same as that with neap tide mixing in this region, and even stronger at some locations (Figure 6d). The ratios, γ and γ_P, follow the variation in residual circulation fairly well with some exceptions (Figure 6e), which will be discussed later.

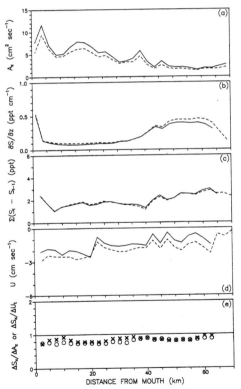

Figure 5. Model results during spring tide (solid line) and neap tide (dashed line) using surface elevation in Figure 4, and constant salinity and freshwater discharge as boundary conditions: (a) the tidal mean vertical viscosity averaged over total depth; (b) the vertical gradient of tidal mean salinity averaged over total depth; (c) the longitudinal gradient of tidal mean salinity integrated over total depth; (d) the tidal mean velocity averaged over the upriver-flowing bottom layer; (e) the ratios, γ ($\times$) and γ_P (O), estimated using vertically averaged quantities. The tidal means are averages over two tidal cycles and the averaging intervals are shown in Figure 4. The negative velocity in (d) indicates upriver flow.

With Time-Varying Boundary Conditions Measured During Summer of 1990

The model was calibrated and verified using the field data collected during summer of 1987 and 1990 [Park et al., 1993; Park and Kuo, 1994]. The model simulation using the 1990 conditions is performed for the time-varying boundary conditions. The surface elevation (Figure 7a) and salinity (Figure 7b) measured at the mouth, and daily varying freshwater discharge rate at the fall line (Figure 7c) are used to force the model. The model results are presented in Figure 8. The two tidal cycle averaging intervals are indicated in Figure 7a for spring and neap tides.

Both the vertical mixing (Figure 8a) and $\partial s/\partial x$ (Figure 8c) are stronger during spring tide than neap tide. Between km 0-20, the residual circulation is stronger during spring tide (Figure 8d) and the ratios, γ and γ_P, are larger than unity (Figure 8e), indicating that the effect of increase in $\partial s/\partial x$ overcomes that of vertical mixing. The water column, however, is slightly more stable between km 0-20 during spring tide than neap tide as shown by slightly higher $\partial s/\partial z$ during spring tide (Figure 8b). This is attributable to the time-varying salinity condition at the mouth (Figure 7b): the surface salinity remains relatively constant over the simulation period while the bottom salinity is about 2.5 ppt higher around spring tide than neap tide.

It is shown in Figure 5 that the fortnightly spring-neap cycle is shorter than the response time scale of salinity distribution so that the enhanced mixing during spring tide cannot

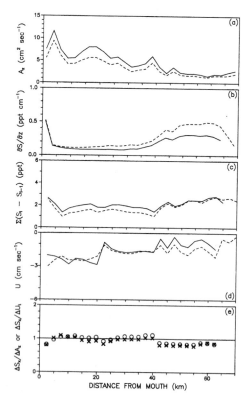

Figure 6. Model results with spring tide amplitude (solid line) and neap tide amplitude (dashed line): all conditions except the surface elevation are the same as Figure 5. See Figure 5 for explanation of (a)-(e).

increase ∂s/∂x in the Rappahannock Estuary. It is also shown in Figure 6 that several months of consistent spring tide mixing is required to affect the salinity distribution and thus ∂s/∂x. The spring and neap tides used to generate the results in Figure 8 are ca. 10 days apart (Figure 7a). Then, what mechanism(s) is responsible for increasing ∂s/∂x in Figure 8c? The difference between the model simulations for Figures 5 and 8 is the boundary conditions: constant conditions for Figure 5 and time-varying conditions for Figure 8. Two model simulations are performed to identify the mechanism(s) responsible for increase in ∂s/∂x in Figure 8c.

A model simulation is performed using the 1990 conditions (Figure 7) except using constant salinity boundary condition at the mouth: the average vertical salinity profile over the simulation period (Figure 7b). The results are presented in Figure 9. Increase in ∂s/∂x during spring tide in Figure 9c is not as large as that in Figure 8c, indicating that most of the increase in ∂s/∂x in Figure 8c is due to the time-varying salinity boundary condition at the mouth. The bottom salinity at the mouth is about 2.5 ppt higher around spring tide than neap tide (Figure 7b). With constant salinity boundary condition, very little increase in ∂s/∂x is caused by increased vertical mixing from neap to spring tide (Figure 9c) because of short time scale of spring-neap cycle. Thus, the effect of increased ∂s/∂x is not large enough to overcome that of increased vertical mixing, resulting in decreased residual circulation during spring tide (Figure 9d). This decrease in circulation also is indicated by the ratios, γ and γp, less than unity with some exceptions (Figure 9e), which will be discussed later. Also note that the stronger ∂s/∂z during spring tide between km 0-20 in Figure 8b, which is attributed to the salinity condition at the mouth, no longer exists in Figure 9b.

Another model simulation is performed using the 1990 conditions (Figure 7) except using constant freshwater discharge at the fall line: the average freshwater discharge over the

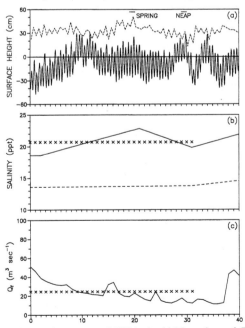

Figure 7. Boundary conditions during summer of 1990 in the tidal Rappahannock River: (a) surface elevation (solid line) and tidal range (dashed line) at the mouth, (b) salinity at bottom (solid line) and surface (dashed line) at the mouth, and (c) freshwater discharge rate (solid line) at the fall line. The horizontal lines in (a) indicate the averaging intervals of two tidal cycle for spring and neap tides. The × marks in (b) and (c) are mean values over the simulation period.

simulation period, 24.6 m³ sec⁻¹ (Figure 7c). The results are presented in Figure 10. For all parameters, very little difference exists between Figures 8 and 10, which indicates the large increase in $\partial s/\partial x$ during spring tide in Figure 8c is not due to the time-varying freshwater discharge. This insensitivity of $\partial s/\partial x$ to the change in freshwater discharge is because of the rather constant freshwater discharge over the simulation period (Figure 7c).

To show the effect of freshwater discharge, two model simulations are performed using the surface elevation in Figure 7a, constant salinity (× marks in Figure 7b) at the mouth, and two constant freshwater discharges at the fall line: one with 100 m³ sec⁻¹ and the other with 10 m³ sec⁻¹. The results from the two model simulations are compared in Figure 11. Increased freshwater discharge enhances the residual circulation by increasing both barotropic flow (U_f in Equations 3 and 8) and baroclinic flow due to increased $\partial s/\partial x$ (Figure 11c). The effect of increased $\partial s/\partial x$ overcomes that of vertical mixing, resulting in enhanced residual circulation (Figure 11d). The ratios, γ and γ_P, are larger than unity throughout the estuarine portion of the river (Figure 11e). The enhanced residual circulation affects vertical mixing positively (arrow D in Figure 1) and negatively by increasing $\partial s/\partial z$ (arrows H and B). Upriver of km 15, the positive and negative effects cancel out, resulting in no change in A_z (Figure 11a). The

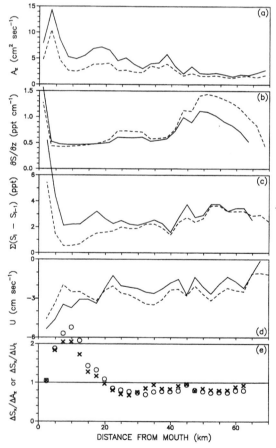

Figure 8. Model results during spring tide (solid line) and neap tide (dashed line) using surface elevation, salinity and freshwater discharge in Fig. 7 as boundary conditions. See Figure 5 for explanation of (a)-(e).

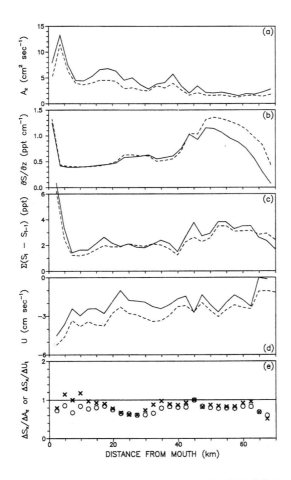

Figure 9. Model results during spring tide (solid line) and neap tide (dashed line) using constant salinity boundary condition (× marks in Figure 7b): all other conditions are the same as Figure 8. See Figure 5 for explanation of (a)-(e).

constant salinity boundary condition fixes the salinity distribution, and thus $\partial s/\partial z$, near the mouth (Figure 11b; also compare Figures 8b and 9b). Little change in $\partial s/\partial z$ near the mouth reduces the negative effect (arrows H and B), which results in the increased vertical mixing near the mouth during high flow (Figure 11a).

Discussion

The ratios, γ and γ_P, generally follow the behavior dictated by the theoretical models, Equations 4 and 9 respectively, with some exceptions. In Figure 6, γ is slightly less than

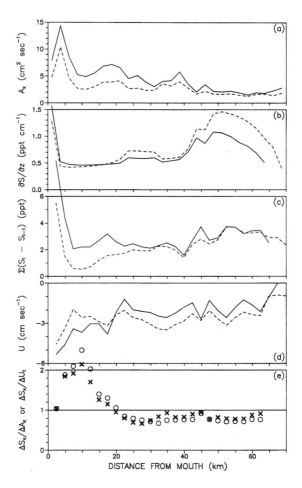

Figure 10. Model results during spring tide (solid line) and neap tide (dashed line) using constant freshwater discharge rate (× marks in Figure 7c): all other conditions are the same as Figure 8. See Figure 5 for explanation of (a)-(e).

unity while the residual circulation is stronger during spring tide around km 20, and γ_P is slightly larger than unity while the residual circulation is weaker during spring tide around km 30. In Figure 9, $\gamma > 1$ while the residual circulation is weaker during spring tide at two segments between km 5-10. These inconsistencies between the residual circulation and the ratios, γ and γ_P, might be attributable to the assumptions that are required in leading to Eq. 4 or 9. Equation 4 is subject to all assumptions made in Hansen and Rattray [1965] including no tidal nonlinearities, steady state, ideal geometry (constant width and depth), no lateral variation in density and flow field, and vertically constant A_z. Another assumption made for Eq. 4 is the constant freshwater discharge rate. Equation 9 is subject to all assumptions used for the theoretical development in Prandle [1985] including no tidal nonlinearities, steady state, linearized equation of motion, no stratification, vertically constant A_z, and A_z

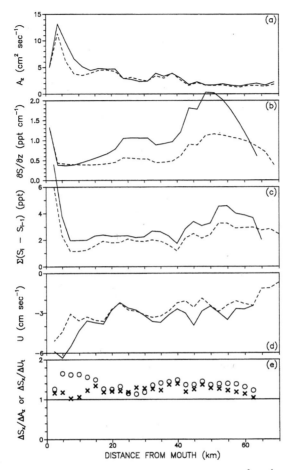

Figure 11. Model results using freshwater discharge at the fall line of 100 m^3 sec^{-1} (solid line) and 10 m^3 sec^{-1} (dashed line) with time-varying surface elevation (Figure 7a) and constant salinity boundary condition (× marks in Figure 7b) at the mouth. See Figure 5 for explanation of (a)-(e).

proportional to U_t. The last assumption that $A_z \propto U_t$ explains the difference between γ (Eq. 5) and γ_P (Eq. 10), and makes the evaluation of γ_P easier since U_t is much easier to measure than A_z. Among the assumptions, lateral homogeneity is not responsible for the inconsistencies in Figures 6 and 9 since the model results themselves are laterally averaged quantities. Despite all the assumptions, the ratios, γ and γ_P, are good indicators of the change in residual circulation as the vertical mixing varies in narrow, weakly nonlinear estuaries (Figures 5, 6, 8, 9 and 10).

In the Rappahannock Estuary, the spring tide occurs ca. 7 to 10 days after neap tide and the time scale for mass response, i.e., salinity distribution, is on the order of months. Because of these different time scales, increase in vertical mixing during spring tide cannot increase $\partial s/\partial x$ (Figure 5c), resulting in reduced residual circulation (Figure 5d). It takes several months of consistent spring tide mixing to affect the salinity distribution and $\partial s/\partial x$, and thus residual circulation (Figure 6). Therefore, the variation in residual circulation as vertical mixing increases is determined by the time scale of variation in vertical mixing relative to the response time of salinity distribution. The longer the increased vertical mixing lasts, the stronger the residual circulation gets. The size of estuaries affects the response time of salinity distribution. For small estuaries, the time scale of mass response is short, which amplifies the resulting increase in $\partial s/\partial x$ as vertical mixing increases.

Using a vertical two-dimensional model, Bowden and Hamilton [1975] conducted numerical experiments for a hypothetical estuary. Result of one experiment with constant salinity at the mouth and constant freshwater discharge indicated that the residual circulation decreased as the tidal amplitude at the river mouth increased. As discussed in Bowden and Hamilton [1975], this model result from the fourth tidal cycle after the model start cannot reflect the mechanism due to change in salinity distribution, and it essentially agrees with Figure 5. Wang and Kravitz [1980] applied a vertical two-dimensional model to the Potomac Estuary. With constant boundary conditions, an arbitrary increase in constant A_z by five times resulted in increase in residual circulation, which they attributed to the increased horizontal density gradient. The present model produces the same result (not shown).

In prototype, in addition to vertical mixing, the boundary conditions including salinity at the river mouth (Figure 9) and freshwater discharge at the fall line (Figure 11) can modify $\partial s/\partial x$ and thus residual circulation. The variation in salinity in the incoming bottom water in response to wind forcing has been frequently observed in many estuaries [Wang and Elliott, 1978; Cannon et al., 1990; Sanford and Boicourt, 1990; Kuo and Park, 1992]. The time scale of wind events tends to be shorter than response time of salinity distribution. Thus, the change in residual circulation due to $\partial s/\partial x$ induced by the variation in vertical mixing, which requires several months of spring tide mixing in the Rappahannock Estuary, might be disguised by the change due to the wind-induced salinity variation in the incoming bottom water, particularly for large estuaries.

Acknowledgments. This paper is contribution number 1990 of School of Marine Science /Virginia Institute of Marine Science, The College of William and Mary, VA. The work was supported by the Hypoxia Program of the Virginia Chesapeake Bay Initiatives.

References

Bowden, K.F., and P. Hamilton, Some experiments with a numerical model of circulation and mixing in a tidal estuary, *Estuarine and Coastal Marine Science, 3*, 281-301, 1975.

Bowden, K.F., *Physical oceanography of coastal waters*, 302 pp., Ellis Horwood Ltd., 1983.

Cannon, G.A., J.R. Holbrook, and D.J. Pashinski, Variations in the onset of bottom-water intrusions over the entrance sill of a fjord, *Estuaries, 13*, 31-42, 1990.

Fischer, K., Salinity intrusion models, in *Mathematical Modelling of Estuarine Physics, Lecture Notes on Coastal and Estuarine Studies*, vol. 1, edited by J. Sündermann and K.-P. Holz, pp. 232-241, Springer-Verlag, 1980.

Geyer, W.R., How does variation in vertical mixing affect the estuarine circulation? (abstract), in *Proceedings of the Twelfth International Estuarine Research Federation Conference*, Hilton Head Island, SC, 43 pp., 1993.

Hansen, D.V., and M. Rattray, Jr., Gravitational circulation in straits and estuaries, *J. of Marine Research, 23*, 104-122, 1965.

Ianniello, J.P., Tidally induced residual currents in estuaries of constant breadth and depth, *J. of Marine Research, 35*, 755-786, 1977.

Ianniello, J.P., Tidally induced residual currents in estuaries of variable breadth and depth, *J. of Physical Oceanography, 9*, 962-974, 1979.

Ianniello, J.P., Comments on tidally induced residual currents in estuaries: dynamics and near-bottom flow characteristics, *J. of Physical Oceanography, 11*, 126-134, 1981.

Ippen, A.T. (ed.), *Estuary and coastline hydrodynamics*, 744 pp., McGraw-Hill Book Company, 1966.

Jay, D.A., Residual circulation in shallow estuaries: shear, stratification and transport processes, in *Residual Currents and Long-Term Transport, Coastal and Estuarine Sciences*, vol. 38, edited by R.T. Cheng, pp. 49-63, Springer-Verlag, 1990.

Jay, D.A., and J.D. Smith, Residual circulation in shallow estuaries 1: highly stratified, narrow estuaries, *J. of Geophysical Research*, *95*, 711-731, 1990.

Kuo, A.Y., B.J. Neilson, and K. Park, A modeling study of the water quality of the upper tidal Rappahannock River. SRAMSOE No. 314, Virginia Institute of Marine Science (VIMS), The College of William and Mary (CWM), VA, 164 pp., 1991.

Kuo, A.Y., and K. Park, Transport of hypoxic waters: an estuary-subestuary exchange, in *Dynamics and Exchanges in Estuaries and the Coastal Zone, Coastal and Estuarine Sciences*, vol. 40, edited by D. Prandle, pp. 599-615, AGU, 1992.

Leonard, B.P., A stable and accurate convective modelling procedure based on quadratic upstream interpolation, *Computer Methods in Applied Mechanics and Engineering*, *19*, 59-98, 1979.

McCarthy, R.K., Residual currents in tidally dominated, well-mixed estuaries, *Tellus*, *45A*, 325-340, 1993.

Officer, C.B., *Physical oceanography of estuaries (and associated coastal waters)*, 465 pp., John Wiley and Sons, Inc., 1976.

Park, K., and A.Y. Kuo, Numerical modeling of advective and diffusive transport in tidal Rappahannock Estuary, Virginia, in *Proceedings of the Third International Conference on Estuarine and Coastal Modeling*, edited by M.L. Spaulding, K.W. Bedford, A.F. Blumberg, R.T. Cheng and J.C. Swanson, pp. 461-474, ASCE, 1994.

Park, K., A.Y. Kuo, and B.J. Neilson, A modeling study of hypoxia and eutrophication in the tidal Rappahannock River, Virginia. SRAMSOE No. 322, VIMS, CWM, VA, 158 pp., 1993.

Prandle, D., On salinity regimes and the vertical structure of residual flows in narrow tidal estuaries, *Estuarine, Coastal and Shelf Science*, *20*, 615-635, 1985.

Pritchard, D.W., Estuarine classification - a help or a hindrance, in *Estuarine Circulation*, edited by B.J. Neilson, A.Y. Kuo and J.M. Brubaker, pp. 1-38, Humana Press, 1989.

Rigter, B.P., Minimum length of salt intrusion in estuaries, *J. of the Hydraulics Division*, ASCE, *99*, 1475-1496, 1973.

Sanford, L.P., and W.C. Boicourt, Wind-forced salt intrusion into a tributary estuary, *J. of Geophysical Research*, 95, 13357-13371, 1990.

Wang, D.-P., and A.J. Elliott, Non-tidal variability in the Chesapeake Bay and Potomac River: evidence for non-local forcing, *J. of Physical Oceanography*, *8*, 225-232, 1978.

Wang, D.-P., and D.W. Kravitz, A semi-implicit two-dimensional model of estuarine circulation. *J. of Physical Oceanography*, *10*, 441-454, 1980.

22

Combined Effects of Buoyancy and Tides Upon Longitudinal Dispersion

Ronald Smith

Abstract

This article focuses upon the fluid mechanics of how salinity modifies the mixing process in partially mixed estuaries. Simplified model equations are used to clarify the interacting effects of tidal oscillations and buoyancy. The maximum value of the longitudinal dispersion coefficient is shown to occur for wider estuaries and be larger when buoyancy effects are significant.

Introduction

Along the minor estuaries in the UK the summer drought of 1976 was aggravated by salinity penetration many tens of kilometres upstream. Farmers who traditionally could rely on extracting fresh water for irrigation from what they regarded as rivers far from the sea, found that they were damaging their crops with brine. By contrast, along the major estuaries such as the Humber and the Thames, the salinity penetration was not very different from any other summer. This marked disparity in salinity response requires both an explanation and quantitative modelling.

The distribution of any solute, such as salt, along a partially mixed estuary can be explained as a balance between the flushing out to sea by the fresh water run-off and the diffusion-like spreading upstream of the solute. In a drought the fresh water run-off is greatly reduced. So, if the longitudinal spreading remained unchanged, the solute (and the solute gradient) would penetrate further upstream. Commonly, there is so much change in salinity that the change in water density becomes dynamically significant: the flow is changed in ways that effect the spreading of solutes. For small estuaries the effective longitudinal diffusivity can increase as the fresh water run-off decreases, giving exaggerated salinity penetration [Smith 1977]. Conversely, in large estuaries both theory [Fischer 1972, Smith 1980] and observation [Williams and West 1974, Uncles and Radford 1980] concur that as the fresh water run-off decreases, so does the effective longitudinal diffusivity. Hence, the low vulnerability of large estuaries to salinity penetration. This paper concerns the transition regime between small and large estuaries i.e. when the mixing time and tidal period are comparable. In terms of estuary widths, this transition regime typically corresponds to widths between 100m and 300 m.

The key ingredient that makes the longitudinal spreading depend upon the run-off is the fact that the transverse flow and the resulting transverse mixing process in a well-mixed estuary is affected by the salinity, as illustrated in Figures 1a,b. An easy way to confirm the existence of the transverse flow is to observe the movement of surface debris or of scum-lines [Nunes and Simpson 1985]. On the ebb, flotsam tends to be carried towards the banks as suggested in Figure 1a. While during flood tide flotsam tends to accumulate towards the middle of the estuary as suggested in

Buoyancy Effects on Coastal and Estuarine Dynamics
Coastal and Estuarine Studies Volume 53, Pages 319-329

Figure 1b. Detailed measurements of the three-dimensional flow and salinity structure are presented by Guymer and West [1991] and agree with the schematical relationship suggested in Figures 1a,b.

Holley, Harleman and Fischer [1970] show that, in the absence of salinity gradients, the longitudinal shear dispersion has a maximum value when the time scale for mixing across the estuary is comparable with the tidal period. This maximum for the longitudinal shear dispersion is about 70 m^{2}s^{-1} and occurs for an estuary width of about 170m. Macqueen [1978] pointed out that in the Thames the observed longitudinal dispersion of heat or salt exceeded the theoretical maximum and occurred for a wider than predicted reach. Macqueen [1978] suggested an empirical combination of the results of Smith[1977] and of Holley, Harleman and Fischer [1970] to account for the buoyancy enhanced mixing across the estuary. This paper provides a quantitative basis for Macqueen's suggestion.

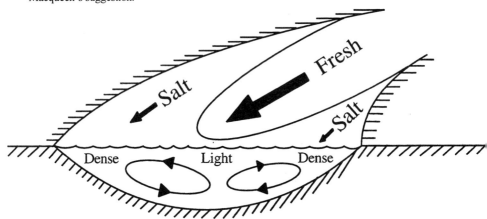

Figure 1a. During the ebb tide the faster flow near the centre of the estuary carries lighter fresh water further seawards and results in density differences across the estuary.

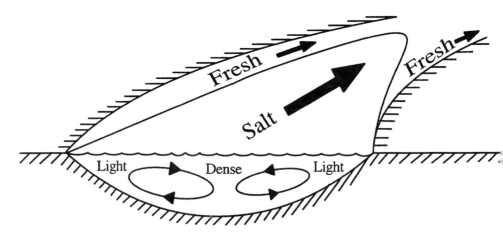

Figure 1b. During the flood tide the density tends to be greater in the centre than at the sides. This drives a density-driven circulation.

Shear Dispersion Mechanism

The way that non-uniform flows dramatically effect the rate of spreading was first explained by Townsend [1951] in a paper communicated to the Royal Society of London by G I Taylor. For a three-dimensional linear flow with arbitrary time-dependence, Townsend [1951] derived exact solutions for the distribution of an initial spot of heat or of solute. Figure 2 gives a graphical version of the way that both shear and diffusion are involved in the transverse dilution process.

If there were neither shear nor diffusion then salt or any other tracer would merely be transported along with the mean flow. A line of small circles and crosses would remain exactly in line. If instead there were shear but no diffusion, then the line of circles and crosses would become increasingly sheared out into a curve whose shape revealed the shape of the velocity profile. Hence, in the upper right-hand part of Figure 2 we infer that the velocity profile is to the left at the (wavy) free surface, to the right deeper down and reduces to zero at the (hatched) bed. At any level within the flow there is zero horizontal spread. It is only by vertically averaging that there is the false illusion of spread which increases linearly with time. When there is diffusion, but no shear, the line of circles and crosses gradually become spread out both vertically and horizontally. The horizontal size increases as the square-root of time (horizontal variance increases linearly with time). Finally, when there is both shear and diffusion the distribution of circles and crosses (or any solute) exhibits some of the features of the pure shear and of the pure diffusion. In the lower right-hand part of Figure 2 we see that near the (wavy) free surface the centroid of the particles is displaced to the left, whilst lower down the centroid becomes displaced to the right. Some of the horizontal spread is inherited from the diffusion (for example a few of the crosses are slightly ahead of where they could have reached by advection alone). Importantly, the vertical migration of the particles combined with the shear gives rise to a considerably augmented horizontal variance.

Townsend [1951] showed that, away from boundaries, as time increases the shear contribution to the variance eventually dominates the diffusive contribution. Initially, the shear dispersion is proportional to the square of the local shear. Hence the much greater horizontal spread of the circles than of the crosses in Figure 2.

Taylor Dispersion Coefficient Across a Shallow Estuary

Eventually the vertical diffusion the circles and crosses, would smooth out any differences in the rate of dispersion at different levels. G I Taylor [1953] showed that beyond this time scale of mixing across the flow, the shear dispersion process becomes diffusive in character. The shear

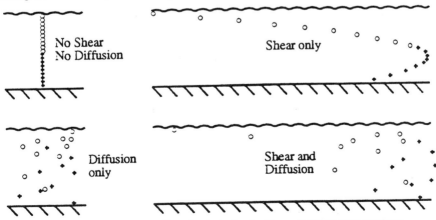

Figure 2: The process of shear dispersion in the density-driven circulation across an estuary.

dispersion can then be characterised by the increase in the effective diffusivity along the flow: the Taylor (or shear) dispersion coefficient. Taylor [1953] calculated the shear dispersion coefficient for laminar pipe flow and verified his results experimentally. What makes G I Taylor's work so important (averaging 80 citations per year) is that the method of calculation is readily adapted to other flows or to encompass additional physics.

For a shallow partially mixed estuary of total depth $(h + \zeta)$, with a cross-flow velocity profile $v(x, y, z, t)$ and vertical diffusivity $\kappa_3(x, y, z, t)$ the eventual Taylor dispersion coefficient $D_2(x, y, t)$ across the estuary is given by the double integral :

$$D_2 = \frac{1}{(h+\zeta)} \int_{-h}^{\zeta} \frac{1}{\kappa_3} \left[\int_{-h}^{z} \left(v - \|v\| \right) dz' \right]^2 dz \ ,$$

(1)

$$\text{with} \qquad \|v\| = \frac{1}{(h+\zeta)} \int_{-h}^{\zeta} v \, dz \ .$$

Hence, the stronger the buoyancy-driven transverse flow $v(x, y, z, t)$ the greater the effective mixing across the estuary. (The co-ordinate system (x, y, z) is assumed to be aligned along, across and vertically with respect to the long shallow estuary.)

The shallowness of the water is required to ensure that the vertical mixing occurs more rapidly than any reversal of the tidal flow i.e. water depths less than 30m. The formula (1) exhibits several standard features of shear dispersion coefficients. There is quadratic dependence upon the velocity shear $(v - \|v\|)$ and upon the water depth $(h + \zeta)$ but inverse dependence upon the mixing κ_3 . The elapsed time necessary for the applicability of the Taylor limit is likewise proportional to $(h + \zeta)^2 / \kappa_3$.

As a particular model of the vertical mixing we use:

$$\kappa_3 = v_{23} = ku_*(h + \zeta)(1 - \sigma)(\sigma + \sigma_*) ,$$

$$\text{where} \qquad \sigma = (h + z) / (h + \zeta) .$$

Here k is von Karman's constant (about 0.4), u_* is the friction velocity, σ is the fractional distance between the bed and the free surface and σ_* is the exceedingly small fractional roughness height. For a smooth bed σ_* can be as small as 0.0001 . In terms of σ_* we define a less tiny parameter ε (about 1/9):

$$\varepsilon = -1 / \ln \sigma_* \ .$$

If salinity concentration s causes a fractional density increase βs, then as a series in ε the transverse velocity has the representation [Smith 1979]:

$$v = \frac{\beta g(h + \zeta)^2}{4ku_*} \partial_y \|s\| \{ 2\sigma - 1 - \varepsilon[1 + \ln(\sigma + \sigma_*)] \}$$

$$+ \|v\| \{ 1 + \varepsilon[1 + \ln(\sigma + \sigma_*)] \} + \cdots \ .$$

By construction the buoyancy term does not contribute to the vertically averaged transverse flow $\|V\|$. The leading terms for the buoyancy and flow contributions to the transverse shear dispersion coefficient are:

$$D_2 = \frac{(h + \zeta)^5}{96k^3 u_*^3} (\beta g \partial_y \|s\|)^2 + \frac{(h + \zeta)^3 \varepsilon \|v\|}{8k^2 u_*^2} \beta g \partial_y \|s\| + 0.4041 \frac{(h + \zeta)\varepsilon^2 \|v\|^2}{ku_*} \ .$$

Only the leading term is present if $\|v\| = 0$. Smith [1979, Figure 6] shows that for the excess lateral spread associated with buoyancy, there is good agreement with the wide range of laboratory experiments conducted by Prych [1970]. Following Fischer [1973] we model the transverse turbulent mixing as being slightly stronger than the vertical mixing:

$$\|\kappa_2\| = 0.15(h + \zeta)u_* \ .$$

Taylor Dispersion Coefficient Along a Narrow Estuary

If the estuary is long and narrow, so that mixing takes place more rapidly than does flow reversal, then qualitatively the shear dispersion process is similar to that schematised in Figure 2. For a narrow, partially-mixed estuary of cross-sectional area $A(x,t)$ with banks at $y_-(x,t)$, $y_+(x,t)$ the counterpart to equation (1) is:

$$E = \frac{1}{A} \int_{y_-}^{y_+} \frac{1}{(h + \zeta)(\|\kappa_2\| + D_2)} \left[\int_{y_-}^{y} (h + \zeta)(\|u\| - \bar{u}) dy' \right]^2 dy \quad . \tag{2a}$$

$$\text{with} \quad \bar{u} = \frac{1}{A} \int_{y_-}^{y_+} (h + \zeta)\|u\| dy \tag{2b}$$

[Smith 1977]. Again, there is quadratic dependence upon the longitudinal current and upon the flow width but inverse dependence upon the mixing. An important feature in equation (2a) is that both turbulent mixing $\|\kappa_2\|$ and transverse dispersion D_2 contribute to the cross-stream mixing. So, the longitudinal shear dispersion coefficient $E(x,t)$ decreases as D_2 increases.

In low river-flow conditions, the salt distribution moves inland and the salinity gradient becomes reduced near the mouth. Consequently, near the mouth there is reduced buoyancy-driven transverse circulation (Figures 1a,b), reduced transverse dispersion D_2 and increased longitudinal dispersion E. It is this increased coefficient of longitudinal spreading as the fresh-water flushing decreases, that makes minor estuaries so vulnerable to salinity penetration [Smith 1977].

Effect of Oscillating Flows Upon Longitudinal Dispersion

If the estuary is sufficiently wide that the tidal flow reverses before there has been mixing across the estuary, then the shear distortion across the estuary reverses and the efficiency of the shear dispersion becomes reduced [Holley, Harleman and Fischer 1970]. A time-dependent longitudinal shear dispersion coefficient can be defined:

$$E = \frac{1}{A} \int_{y_-}^{y_+} (h + \zeta)(\|\kappa_2\| + D_2)(\partial_y G)^2 dy \tag{3}$$

[Smith 1985]. The centroid displacement function $G(x, y, t)$ satisfies the transverse dispersion equation

$$\partial_t\{(h + \zeta)G\} + \partial_y\{(h + \zeta)\|v\|G\} - \partial_y\{(h + \zeta)(\|\kappa_2\| + D_2)\partial_y G\} \tag{4a}$$

$$= (h + \zeta)(\|u\| - \bar{u})$$

$$\text{with} \quad (h + \zeta)(\|\kappa_2\| + D_2)\partial_y G = 0 \quad \text{on} \quad y = y_-, y_+ . \tag{4b}$$

For narrow (less than 100m) estuaries we can neglect the ∂_t term in equation (4a). An integration with respect to y, then yields an integral expression for $\partial_y G$. Substituting into equation (3), we can recover the result (2a).

The tidally averaged longitudinal dispersion ceases to increases quadratically with the flow width and instead reaches a maximum when the tidal and mixing time scales are comparable [Holley, Harleman and Fischer 1970]. In this regime there can be noticeable time lags between the centroid displacement G and the tidal current $\|u\|$.

In terms of $G(x,y,t)$, the solute concentration can be decomposed:

$$\|c\| = \overline{c} - G\partial_x\overline{c} + \cdots .$$

Thus a forwards displacement G in the faster-moving central part of an estuary gives a local concentration corresponding to the cross-sectionally averaged concentration $\overline{c}$ slightly upstream. When the solute is salt we denote the vertically and cross-sectionally averaged concentrations by $\|s\|$ and $\overline{s}$. We recall that it is the salinity differences across the estuary that drive the secondary flows across the estuary, as illustrated in Figures 1a,b,

Eigenfunction Expansion

To focus our attention upon the role of the salinity gradient $\partial_x\overline{s}$ and of the associated secondary flow, the illustrative example has as few other features as possible. The depth $h(y)$ and the flow properties do not vary along the estuary. Also, the tidal elevation ζ is regarded as small relative to the depth. The simplified equation (4a) for the centroid displacement function is:

$$h\partial_t G - \partial_y\left\{\left[0.15u_* + \frac{(\beta g\partial_x\overline{s})^2(h^2\partial_y G)^2}{96k^3u_*^3}\right]h^2\partial_y G\right\} = h(\|u\| - \overline{u}) . \tag{5a}$$

with

$$u_* h^2\partial_y G = 0 \; on \; y = y_-, y_+ . \tag{5b}$$

On the hypothesis that the profile of the friction velocity varies more in magnitude than in shape, we introduce the eigenvalue problem

$$\frac{d}{dy}\left\{\langle u_*\rangle h^2\frac{d\psi_n}{dy}\right\} + \lambda_n h\psi_n = 0 \; \text{with} \; \frac{1}{A}\int_{y_-}^{y_+} h\psi_n^2 dy = 1 .$$

The angle brackets $\langle\cdots\rangle$ indicate the tidal average value. The zero mode is $\psi_0 = 1$ with $\lambda_0 = 0$. For comparison, Chebyshev polynomials of the second kind $\psi_n = U_n(y/B)$ satisfy:

$$\frac{d}{dy}\left\{\left(1-\frac{y^2}{B^2}\right)^{3/2}\frac{d\psi_n}{dy}\right\} + \frac{n(n+2)}{B^2}\left(1-\frac{y^2}{B^2}\right)^{1/2}\psi_n = 0 .$$

The corresponding depth and friction velocity profiles would be

$$h = H\frac{4}{\pi}\left(1-\frac{y^2}{B^2}\right)^{1/2} , \quad u_* = U_*\frac{4}{\pi}\left(1-\frac{y^2}{B^2}\right)^{1/2} ,$$

with average depth H, total width $2B$ and reference friction velocity $U_*(t)$.

The velocity and centroid displacement can be represented as series:

$$u = \overline{u} + \sum_{n=1}^{\infty} u_n\psi_n , \quad G = \sum_{n=1}^{\infty} G_n\psi_n .$$

The ψ_n component of equation (14a) gives an equation for the coefficient G_n:

$$\frac{dG_n}{dt} + 0.15\lambda_n\xi_n G_n + \frac{(\beta g\partial_x\overline{s})^2}{96k^3 A}\int_{y_-}^{y_+}\frac{h^6}{u_*^3}\left(\frac{dG}{dy}\right)^3\frac{d\psi_n}{dy}dy$$

$$= u_n + \frac{0.15}{A}\int_{y_-}^{y_+}\left(\xi_n\langle u_*\rangle - u_*\right)h^2\frac{dG}{dy}\frac{d\psi_n}{dy}dy \tag{6}$$

Although the eigenfunction ψ_n is independent of time, the turbulence and decay rate are time-dependent. The parameter $\xi_n(t)$ will be selected to account for that time-dependence as it effects the n'th mode.

To eliminate the diagonal G_n contribution to the right-hand side of equation (6), we introduce a weight factor $W_n(y)$ and select $\xi_n(t)$

$$W_n(y) \propto h\left(\frac{d\psi_n}{dy}\right)^2 \quad , \quad \xi_n = \frac{\overline{u_* W_n}}{\langle \overline{u_* W_n} \rangle} .$$

Hence the time-dependent decay rate for the n'th mode is particularly sensitive to the strength of the turbulence in that part of the estuary where the n'th mode is changing rapidly.

We can achieve dramatic simplification if we truncate the series for G with just $n=2$:

$$\frac{dG_2}{dt} + 0.15\lambda_2 \frac{\overline{u_* W_2}}{\langle \overline{u_* W_2} \rangle} G_2 + G_2^3 \frac{(\beta g \partial_x \overline{s})^2}{96 k^3 A} \int_{y_-}^{y_+} \frac{h^6}{u_*^3}\left(\frac{d\psi_2}{dy}\right)^4 dy = u_2 . \quad (7)$$

The formula (2a) for the shear dispersion coefficient becomes:

$$E = 0.15\lambda_2 \frac{\overline{u_* W_2}}{\langle \overline{u_* W_2} \rangle} G_2^2 + G_2^4 \frac{(\beta g \partial_x \overline{s})^2}{96 k^3 A} \int_{y_-}^{y_+} \frac{h^6}{u_*^3}\left(\frac{d\psi_2}{dy}\right)^4 dy . \quad (8)$$

The example of Chebyshev polynomials of the second kind has

$$\psi_2 = 4(y^2 / B^2) - 1 , \quad \frac{W_2}{\overline{W_2}} = \frac{15\pi}{8}\left(1 - \frac{y^2}{B^2}\right)^{1/2}\frac{y^2}{B^2} ,$$

with $\qquad \dfrac{\overline{u_* W_2}}{\overline{W_2}} = U_* \qquad$ and $\qquad \lambda_2 = \dfrac{8\langle U_* \rangle H}{B^2} .$

For the longitudinal velocity we take a square-root profile:

$$\|u\| = \overline{u}\,\frac{3\pi}{8}\left(1 - \frac{y^2}{B^2}\right)^{1/2} .$$

The coefficients in equations (7,8) can be evaluated:

$$\frac{dG_2}{dt} + 1.2\frac{U_* H}{B^2} G_2 + G_2^3 \frac{32(\beta g \partial_x \overline{s})^2 H^5}{\pi^2 k^3 B^4 U_*^3} = -\frac{\overline{u}}{5} , \quad (9a)$$

$$E(t) = 1.2\frac{U_* H}{B^2} G_2^2 + G_2^4 \frac{32(\beta g \partial_x \overline{s})^2 H^5}{\pi^2 k^3 B^4 U_*^3} . \quad (9b)$$

Once the effective friction velocity $U_*(t)$ and the bulk tidal velocity $\overline{u}(t)$ are specified, it is merely a matter of integration to calculate G_2 and hence the shear dispersion coefficient $E(t)$.

Illustrative Example

The simplicity of equation (9a) makes it easy for us to identify appropriate non-dimensional measures Ω and S for the tidal frequency ω and salinity gradient $\partial_x \overline{s}$

$$\Omega = \frac{\omega B^2}{\langle U_* \rangle H} \quad , \quad S = \frac{\beta g \partial_x \overline{s} H B}{k^2 \langle U_* \rangle^2} \quad \text{with} \quad G_2 = \frac{B^2}{H}\hat{G}_2 \quad , \quad E = \hat{E}\frac{\langle U_* \rangle^2}{\omega} .$$

A large value of Ω signifies that the tide reverses frequently in a mixing time i.e. the estuary is wide. Large values of S signify that the buoyancy-driven transverse flow strongly effects the mixing. Non-dimensional counterparts of equations (9a,b) are

$$\Omega \frac{d\hat{G}_2}{dT} + 1.2\left(\frac{U_*}{\langle U_*\rangle}\right)\hat{G}_2 + \hat{G}_2^3 S^2 \frac{32k}{\pi^2}\left(\frac{\langle U_*\rangle}{U_*}\right)^3 = -\frac{1}{5}\left(\frac{\bar{u}}{\langle U_*\rangle}\right) \quad , \qquad (10a)$$

$$\hat{E}(T) = \Omega\left[1.2\left(\frac{\langle U_*\rangle}{U_*}\right)\hat{G}_2^2 + \hat{G}_2^4 S^2 \frac{32k}{\pi^2}\left(\frac{\langle U_*\rangle}{U_*}\right)^3\right] \quad , \quad \text{with} \quad \omega t = T . \qquad (10b,c)$$

For illustration, we take the friction velocity to be constant and the tidal velocity to be sinusoidal:

$$U_* = \text{constant} \quad , \qquad \text{and} \qquad \bar{u} = 10U_* \cos(T) . \qquad (11a,b)$$

$$\partial_x \bar{s} = 0, \ 0.9, \ 1.8 \ \%o\,\text{km}^{-1} .$$

Figure 3a shows the dimensionless shear dispersion coefficient for S=0, 0.5, 1 when the estuary is narrow $\Omega = 0.5$. An appropriate specification of such an estuary might be:

$$B = 35\text{m} , \ H = 3.5\text{m} , \ \omega = 1.5\times10^{-4}\text{s}^{-1} , \ \langle U_*\rangle = 0.1\text{ms}^{-1} \quad ,$$

We note that on average the salinity reduces the dispersion and that the phase of the dispersion only lags slightly behind the phase of the square of the tidal velocity. The increased cross-sectional mixing associated with salinity can be characterised as reducing the effective estuary width $2B$ both as regards the phase lag and the magnitude of the dispersion. Figure 3b gives the corresponding shear dispersion results for a wide estuary $\Omega = 2$

e.g . $\qquad B = 140\text{m} , \ H = 14\text{m} , \ \omega = 1.5\times10^{-4}\text{s}^{-1} , \ \langle U_*\rangle = 0.1\,\text{ms}^{-1} \quad ,$

$$\partial_x \bar{s} = 0, \ 0.06, \ 0.12 \ \%o\,\text{km}^{-1} \quad .$$

In keeping with the ideas of Macqueen [1978], the salinity now tends to increase the dispersion. Also, we observe that in the absence of buoyancy the phase is approaching orthogonality to the phase of the square of the tidal velocity. Again, salinity can be characterised as reducing the effective estuary width $2B$ as regards the phase lag.

Fourier Series

If, in the example (11a,b) we sought Fourier series expansions:

$$\hat{G}_2 = \sum_{n=0}^{\infty} a_{2n+1} \cos((2n+1)T) + \sum_{n=0}^{\infty} b_{2n+1} \sin((2n+1)T) ,$$

then the cosine and sine components of equation (10a) give the component equations

$$(2n+1)\Omega b_{2n+1} + 1.2a_{2n+1} + 1.297\frac{S^2}{\pi}\int_0^{2\pi}\hat{G}_2^3 \cos((2n+1)T)dT = -2\delta_{n,0} , \qquad (12a)$$

$$-(2n+1)\Omega a_{2n+1} + 1.2b_{2n+1} + 1.297\frac{S^2}{\pi}\int_0^{2\pi}\hat{G}_2^3 \sin((2n+1)T)dT = 0 . \qquad (12b)$$

Again, we can achieve dramatic simplification if we retain just the a_1 , b_1 terms:

$$a_1 = \frac{-2\{1.2 + 0.973S^2(a_1^2 + b_1^2)\}}{\{1.2 + 0.973S^2(a_1^2 + b_1^2)\}^2 + \Omega^2} \quad , \qquad (13a)$$

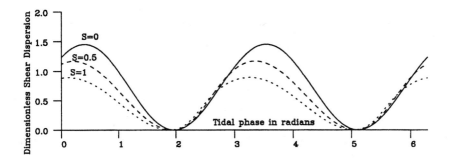

Figure 3a: Time dependence of the shear dispersion coefficient for a narrow estuary.

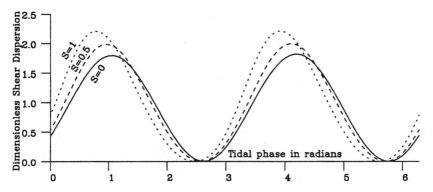

Figure 3b: Time dependence of the shear dispersion coefficient for a wide estuary.

$$b_1 = \frac{2\Omega}{\{1.2 + 0.973S^2(a_1^2 + b_1^2)\}^2 + \Omega^2} \quad , \tag{13b}$$

where
$$a_1^2 + b_1^2 = \frac{4}{\{1.2 + 0.973S^2(a_1^2 + b_1^2)\}^2 + \Omega^2} \quad . \tag{13c}$$

The equation (13c) can also be written as a cubic for $a_1^- + b_1^-$.

The corresponding drastic simplification for the tidally averaged shear dispersion can be written:

$$\langle \hat{E} \rangle = \frac{2.4\Omega_M}{1.44 + \Omega_M^2} \quad \text{with} \quad \Omega_M = \frac{1.2\Omega}{1.2 + 0.973S^2(a_1^2 + b_1^2)} \quad . \tag{14}$$

Macqueen [1978] suggested that such a modified frequency ratio might replicate the combined effects of buoyancy and tides (hence the M subscript). The maximum value is 1 when $\Omega_M = 1.2$. i.e. when the buoyancy augmented mixing rate equals the tidal frequency. In the absence of buoyancy equation (13) can be identified as the leading term in the infinite series given by Holley, Harleman and Fischer [1970]. For the illustrative examples in the previous section, with an $H:B$ aspect ratio1:10, $\Omega = 1.2$ would correspond to an estuary width $2B = 168$m and a maximum longitudinal shear dispersion of 66.6 m^2s^{-1}.

Unfortunately, equation (13) fails to predict Macqueen's[1978] observation that the maximum value of the shear dispersion can be increased by salinity. The inclusion of additional Fourier harmonics (or more eigenfunctions ψ_n) allows a more complete representation of buoyancy. For example, in equations (12a,b) the cubic non-linearity involving a_1 and b_1 only drives the third harmonics:

$$a_3 = \frac{0.973S^2\{1.2a_1(3b_1^2 - a_1^2) + 3\Omega b_1(3a_1^2 - b_1^2)\}}{3[1.44 + 9\Omega^2]} \quad,$$

$$b_3 = \frac{0.973S^2\{-1.2b_1(3a_1^2 - b_1^2) + 3\Omega a_1(3b_1^2 - a_1^2)\}}{3[1.44 + 9\Omega^2]} \quad.$$

The presence of such $\cos(3T)$ and $\sin(3T)$ contributions can be inferred from the slight asymmetry of the dispersion curves in Figures 3a,b. The extended equation for the tidally averaged shear dispersion is

$$\frac{2}{\Omega}\langle \hat{E} \rangle = (a_1^2 + b_1^2)\{1.2 + 0.973S^2(a_1^2 + b_1^2) + 3.89S^2(a_3^2 + b_3^2)\}$$

$$+ (a_3^2 + b_3^2)\{1.2 + 0.973S^2(a_3^2 + b_3^2)\}$$

$$+ 1.297S^2[b_3b_1(3a_1^2 - b_1^2) - a_3a_1(3b_1^2 - a_1^2)] \quad.$$

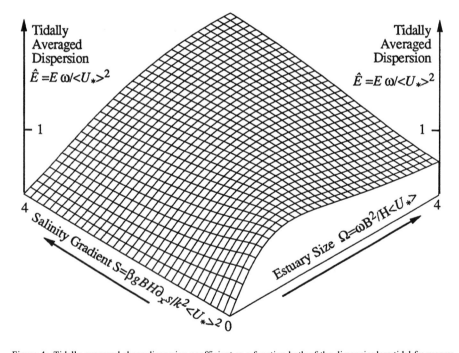

Figure 4: Tidally averaged shear dispersion coefficient as a function both of the dimensionless tidal frequency and of the dimensionless longitudinal salinity gradient. The frequency increases in the NE direction and the salinity gradient increases in the NW direction.

The maximum value now exceeds 1. Figure 4 gives a perspective plot of the tidally averaged shear dispersion coefficient $<\hat{E}>$ for S and Ω both ranging between 0 and 4. It is noticeable how the buoyancy decreases the dispersion for narrow estuaries (with Ω less than 1.2) and increases the dispersion for wide estuaries (with Ω greater than 1.2).

Concluding Remarks

One of the more important classifications for solute dispersion along estuaries is that of Holley, Harleman and Fischer [1970] which uses the cross-estuary mixing time. In the absence of secondary circulation, the longitudinal dispersion of solutes has its maximum when the cross-estuary mixing time equals the tidal frequency (estuary widths of about 170m). The present article modifies that classification to include the buoyancy-driven secondary circulation. The cross-estuary shear dispersion augments the cross-estuary mixing and it is not until wider estuaries that the mixing time equals the tidal frequency. The resulting maximum for the longitudinal dispersion is larger than in the absence of salinity.

Acknowledgement . My attendance at this conference was funded by the Royal Society.

References

Fischer, H. B., Mass transport mechanisms in partially stratified estuaries, *Journal of the Hydraulics Division, ASCE, 53*, 691-687, 1972.

Fischer, H. B., Longitudinal dispersion and turbulent mixing in open channel flow, *Ann. Rev. Fluid Mech., 5*, 59-78, 1973.

Guymer, I., and J. R. West, Field studies of the flow structure in a straight reach of the Conwy Estuary, *Estuarine, Coastal Shelf Sci., 32*, 581-596, 1991.

Holley, E. R., D. R. F. Harleman, and H. B. Fischer, Dispersion in Homogeneous Estuary Flow, *Journal of the Hydraulics Division, ASCE, 96*, 1691-1709, 1970.

Macqueen, J. F., On factors affecting longitudinal mixing in tidal waterways, *Central Electricity Laboratories*, No. RD/L/N127/78, 1978.

Nunes, R. A., and J. H. Simpson, Axial convergence in a well-mixed estuary, *Estuarine, Coastal Shelf Sci., 20*, 637-649, 1985.

Prych, E. A., Effects of density differences on lateral mixing in open channel flow, *Keck Lab. Hydraul. Water Res. Calif. Inst. Tech.* Rep. KH-R-21, 1970.

Smith, R., Long-term dispersion of contaminants in small estuaries, *J. Fluid Mech., 82*, 129-146, 1977.

Smith, R., Buoyancy effects upon lateral dispersion in open-channel flow, *J. Fluid Mech., 90*, 761-779, 1979.

Smith, R., Buoyancy effects upon longitudinal dispersion in wide well-mixed estuaries, *Phil. Trans. Roy. Soc. London A, 296*, 467-496, 1980.

Smith, R., When and where to put a discharge in an oscillatory flow, *J. Fluid Mech., 153*, 479-599, 1985.

Taylor, G. I., Dispersion of soluble matter in solvent flowing slowly through a tube, *Proc. Roy. Soc. London A, 219*, 186-203, 1953

Townsend, A. A., The diffusion of heat spots in isotropic turbulence, *Proc. Roy. Soc. London A, 209*, 418-430, 1951.

Uncles, R. J., and P. J. Radford, Seasonal and spring-neap tidal dependence of axial dispersion coefficients in the Severn - a wide, vertically mixed estuary, *J. Fluid Mech., 98*, 703-726, 1980.

Williams, D. J. A., and J. R. West, Salinity distribution in the Tay estuary, *Proc. Roy. Soc. Edinburgh B, 75*, 29-39, 1974.

23

The Onset and Effect of Intermittent Buoyancy Changes in a Partially Stratified Estuary

Roy E Lewis and Jonathan O Lewis

Abstract

Vertical profiling of an estuary has illustrated the variety of changes to shear and stratification which can occur on ebb and flood tides. In particular, a series of intermittent mixing events were observed on the ebb tide, culminating in an abrupt transition to a state of near homogeneity. The intrusion of the saline wedge on the flood tide damped out the interfacial mixing but, despite the high degree of stratification, a few vigorous mixing events were observed. Detailed analyses of the results suggest that the ebb tide mixing was due to bed generated turbulence assisted by Kelvin-Helmholtz billows on the pycnocline, this latter process being associated with generally low values for the bulk Richardson number. The rapid transition to a fully mixed state towards the end of the ebb was apparently triggered by the breakdown of the density interface by shear instabilities. The flood tide mixing was generally attributed to Kelvin-Helmholtz instabilities on the upper surface of the saline intrusion, but major upward movements of salt are more likely to have been associated with rotors or internal hydraulic jumps. A mathematical model was used to show that the potential energy change due to mixing on the ebb tide was associated with a maximum vertical mixing coefficient of $0.0020 \text{ m}^2 \text{ s}^{-1}$.

Introduction

Echo-sounder observations in the Tees estuary, which is a partially stratified system, have shown that appreciable changes in density structure can occur during a tidal period [Dyer and New, 1988; New et al. 1986]. These changes have been associated with the advection of mixed water though the estuary and with abrupt mixing events arising from local instabilities. Observations of velocity, salinity and temperature at anchored stations along the estuary axis have revealed how different salt flux components contribute to the salt balance [Lewis and Lewis, 1983], and how the internal shear stress varies during a tide [Lewis and Lewis, 1987]. However, these measurements have generally been taken at discrete one metre intervals which means that it has taken about 20 minutes to profile the full depth. As significant changes in density structure are produced by mixing events on this timescale, such observations are of limited value in determining the causes and effects of mixing.

Instrument packages have been specifically developed for investigating detailed changes in estuary structure. These have revealed highly time dependent mixing during the tidal period [Partch and Smith, 1978; Pietrzak et al., 1991], and indicated that shears resulting from the internal wave field make a significant contribution to the active mixing [Geyer and Smith, 1987]. Continuous records over a number of tidal cycles have been obtained using anchored instruments set at a number of depths through the water column. Results from such

Buoyancy Effects on Coastal and Estuarine Dynamics
Coastal and Estuarine Studies Volume 53, Pages 331-340
Copyright 1996 by the American Geophysical Union

observations have shown how mixing conditions change over the spring-neap tidal cycle and the contribution to mixing from the wind stirring [Sharples et al., 1994].

To gain more insight into mixing processes in the Tees estuary, a sampling device was developed which measured velocity and density over the full depth of the water column at a fixed position. Preliminary results obtained with this instrument revealed instances in which conditions rapidly changed from stratified to mixed at the measuring point. This paper describes these findings and determines the flow conditions at which these events occur, drawing comparisons with similar observations in other estuaries. To assess the relative significance of the rapid changes in density structure, the potential energy anomaly was computed throughout the tidal period for the two occasions on which measurements were made.

Field Investigation

The profiling device deployed in the Tees estuary consisted of an electromagnetic flowmeter (Valeport Series 800) and a CTD (Valeport Series 600 MK2) attached to a trolley. An aluminium beam was set vertically in the bed of the estuary alongside an anchored vessel, and the trolley was attached to this beam so that it could be lowered and raised at any required speed (Figure 1). In this way velocity profiles could be obtained over the full water depth without any significant disturbance by the beam or the vessel which logged the results. The accuracy of the e/m flowmeter was +/- 0.02 m s^{-1}, with a resolution of 10^{-3} m s^{-1}, and the accuracies of the salinity and temperature sensors were +/- 0.05 and +/- 0.1 °C respectively.

The profiler was deployed in Billingham Reach, which is a canalised section of the Tees estuary in north-east England, on 24 and 25 August 1993 (Figure 2). This is a straight section of the Tees in which considerable internal structure has been noted before [New et al., 1986). The array was set in the river bed to one side of the main channel where the depth was approximately 7 m at high water. The trolley was lowered down the beam at about 0.05 m s^{-1} so that, over the two second interval between successive recordings by the device, the probes had moved though a vertical distance of approximately 0.1 m. Profiles were obtained at 15minute intervals over the full 12.5 hour period of the semi-diurnal tidal cycle on two successive days. Details of the times of measurement, tidal ranges and freshwater flows are give in Table 1. The annual average freshwater flow of the River Tees is 20 m^3 s^{-1} and a typical dry weather flow is 3 m^3 s^{-1}; the flow of about 7 m^3 s^{-1} at the time of the survey would have been high enough to induce significant stratification in Billingham Reach.

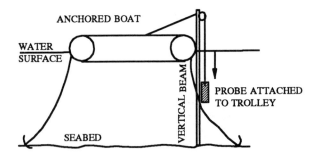

Figure 1. Diagram of profiling device

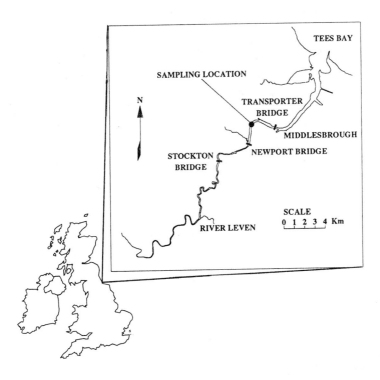

Figure 2. Location Tees estuary and sampling position.

Table 1 Tidal ranges and river flows during the deployment of the profiling system.

DATES (PERIOD OF OBSERVATIONS)		24/8/93 (08.00 - 20.30 BST)		25/8/93 (06.00 - 19.00 BST)	
		TIME (hrs mins)	HEIGHT (m)	TIME (hrs mins)	HEIGHT (m)
LW		02.59	1.4	04.00	1.7
HW		09.13	5.0	10.22	4.6
LW		15.40	1.4	16.45	1.8
HW		21.47	4.6	22.56	4.4
RIVER FLOW	(m^3/s)	6.9		6.5	

Results and Analysis

Velocity and Salinity Contours

The results obtained with the profiling device are exemplified by the velocity and salinity contour plots shown in Figure 3. These illustrate the variation in these parameters throughout the tidal period at the fixed position, and indicate the change in total water depth. In Figure 3

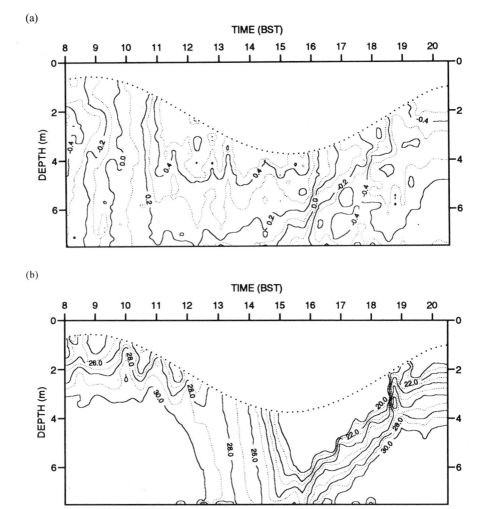

Figure 3. Tidal variation in velocity and salinity - Tees estuary (24/8/93)

(a), the closer the separation of the velocity contours over a selected depth interval, the stronger the current shear.

The density of the waters of the Tees estuary is dominated by salinity as temperature variations are relatively small. Thus the isohalines shown in Figure 3 (b) can be taken as indicative of the changes in density structure which occurred during the tidal cycle. This plot and that obtained on the following day displayed similar features at corresponding times in the tidal cycle; for example, stratification tended to break down during the second half of the ebb tide and then reform towards low water due to the saline intrusion. On both sampling days marked vertical incursions of saline water from the lower to the upper layer were recorded on the flood tide, as exemplified by the feature at 18.45 on 24/8/93.

Stability and Mixing Parameters

To investigate the possible causes of instability in the Tees, numbers which are believed to characterise the mixing were computed. Assuming two distinct layers of different density, bulk Richardson numbers Ri_B were computed in analogous way to the computation of internal Froude numbers by Geyer and Farmer [1989]. The value of Ri_B was computed from the expression

$$\frac{1}{Ri_B} = (\frac{1}{Ri_{B1}} + \frac{1}{Ri_{B2}})$$

where

$$Ri_{Bi} = \frac{g' h_i}{u_i^2}$$

u_1, u_2 and h_1, h_2 being the velocities and depths of the upper and lower layers, and g' is the reduced gravitational acceleration.

Figure 4 presents a plot of the computed values for the bulk Richardson number through the tidal period on 24/8/93; a logarithmic scale is used to allow for the range in values. The depth of the interface was taken as the point where the ratio of the measured density gradients above and below became greater than 1.5. Also included on the plot is the variation in the potential energy anomaly φ as computed from

$$\phi = \frac{g}{H} \int_0^H (\rho - \rho_m) z \, dz$$

where H is the total depth, ρ is the local density and ρ_m is the depth mean density [Simpson et al., 1990].

To define the local stability of the flow, gradient Richardson numbers Ri were computed from

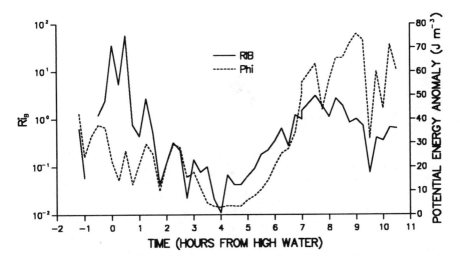

Figure 4. Tidal variation in bulk Richardson number and potential energy anomaly - Tees estuary (24/8/93).

$$Ri = \frac{g}{\rho} \frac{\partial\rho/\partial z}{(\partial u/\partial z)^2}$$

With the profiling technique employed, values for current, salinity and temperature were measured at 2 second intervals. The electromagnetic current meter filtered velocity fluctuations at higher frequencies than 0.8 seconds and hence the averaging time for any shear determination was approximately 3 seconds. Typical internal wave periods in the Tees estuary range from 44 to 126 secs [New et al., 1986], suggesting that shear measured with the profiler would have been influenced by internal waves. Geyer and Smith [1987] indicated that internal waves within the pycnocline could provide an additional component of shear to that arising from the mean flow and consequently increase the likelihood of instability in proportion to the intensity of the internal waves in the system.

A contour plot of gradient Richardson numbers for 24/8/93 is presented in Figure 5. As local shears and stratification are likely to change considerably over the 15 minute profiling interval, contouring can only be justified in that it gives an overall impression of the varying field of stability. The plot reveals the relatively patchy nature of the Ri distribution, with darker regions indicating high stability or low shear. It can be seen that the lighter regions, corresponding to low stability or high shear, occur in the near bed waters throughout the tide and over much of the water column at the end of the ebb tide.

Mixing Events and Conditions for their Onset

Before attempting to understand the causes of the observed changes in density structure, it is preferable to select certain features which may be indicative of mixing events. In examining data from a single fixed position there is always an element of doubt as to whether the structural changes have occurred locally or whether they represent the simple advection of water past the measuring point from a remote location where a mixing event has happened. Advection has been suggested to explain mixed zones in the Tees [New et al., 1987]. However, as can be seen from Figure 3 (a), simple advection is unlikely on the ebb because shearing flow is nearly always present, and would tend to stratify even an initially well mixed

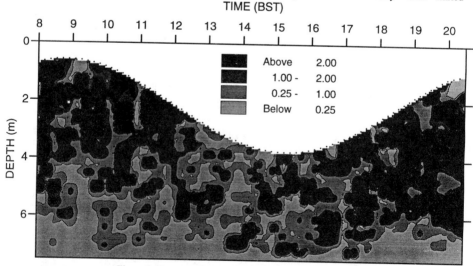

Figure 5 Contours of gradient Richardson number (24/8/93)

Table 2 Bulk Richardson numbers prior to mixing events

DATE		24/8/93			25/8/93	
	Time BST (hrs mins)	Time from HW or LW (hrs mins)	Ri_B	Time BST (hrs mins)	Time from HW or LW (hrs mins)	Ri_B
EBB	10.30	HW+1.15	1.2	12.00	HW+1.45	1.1
	11.30	HW+2.15	0.2			
	12.30	HW+3.15	0.1			
FLOOD	18.30	LW+2.45	0.6	06.45	LW+2.45	0.9
				08.00	LW+4.00	0.8
				09.30	LW+5.30	0.5

patch as it is carried downstream. Furthermore, it was observed that the surface salinities tended to increase during the ebb (Figure 3 (b)), implying that mixing was occurring locally and overcoming the tendency for advection to increase the degree of stratification.

To allow for the finite time it takes for mixing to occur, it is reasonable to presume that critical conditions for the onset of mixing occur just before a change in density structure is observed. So for example, the increase in surface salinities at 11.00 on 24/8/94 is liable to have been triggered by the flow conditions at 10.30 (Figure 3 (b)). On this basis, the times of commencement of ebb and flood tide events were tabulated and compared with the corresponding bulk Richardson numbers shown in Figure 4. As shown in Figure 4, the Ri_B values vary appreciably and the numbers given in the Table 2 are averages of three Ri_B values 15 minutes before, at and 15 minutes after the estimated time at which flow conditions became critical.

It can be seen from Table 2 that the listed Ri_B values were in the range between 0.1 to 1.2. A value of 1.0 for the bulk Richardson number is the boundary between subcritical and supercritical flow for a two-layer system.

Interpretation of Results

Mixing Mechanisms During a Tide

To date mixing in stratified or partially stratified estuaries has been ascribed to three principal causes - bed generated turbulence, interfacial turbulence due to breaking or arrested internal waves, and interfacial turbulence produced by shear instabilities. Bed generated turbulence is typified by erosion of the pycnocline from below and is usually the dominant factor in mixing in shallow estuaries. A prerequisite for wave generated mixing is that interfacial waves with sufficient energy be present in the system, and this in turn depends on a the degree of stratification and the form of the estuary topography. Shear instabilities of the Kelvin-Helmholtz type are most often considered important and, as the name implies, they require strong shear across a density interface; such shears are believed to be assisted by internal wave shear [Geyer and Smith, 1987].

The flow can be characterised by bulk and gradient Richardson numbers so that together with direct and echo-sounder observations of the way the pycnocline alters, it is possible to infer which mechanisms might be responsible for the changes in salinity distribution illustrated in Figure 3 (b).

The mixing event at about 10.30 was triggered when Ri_B was 1.2 and the minimum Ri value of about 1.0 was measured at 1.1 m, which lay within the pycnocline. The flow conditions favour the formation of Kelvin-Helmholtz instabilities which could produce an overturning which carries saline water up to the surface. The occurrence of K-H instabilities depends on there being sufficient shear across the density interface. As the flow was still just subcritical, the influence of bed drag on interfacial shear may have been limited [Geyer and

Farmer, 1989], but internal waves could have provided additional shear. Evidence from earlier surveys in the Tees suggests that the flow of water over the irregular bottom topography on the flood tide results in the formation of internal waves at the pycnocline [New and Dyer, 1988]. These waves travel downstream during the early ebb, becoming dissipated after about 1 hour.

The next event at 11.30 occurred when Ri_B was about 0.2, implying that the flow conditions were supercritical. Interestingly, intense mixing on the ebb tide has been observed in Billingham Reach when the internal Froude number was 2.1, which corresponds to a bulk Richardson number of about 0.2 [New et al., 1987]. Gradient Richardson numbers down to a depth of about 4 m were above 3.0 at this time, but reduced to less than 1.0 just below this depth. The strengthened shear across the density interface in the supercritical flow would promote K-H shear instabilities, even in the absence of shears due to internal waves.

The 11.30 mixing event appeared to be complete by about 12.15 as the depth of the pycnocline started to increase. However, by 12.30 the isohalines in the near surface waters had moved upwards and the deeper isohalines became steeply directed downwards to the bed, apparently due to the vertical transport of salt and the seaward advection of the salt wedge. At 12.30 Ri_B was approximately 0.1, meaning that the flow was distinctly supercritical and shear across any density interface would be greatly enhanced by bottom friction. This appears to have permitted shear instabilities to continue, thus destroying the stability of the upper layer and allowing bed generated turbulence to complete the process of homogenisation. From 12.30 until the start of the flood tide the gradient Richardson numbers were generally about 0.25, suggesting that a state of marginal stability had been reached.

The importance of current speed was emphasised in an alternative explanation of similar vigorous mixing events observed in the Duwamish estuary [Partch and Smith, 1978]. They proposed that the mixing resulted from a 'blocking flow' in which internal waves were unable to travel upstream against the ebb current [Turner, 1973;p 66]. Thus, under a weak flow faster moving and more energetic internal waves could travel upstream against the flow, and only the slower waves would produce a hydraulic jump, leading to relatively weak turbulent mixing. By contrast, when the current is stronger, hydraulic jumps would be associated with faster moving internal waves and the turbulence generated would be more energetic. This explanation could also account for the mixing event triggered at 11.30 developing into a complete homogenisation of the water column. However, observations in the Tees show that vigorous mixing at about mid ebb tide occurs over a considerable length of the estuary and it appears unlikely that internal waves would always be present over such a distance. Furthermore, the energy in the waves has been shown to be insufficient to account for the degree of mixing observed [New et al., 1987].

Very different processes occur on the flooding tide. Turbulence appears to be greatly reduced by low water and, as observed in laboratory experiments [Linden and Simpson, 1988], a distinct gravity flow develops as a result of frontogenesis. The resulting landward intrusion of saline water near the bottom results in a very marked pycnocline. The low Ri values early on the flood tide imply that Kelvin-Helmholtz instabilities could have formed on the saline intrusion, leading to small prominences on the density interface as have been observed in laboratory experiments [Simpson, 1987].

Nearly 3 hours into the flood tide, a large upward movement of salt occurred over a distance of at least 1 km. The incursion of saline water into the brackish layer had a typical timescale of about 25 minutes, and was only detected by two successive profiles at the 15 minute sampling frequency. This gave an estimated length scale for the incursion of about 1.0 km. The Billingham Reach is only about 2.0 km long so the upward transfer of salt by these features was happening over about half the reach length. It is also interesting to note from Table 2 that the two principal incursions of salt from the lower to the upper layer during the survey both occurred at 2 hours and 45 minutes after predicted low water. The incursions appeared to resemble the larger billows which have been observed to form on the backs of saline intrusions in laboratory investigations [Simpson, 1987]. As the flow had become interfacially supercritical, these events could have had their origin in an internal hydraulic jumps but more evidence is needed to substantiate such an argument.

Effect of Mixing Processes

The recent observations support the mechanisms outlined for mixing in the estuary of the Fraser River, which is a much larger system than the Tees [Geyer and Farmer, 1989]. However, in the shallower waters of the Tees, the effect of mixing on the salinity structure is even more marked. Particular interest centres on the vigorous mixing on the ebb tide which is strong enough to overcome the tendency for the shear of fresher water over the lower saline layer to induce stratification. This mixing on the ebbing tide has been investigated using a 2-D, laterally averaged, model of the Tees estuary.

The model simulated the changing density structure along the full length of the estuary, some 44 km, from the tidal limit to the sea. The equations of motion and continuity were solved using a finite difference technique at each mesh point on a grid with 90 horizontal points and 13 vertical points. Allowance was made for the effect of stability on the vertical transfer of momentum and mass using a damping expression of the form suggested by Munk and Anderson [1948]. The model was calibrated for conditions in the Tees estuary by comparing predictions with velocity and salinity data obtained at locations along the seaward half of the system [Lewis and Lewis, 1983].

The model was run with weak and strong ebb tide mixing conditions to predict the tidal variation in salinity at the position employed in the profiling studies. Figure 6 presents the values for the potential energy anomaly φ derived from these results together with the values of φ from the observed data. It can be seen that with weak mixing the estuary would be expected to be much more stratified than was actually found. However, by increasing the base value (i.e. prior to allowance for damping) for the vertical mixing coefficient on the ebb by a factor of 5.0, corresponding to a value of 0.0020 m^2 s^{-1}, a much better match with observations was obtained. The effect on φ of the flood mixing associated with the relatively short-lived incursions of salt into brackish layer were not so well defined by this model

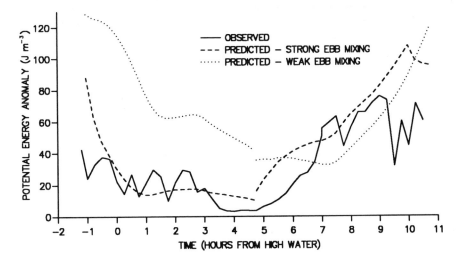

Figure 6.   Observed and predicted values for the potential energy anomaly φ..

The overall effect of the intrusion of saline water on the flood tide and strong mixing on the ebb is to bring salt into the Tees estuary to counter the loss to sea by the net river flow [Lewis and Lewis, 1983]. Although the results described in this paper suggest that mixing is due to a number of different mechanisms, the energy changes associated with these processes are part of an overall energy budget which ensures that a salt balance is maintained.

References

Geyer, W.R., and J.D. Smith, Shear instability in a highly stratified estuary, *J Phys Ocean, 17*, 10, 1668-1679, 1987.

Geyer, W.R., and D.M. Farmer, Tide-induced variation of the dynamics of a salt wedge estuary, *J Phys Ocean, 19*, 8, 1060-1072, 1989.

Lewis, R.E., and J.O. Lewis, The principal factors contributing to the flux of salt in a narrow, partially stratified estuary, *Estuarine, Coastal and Shelf Science, 16*, 599-626, 1983.

Lewis, R.E., and J.O. Lewis, Shear stress variations in an estuary, *Estuarine, Coastal and Shelf Science, 25*, 621-635, 1987.

Linden, P.F., and J.E. Simpson, Modulated mixing and frontogenesis in shallow seas and estuaries., *Continental Shelf Research, 8*, 10, 1107-1127, 1988.

Munk, W., and E.R. Anderson, Notes on a theory of the thermocline, *J. Marine Research, 7*, 276-295, 1948.

New, A.L., K.R. Dyer, and R.E. Lewis, Predictions of the generation and propagation of internal waves and mixing in a partially stratified estuary, *Estuarine, Coastal and Shelf Science, 22*, 199-214, 1986.

New, A.L., K.R. Dyer, and R.E. Lewis, Internal waves and intense mixing periods in a partially stratified estuary, *Estuarine, Coastal and Shelf Science, 24*, 15-33, 1987.

New, A.L., and K.R. Dyer, Internal waves and mixing in stratified estuarine flows, in *Physical Processes in Estuaries,* edited by J. Dronkers and W. van Leussen, Springer-Verlag, Heidelberg, 1988.

Partch, E.N., and J.D. Smith, Time dependent mixing in a salt wedge estuary, *Estuarine, Coastal and Shelf Science, 6*, 3-19, 1978.

Pietrzak, J.D., C. Kranenburg, G. Abraham, B. Kranenborg, and A. van der Wekken, Internal wave activity in the Rotterdam Waterway, *J Hydraul Eng, 117*, 6, 738-757, 1991.

Sharples, J., J.H. Simpson, and J.M. Brubaker, Observations and modelling of periodic stratification in the Upper York River estuary, Virginia, *Estuarine, Coastal and Shelf Science, 38*, 301-312, 1994.

Simpson, J.E., Gravity Currents: In the Environment and the Laboratory, Ellis Horwood, Chichester, 1987.

Simpson, J.H., J. Brown, J. Matthews, and G. Allen, Tidal straining, density currents and stirring in the control of estuarine stratification, *Estuaries, 13*, 125-132, 1990.

Turner, J.S., Buoyancy Effects in Fluids, Cambridge University Press, Cambridge, 1973.

24

The Influence of Buoyancy on Transverse Circulation and on Estuarine Dynamics

Job Dronkers

Abstract

Transverse circulation plays an important role in the dynamics of mixing and sedimentation in partially and well mixed coastal plain estuaries [Dyer, 1989]. Buoyancy induced reduction of vertical momentum exchange causes a substantial increase of transverse secondary currents. Measurements of the current structure in the Volkerak estuary reveal transverse secondary currents in the bends of the meandering main channel which are significantly larger than one might expect in homogeneous flow conditions. These transverse currents increase longitudinal dispersion by enhancing exchange of water masses over the tidal shear layer at the channel boundaries. In the landward range of the Volkerak estuary strong lateral density gradients build up during ebb. As a consequence the topographically induced transverse circulation is reversed at the inner channel bend by a density current entering the channel from the tidal flat. This affects the channel-flat morphology; the characteristic asymmetry of the inner and outer channel bend slopes is absent. The corresponding down-slope transport of sediment counteracts tidal flat sedimentation. When comparing the topographical characteristics of tidal basins in the Wadden Sea, which are subject to identical tidal forcing but to different fresh water inflow, it is found that the tidal flats are highest in the basin with the weakest buoyancy influence.

Introduction

Many large estuarine systems possess a complex morphological structure which is generated by the interaction of tidal currents with topography. This is the case, in particular, for coastal plain estuaries, which are characterized by one or several flood deltas with numerous shoals and tidal flats, and a branched, winding channel system. Examples are Chesapeake Bay, San Francisco Bay, Wadden Sea, Baie d'Arcachon, Thames, Severn and many others. These estuaries receive a fresh water input which may be considerable, but which is yet much less than the tidal discharge through the estuarine inlet. The major part of the estuary can be described as "well mixed": the salinity difference over the vertical is on the same order of magnitude or smaller than the salinity difference over the channel width, and there is no sharp interface between fluid layers.

Still buoyancy effects are present and play a significant role in the hydrodynamic behaviour of these estuaries, as will be discussed in this paper. In the sixties it was recognized, in particular owing to the work of Pritchard [1967], that mixing processes prevent the establishment of a stationary salt wedge and maintain a density gradient along the channel axis. This lolngitudinal density gradient induces a depth dependent pressure gradient

Buoyancy Effects on Coastal and Estuarine Dynamics
Coastal and Estuarine Studies Volume 53, Pages 341-356

which cannot be balanced by landward water level set up and therefore drives a circulation with landward velocity in the lower part of the vertical and seaward velocity in the upper part. This so-called estuarine circulation may possess a 3-dimensional character as a result of density differences and frictional differences between the deeper channel and the shallower channel banks [Fischer, 1972; Smith, 1980]. In estuaries without pronounced topographical structures, such as flood deltas, estuarine circulation constitutes the major secondary current pattern. It strongly influences salt intrusion, dispersion and flushing of contaminants and the residual transport of suspended sediments.

In the well-mixed estuaries described earlier the situation is different. Secondary currents are due not only to horizontal density gradients, but also to the interaction of tidal currents with topography. The larger scale of estuarine circulation with respect to topographical eddies would enhance its role in large scale dispersion and residual transport, but this is counteracted by tidal mixing. As shown by Fischer [1972], if the time scale for vertical mixing is short with respect to the tidal period then the longitudinal dispersion resulting from the vertical estuarine circulation is strongly reduced. This is confirmed by measurements in well mixed estuaries, by decomposition of the residual salt transport into different components [Fischer et al., 1979; Dronkers and van de Kreeke, 1986].

Neglecting buoyancy effects in well mixed estuaries is too strong a simplification, however. The reason is that buoyancy affects the hydrodynamics in still another way, which is the damping of turbulent water motions. This effect has been studied since a long time already. It means that the diffusion of momentum through the water column is reduced. The damping function, which is the ratio of vertical momentum diffusion coefficients ε and $\varepsilon_{neutral}$ for buoyant and neutral fluids, is a decreasing function of the local Richardson number

$$Ri = -\frac{g}{\rho}\frac{\partial r/\partial z}{(\partial u/\partial z)^2}$$

Ri represents the ratio of buoyant potential energy and kinetic energy potentially available for mixing; for well mixed fluids Ri < 1. From recent experiments [Uittenbogaard, 1993] it appears that the damping of turbulent diffusion with increasing Ri-number is stronger than indicated by the well known Munk-Anderson relation, see Figure 1. As a consequence, there is less cohesion between the water motions in the upper and lower parts of the vertical in a well mixed buoyant flow than in a similar neutral flow; small scale dissipative fluid motions are suppressed, while large scale organized fluid motions are enhanced. This is

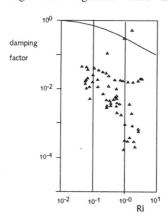

Figure 1. Munk-Anderson damping function of turbulent vertical diffusion of momentum by buoyancy (solid line) and the ratio of the diffusion coefficients for buoyant and neutral flows obtained from recent flume experiments [Uittenbogaard, 1993].

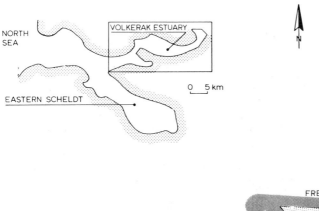

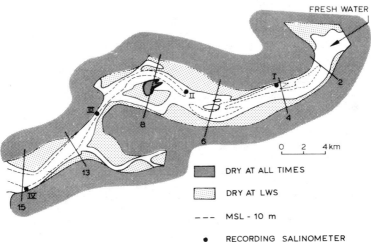

Figure 2. Bathymetry of the Volkerak estuary and location of measuring sections.

illustrated by the occurrence of large internal wave motions in well mixed estuaries [Uittenbogaard, 1995; Dyer, 1982]; such internal waves have been observed also in the Volkerak estuary.

In this paper the secondary currents measured in the Volkerak estuary (part of the Rhine-Meuse-Scheldt delta) will be discussed. Particular attention will be given to secondary currents resulting from the interaction of tidal currents with topography. It will be shown that these secondary currents are strongly enhanced by the influence of buoyancy, due to the damping of turbulent diffusion of momentum and the reduction of internal friction. This phenomenon has been described earlier by Dyer [1989] and by Geyer and Signell [1993] in comparable tidal environments. These secondary currents contribute to the transverse mixing in the estuary. Transverse mixing, in turn, is important for the longitudinal dispersion by lateral shear. In this indirect way buoyancy affects longitudinal dispersion; in well mixed flood plain estuaries this mechanism can be more important than estuarine circulation. Secondary currents also affect the estuarine morphology, in particular, the shape of channel cross-sections and the accretion of tidal flats. Some field evidence is presented on morphological characteristics which may be attributed to the influence of buoyancy effects on secondary currents.

The field campaign in the Volkerak estuary

The analysis reported in this paper mainly relies on field data collected during an intensive measuring campaign in the Volkerak estuary in 1977. Several studies on estuarine transport processes based on this field campaign have been published earlier [Dronkers and Van de Kreeke, 1986; Van de Kreeke and Robaczewska, 1989].

The Volkerak estuary is the northern branch of the Eastern Scheldt basin. It is a well mixed estuary with winding channels and large tidal flats bordered by sea dikes. Characteristic length, width and depth dimensions are respectively 35,000 m, 2000 m, and 10 m. At the head the estuary is closed by a dam with locks and sluices. A fresh water inflow through the locks and sluices is kept constant at 50 m³/s. This fresh water inflow is very small compared to the maximum tidal discharge through the mouth of the estuary of about 30,000 m³/s. The tidal range is between 3 and 4 m with hardly any neap-spring variation. The bathymetry is shown in Figure 2, where also the location of the measuring sections is indicated. In each of these sections (which are numbered 2, 4, 6, 8, 13 and 15) the salinity and current velocity have been measured every 15 minutes during two tidal cycles, in 5 or 6 verticals at 4 to 7 depths.

In Figure 3 the salinity distribution in a longitudinal section of the estuary is shown at high and low water slack (HWS and LWS). Salinity gradients are strongest at LWS, but no distinct salt layer is formed. At the estuarine mouth (section 15) vertical salinity gradients are nearly absent. Figure 4 displays the surface salinity distribution at HWS and LWS. In these figures small fronts which are recorded by continuous sampling are smoothed out. At high water, and during most of the flood period, there is no strong lateral salinity gradient. This contrasts with the ebb period, during which a strong salinity gradient is built up between the main channel and the tidal flats, especially in the landward part of the estuary.

The tidally averaged vertical profiles of salinity (C_V) and current velocity (U_V) along the channel axis are showin in Figure 5 for all mid-channel measuring stations. Throughout the estuary an estuarine circulation is present. In the upper ranges of the estuary the average bottom-surface salinity difference is in the order of 2 ppt; in the lower range the water column is almost completely mixed. Figure 6 shows the tidally averaged transverse profile of salinity (C_H) and current velocity (U_H). The transverse distribution of the residual current is mainly caused by tide-topography interactin and can be explained by residual vorticity built up due

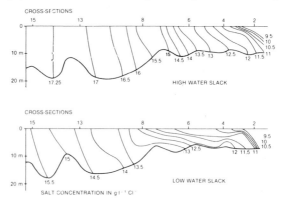

Figure 3. Chlorosity distributions (g/l Cl) in a section along the channel axis at high and low slack water. Salinity ≈ 1.8 * Chlorosity.

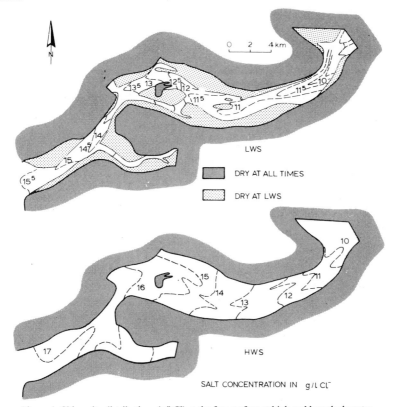

Figure 4. Chlorosity distributions (g/l Cl) at the free surface at high and low slack water.

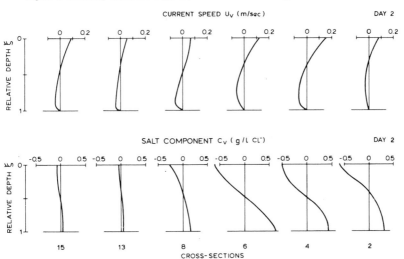

Figure 5. Vertical profiles of the tide-averaged velocity (m/s) and chlorosity (g/l Cl) deviations in the middle of the channel at all measuring sections.

to channel curvature and bottom friction (similar to headland circulation). As a result distinct flood and ebb channels are formed along channel meanders.

The strength of horizontal circulation is approximately the same throughout the estuary, but only in the upper range a distinct correlation with the transverse salinity distribution appears.

The analysis of salt intrusion mechanisms in the Volkerak estuary by Dronkers and Van de Kreeke [1986] yielded the following results:

- vertical gravitational circulation accounts for approximately half of the upstream salt transport in the landward part of the estuary (sections 2,4,6); in the seaward part (sections 13,15) the contribution is insignificant.

- the contribution of horizontal circulation is manifest in the central part of the estuary; elsewhere the contribution of horizontal exchange processes associated with residual eddies or oscillating velocity shear is lumped into the tidal pumping transport term. This is due to a strong longitudinal variation of horizontal exchange processes; correlation between horizontal fluctuations in velocity and salinity therefore tend to disappear when considering a fixed reference frame instead of a frame moving with the average tidal flow. The tidal pumping transport term is particularly important in the seaward part of the estuary (sections 13,15) and provides most of the upstream salt transport required to maintain the salt balance in the estuary.

The measuring sections 6, 8 and 15 coincide with bends in the main tidal channel. Cross section 15 displays the characteristic shape of a steep outer bend and a shallow inner bend,

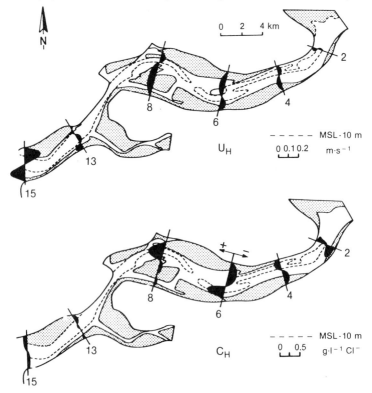

Figure 6. Transverse profiles of the depth- and tide-averaged velocity (m/s) and chlorosity (g/l Cl) deviations.

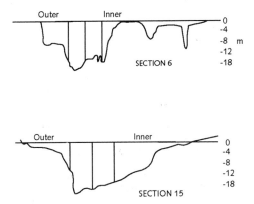

Figure 7. Depth profiles of sections 6 and 15, situated at channel bends.

see Figure 7. The inverse seems to hold for the upstream cross-section 6; here the inner bend is steeper than the outer bend. The situation is complicated by the presence of small parallel channels cutting through the tidal flat at the inner bend. The transverse circulation for each section is shown in Figures 8 and 9 at maximum flood and maximum ebb flow based on current velocities measured in three verticals of each section. The vertical profiles of the streamwise current velocity are shown in the same Figure. Wind effects can be neglected, as the measurement took place during very calm weather conditions. On the other hand an aliasing effect of internal waves may be expected, as the sampling frequency is lower than the internal wave frequency; this effect probably does not exceed a few cm/s.

Figures 8 and 9 show that at maximum ebb a strong vertical variation is present both in the streamwise and the transverse current profiles. The latter will be discussed more in detail in the next section. Measuring section 8 is left out of consideration. The reason is a strong local disturbance of the morphology due to the construction of a sluice complex on the tidal flat.

Transverse circulation in channel bends

In this section we will discuss the interaction of the tidal current with topography which drives the transverse circulation in channel bends. The channel bend topography imposes a torque on the main tidal flow to which the flow adapts gradually when following the channel bend [Kalkwijk & De Booij, 1986]. It is assumed that the channel bend is sufficiently smooth such that locally a radius of curvature R can be defined for the depth averaged tidal flow. The transverse (radial) component of the momentum equation then reads

$$\frac{\partial u_n}{\partial t} - \frac{u_s^2}{R} + f u_s + g\frac{\partial \eta}{\partial n} + \frac{\partial p_\rho}{\partial n} + \frac{\partial}{\partial z}\frac{\tau_n}{\rho} = 0 \qquad (1)$$

Here n = transverse coordinate
    u_n = transverse velocity, u_s = streamwise velocity,
    h = total water depth, η = water level,

$$\frac{\partial p_\rho}{\partial n} = \frac{g}{\rho}\int_z^h \frac{\partial r}{\partial n}dz' \quad = \text{buoyancy induced transverse pressure gradient.}$$

We will consider only tidal stages at which the time derivative $\partial u_n/\partial t$ can be neglected. It is assumed that the shear stress τ_n is due to vertical diffusion of momentum and that it can be parametrized by means of a momentum diffusion coefficient $\varepsilon(z,t)$. The lateral surface inclination can be eliminated from Eq.1 by integration over the depth. Then the following expression for the transverse velocity component u_n results:

$$u_n = \int_0^z \frac{dz'}{\varepsilon(z')} \int_{z'}^h dz'' \left[\underbrace{\frac{u_s^2 - \overline{u_s^2}}{R}}_{A} - \underbrace{f\left(u_s - \overline{u_s}\right)}_{B} - \underbrace{\frac{\partial p_\rho}{\partial n} + \frac{\overline{\partial p_\rho}}{\partial n}}_{C} - \frac{\tau_n}{\rho h} \right] \tag{2}$$

The driving forces for the transverse current are the imbalance between local and depth averaged centrifugal acceleration, (A), the imbalance between local and depth averaged Coriolis acceleration, (B) and the imbalance between local and depth averaged transverse baroclinic pressure gradient (C).

In mid-latitude regions ($f \approx 10^{-4}$) the effect of earth rotation can be neglected with respect to the centrifugal term if the radius of curvature of the channel bend is a few kilometers or less. In the Volkerak estuary most channel meanders have sufficient curvature to justify the neglect of Coriolis, in a first approximation. The transverse flow due to the centrifugal forces will induce an inclination of the pycnoclines such that the baroclinic pressure gradient counteracts the local imbalance in the centrifugal acceleration. If stratification is sufficiently strong an equilibrium between the different forces may establish, and transverse circulation will be

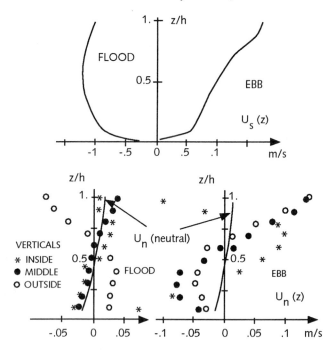

Figure 8. Section 6. Vertical profiles of the streamwise velocity at maximum ebb and maximum flood current at the centre of the channel. Vertical profiles of the transverse velocity in different verticals at maximum ebb and maximum flood current, compared to the theoretical profile in homogeneous channel bend flow.

weak [Chant et al., 1996]. In well mixed estuaries like the Volkerak this is not the case: tilting of the pycnoclines would yield a transverse pressure gradient which is almost an order of magnitude smaller than the centrifugal term. The transverse density gradient is mainly due to transverse shear in the streamwise tidal current.

During ebb the vertical imbalance in centrifugal acceleration is quite large, see Figures 8 and 9. From Figure 6 the transverse baroclinic pressure gradient can be estimated; this pressure gradient is of minor importance in Eqs.1 and 2, except near the channel boundaries in the upper part of the estuary. At the tidal stage of maximum ebb the time derivative $\partial u_n/\partial t$ is an order of magnitude smaller and can be neglected too.

During flood the vertical imbalance in centrifugal acceleration is small. The transverse baroclinic pressure gradient in Eq.2 can be of the same order or even larger. The neglect of the time derivative $\partial u_n/\partial t$ at maximum flood is not entirely justified, but has no consequences for the following qualitative discussion.

Eq.2 demonstrates the influence of turbulence damping on the lateral velocity, explicitly by the presence of $1/\varepsilon$ in the integral and implicitly by the dependence of the streamwise velocity shear (see Eq. 2 [A]) on ε. The theoretical transverse current velocity distribution which occurs in homogeneous channel bend flow [Rozovskii, 1957] has been indicated as a reference in the Figures 8 and 9. In the homogeneous case the streamwise velocity profile is assumed to follow a logarithmic law and the bottom drag coefficient is taken equal to 3.10^{-3}. In the field situation the streamwise velocity u_s differs strongly from a logarithmic profile. This is due not only to the damping of turbulence, but especially to the longitudinal baroclinic pressure gradient. During ebb the vertical shear is increased while during flood it is decreased. This effect is most pronounced in the upper reach of the estuary. Following Eq.2 a strong transverse circulation would be expected during ebb and a much weaker circulation during flood. For the ebb flow this agrees with the measurements. During flood, however, the transverse circulation is also stronger than in homogeneous flow. This shows that damping of turbulence must be stronger during flood than during ebb, even if vertical stratification is less. This can be explained by the absence of strong velocity shear and turbulence production, leading to larger Richardson numbers for flood (order 0.2 to 2) than for ebb (order 0.05 to 0.2).

In section 6, the transverse circulation is reversed at the outer bend during flood and at the inner bend during ebb, see Figure 10. This reversal should most probably be attributed to buoyancy effects. During flood the streamwise velocity is constant or even slightly decreased in the upper part of the vertical. Transverse currents are mainly driven by the transverse baroclinic pressure gradient. The salinity at the center of the main channel is higher than at the channel slopes. The result is a an axially convergent circulation. Such a circulation has been observed also in other well mixed estuaries [Nunes and Simpson, 1985]. During ebb a strong salinity gradient builds up between the surface flow in the channel and the tidal flat at the inner bend, see Figure 4. This higher density of the water stored on the tidal flat produces a density current down the slope of the tidal flat into the channel. This transverse density current reverses in the upper part of the vertical the transverse circulation related to channel bend topography, as can been seen in Figure 8. A similar observation has been reported by Huzzey [1988] in the York River.

In section 15 the transverse channel bend circulation is not strongly altered by the transverse baroclinic pressure gradient, see Figure 11. Also the cross-section exhibits the characteristic shape of channel bend flow.

Whatever be the cause of the transverse circulation, the strength of the circulation is remarkable. It illustrates the important influence of turbulence damping on the vertical flow structure. The magnitude of the transverse circulation is such that transverse mixing is strongly enhanced. Water masses experience an important cross channel displacement during one tidal cycle, especially if vertical mixing is not very fast. In the next section the effect on longitudinal dispersion is discussed.   Arrows correspond to the measured transverse velocities of Figure 9.

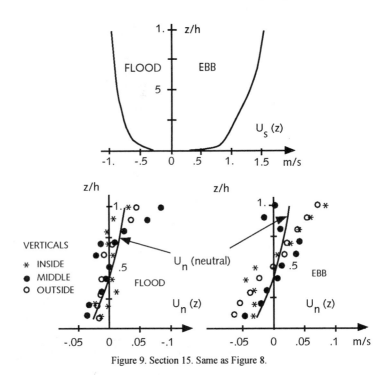

Figure 9. Section 15. Same as Figure 8.

Implications for longitudinal dispersion

In the literature many mechanisms are proposed which may contribute to the mixing and flushing of estuarine waters. These mechanisms have been categorized by Fischer et al. [1979], who distinguished between processes acting more or less uniformly along the estuarine axis, and processes acting locally, but whose influence is advected along with the tidal current. This latter category is designated as "tidal pumping". This distinction is artificial in a physical sense, because the processes involved are not fundamentally different; the distinction is useful, however, for the present discussion. The processes acting more or less uniformly along the estuary are divided in two categories: large scale circulation in vertical or horizontal planes along the estuarine axis and oscillatory shear in vertical or lateral directions. Here again the distinction is somewhat artificial because of the interaction which exists between these processes.

Large scale circulation

The contribution of large scale circulation to longitudinal dispersion is important if cross sectional mixing is slow. The time scale for vertical mixing is in general shorter than for lateral mixing; the dispersive effect of large scale horizontal circulation should therefore be stronger than for vertical circulation. It appears that this is not true for the landward part of the Volkerak estuary [Dronkers and Van de Kreeke, 1986]. The reason lays in the strong spatial variation of horizontal circulation related to channel meandering; the net displacements due to these residual circulation do not simply add up but should be described by random walk [Zimmerman, 1986] or by chaotic diffusion processes [Ridderinkhof and Zimmerman, 1992].

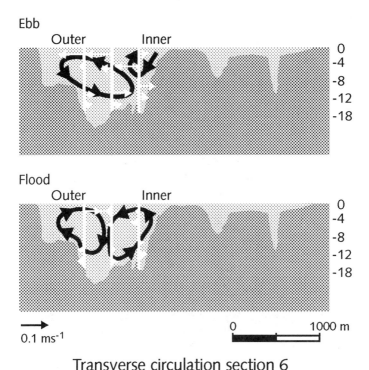

Transverse circulation section 6

Figure 10. Transverse circulation at maximum flood and maximum ebb in section 6. The white arrows correspond to the measured transverse velocities of Figure 8.

Oscillatory shear

The contribution of oscillatory shear is most important if the time scale of cross sectional mixing is in the order of the tidal period [Fischer et al., 1979]. The time scale for vertical mixing is generally much shorter than a tidal period. As a consequence, vertical oscillatory shear does not strongly contribute to longitudinal dispersion, although the interaction with gravitational circulation cannot be ignored. In contrast with mixing over the depth, mixing of water masses over the width of the estuary in general takes many tidal periods; for this reason transverse oscillatory shear is sometimes not considered as a major mechanism for longitudinal dispersion. However, in most coastal plain estuaries, like the Volkerak, longitudinal dispersion depends more on transverse mixing over the width of the shear layer at the channel boundaries than on mixing over the entire width of the estuary. The pertinent transverse mixing time therefore is much shorter. In the Volkerak estuary 20 to 30 % of the water is stored on tidal flats and in retention zones along shoreline irregularities. These water masses lag behind the flow in the channel, both during ebb and flood. The typical width of the boundary shear layer is in the order of several hundred meters and the magnitude of transverse circulation is in the order of 0.05 m/s. The time scale for exchange of water masses over the boundary shear layer is therefore of the same order as the ebb or flood duration. Transverse circulation in a partially or well mixed meandering estuary thus provides optimum efficacy for longitudinal dispersion by oscillatory shear. The importance of this dispersion process has already been recognized in literature [Okubo, 1973; Dronkers and Zimmerman, 1982; Geyer and Signell, 1989].

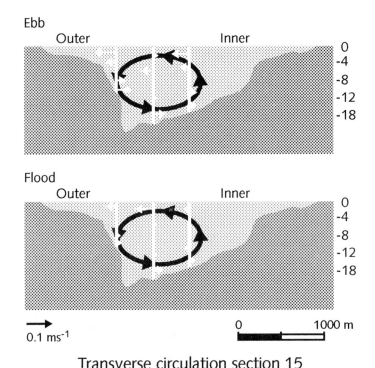

Transverse circulation section 15

Figure 11. Transverse circulation at maximum flood and maximum ebb in section 15. The white arrows correspond to the measured transverse velocities of Figure 8.

Strengthening of transverse circulation by buoyancy effects may considerably increase longitudinal dispersion, especially in coastal plain estuaries. This effect may provide an explanation for the observed increase of longitudinal dispersion in well mixed estuaries when fresh water inflow is increased [Helder and Ruardij, 1983].

Tidal flats or flow separation zones often coincide with channel bends. If the spatial scales of retention zones and secondary current structures are smaller than the tidal excursion no strong correlation between tidal velocity and salinity anomalies may show up in a fixed cross-section of the estuary. In the Volkerak estuary most of the dispersive transport is lumped in the tidal pumping term, as shown by Dronkers and van de Kreeke [1986].

Morphological implications

The transverse circulation in a meandering channel system is generated as a response of the tidal flow to topography. However, the topography in turn is generated as a response of the estuarine bed to the structure of the tidal flow. The meandering of the channel develops through scouring of the outer bend by the tidal current and subsequent transport of the scoured material to the inner bend by secondary currents. Secondary currents, such as headland eddies and transverse circulation, are both result and cause of channel meandering. Alteration of transverse circulation by buoyancy effects thus influences the morphodynamic system and thereby its morphological characteristics.

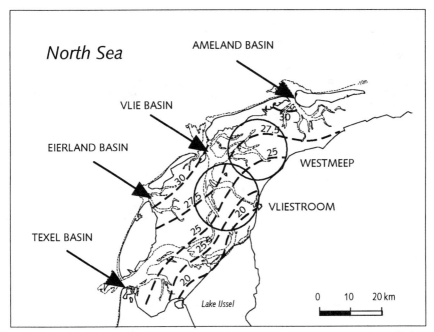

Figure 12. Map of the western Wadden Sea with tidal basins of Texel, Eierland, Vlie and Ameland, and average winter salinity distribution.

In the Volkerak estuary the relationship between transverse circulation and channel morphology is manifest. In channel sections where the direction of the transverse circulation is not altered by buoyancy effects (for example, section 15) the cross-section displays the characteristic asymmetric shape of a steep sloping outer channel bend and a gently sloping inner channel bend. In channel sections were the transverse circulation is reversed (for example, section 6) this asymmetry is absent. In this case a strong density current enters the channel from the tidal flat situated at the inner channel bend. This density current causes scouring of the inner channel bend and a downward transport of sediment.

In the normal situation transverse channel bend circulation stimulates the accretion of tidal flats. By strengthening and modifying transverse circulation buoyancy affects tidal flat accretion. Buoyancy affects estuarine sediment transport also by generating estuarine circulation, causing an increase of near bottom flood currents and a decrease of near bottom ebb currents. Altogether it is not easy to predict, even in a qualitative sense, what will be the long term effect of changing fresh water inflow and buoyancy on estuarine morphodynamics. A general answer requires more refined estuarine transport models than presently available. A clue, although with restricted general validity, can be obtained from field observations.

In the Volkerak estuary tidal flats appear to be lower in those regions where buoyancy effects are stronger. This holds in particular for the tidal flats situated at the inner channel bends of sections 6 and 15 (see Figure 7). Whether buoyancy effects are responsible cannot be assessed; tidal and topographical conditions may play a role as well.

For other field evidence we direct our attention to the Wadden Sea, and in particular to the Vliestroom, the Westmeep and the Ameland tidal systems. These tidal basins are similar in size, they experience almost identical tidal forcing and wave influence and possess similar sedimentary structure. The major difference is the fresh water inflow, which is significant (although two orders of magnitude smaller than the tidal discharge) for the Vliestroom and Westmeep tidal systems, and nearly absent for the Ameland tidal system (see Figure 12). The three tidal basins can be characterised as well-mixed meandering coastal plain estuaries, similar to the Volkerak estuary.

Two morphologic characteristics of the different tidal basins are compared. In the first place the ratio V/A of inlet cross-section to tidal prism is considered. This quantity represents an empirical measure for the morphological equilibrium state of the tidal inlet. Differences in the V/A ratio are indicative of differences in sediment transport conditions at the inlet, due to bed material, tides, waves and buoyancy [Gao and Collins, 1994]. A difference in V/A ratio for the Wadden Sea tidal basins might be related to the last factor. In fact, no significant difference in V/A ratio shows up from the data for the Vliestroom, Westmeep and Ameland tidal basins, see table 1.

Table 1. Comparison of morphological characteristics of two Vlie sub-basins (Vliestroom and Westmeep) with the Ameland basin

	Vliestroom basin	Westmeep basin	Ameland basin
Tidal volume V (10^6 m^3)	380	321	450
Inlet cross-section A (10^3 m^2)	25	19	29
Ratio V/A (10^3 m)	15	17	15.5
Hydraulic radius inlet (m)	14.4	12	14.7

The similarity of inlet equilibrium characteristics may be explained by the fact that buoyancy effects are much weaker near the inlet than in the landward part of the tidal basins. Therefore also another morphological characteristic is compared, involving the inland morphological structure. In Figure 13 the hypsometric curves are shown for the different Wadden Sea tidal basins. Here a striking difference appears: the average level of tidal flats in the Ameland basin is higher than in the Vliestroom and Westmeep basins. The Volkerak study suggests that alteration of transverse circulation driven by buoyancy effects may play a role. It is expected that in the buoyancy influenced basins Vliestroom and Westmeep reversal of the up-slope transverse circulation at channel bends occurs, similar as observed in the Volkerak estuary. This provides a mechanism for countering the accretion of tidal flats. Further investigations are undertaken to analyse this hypothesis in more detail.

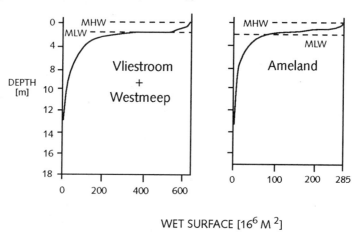

WET SURFACE [16^6 M ^{2}]

Figure 13. Hypsometric curves of the Vlie and Ameland tidal basins, displaying the wet surface as a function of depth.

Conclusions

The present study stresses the importance of buoyancy effects on the dynamics of well mixed estuaries. Particular emphasis is given to the role of transverse circulation, and to the modification of transverse circulation due to buoyancy effects. Even if density stratification is small, the magnitude of transverse circulation is significantly strengthened. As a consequence lateral exchange of water masses over the tidal shear layer at the channel boundaries is increased. This enhances tidal flushing and provides an explanation for the significant increase of longitudinal dispersion with increasing fresh water inflow which is observed in well mixed estuaries. The strengthening of transverse circulation also influences morphodynamic processes. Buoyancy effects may reverse the topographically induced transverse currents in the bends of a meandering tidal channel. In that case the characteristic asymmetry of the channel bend cross-section is absent; the inner channel bend is scoured by density driven flow from the adjacent tidal flat. Some evidence is presented that this density driven transverse flows counteracts the accretion of tidal flats.

References

Bruun, P., *Stability of tidal inlets*, 510 pp.,Elsevier, Amsterdam, 1978.

Chant, R. J., R. E. Wilson and J. Brubaker, Interaction between tidally driven eddy and stratification, *this volume*, 1996.

Dronkers, J. and J. Van de Kreeke, Experimental determination of salt intrusion mechanisms in the Volkerak Estuary, *Netherlands Journal of Sea Research*, 20, 1-19, 1986.

Dronkers, J. and J. T. F. Zimmerman, Some principles of mixing in tidal lagoons, *Oceanologica Acta SP*, 107-117, 1982.

Dyer, K. R., Mixing caused by lateral internal seiching within a partially mixed estuary, *Estuarine and Coastal Shelf Science*, 15, 443-457, 1982.

Dyer, K. R., Estuarine flow interaction with topography, lateral and longitudinal effects, in *Estuarine Circulation*, edited by B.J.Neilson et al., pp. 39-59, Humana Press., Clifton, New Yersey, 1989.

Fischer, H. B., Mass transport mechanisms in partially stratified estuaries, *Journal of Fluid Mechanics*, 53, 671-687, 1972.

Fischer, H. B., E. J. List, R. C. Y. Koh, J. Imberger and N. H. Brooks, *Mixing in inland and coastal waters*, 483 pp., Acadamic Press, New York, 1979.

Gao, S. and M. Collins, Tidal inlet equilibrium, in relation to cross-sectional area and sediment transport patterns, *Estuarine, Coastal and Shelf Science*, 38, 157-172, 1994.

Geyer, W. R. and R. P. Signell, A reassessment of the role of tidal dispersion in estuaries and bays, *Estuaries*, 15, 97-108, 1992.

Geyer, W. P. and R. P. Signell, Three-dimensional tidal flow around headlands, *Journal of Geophysical Research*, 98C, 955-966, 1993.

Helder, W. and P. Ruardij, A one-dimensional mixing and flushing model of the Ems-Dollard Estuary: calculation of time-scales at different river discharges, *Netherlands Journal for Sea Research*, 17, 293-312, 1983.

Huzzey, L. M., The lateral density distribution in a partially mixed estuary, *Estuarine, Coastal and Shelf Science*, 26, 351-358, 1988.

Kalkwijk, J. P. T. and R. Booij, Adaptation of secundary flow in nearly horizontal flow, *Journal of Hydraulic Engineering*, 24, 19-37, 1986.

Kreeke, J. Van de and K. Robazcewska, Effect of wind on the vertical circulation in the Volkerak Estuary, *Netherlands Journal of Sea Research*, 23, 239-253, 1989.

Nunes, R. A. and J. H. Simpson, Axial convergence in a well-mixed estuary, *Estuarine, Coastal and Shelf Science*, 20, 637-649, 1985.

Okubo, A., Effect of shoreline irregularities on streamwise dispersion in estuaries and other embayments, *Netherlands Journal of Sea Research*, 6, 213-204, 1973.

Pritchard, D. W., Observations of circulations in coastal plain estuaries, in *Estuaries*, edited by Lauff, pp. 37-44, A.A.A.S.Publication 83, Washington, 1967.

Ridderinkhof, H. and J. T. F. Zimmerman, Chaotic stirring in a tidal system, *Science*, 258, 1107-1111, 1992.

Rozovskii, I. L., Flow of water in bends of open channels, *Academy of Sciences of the Ukrainian SSR*, 233 pp., Kiev, 1957.

Smith, R., Buoyancy effects upon longitudinal dispersion in wide well mixed estuaries, *Philo-sophical Transactions of the Royal Society*, A296, 467-496, 1980.

Uittenbogaard, R. E., Testing some damping functions for mixing-lenght turbulence models, *Rep. Z721*, 44 pp., Delft Hydraulics, 1993.

Uittenbogaard, R. E., The importance of internal waves for mixing in a stratified estuarine tidal flow, Ph.D. Thesis, Delft University of Technology, 1995.

Zimmerman, J. T. F., The tidal wihrlpool: a review of horizontal dispersion by tidal and residual currents, *Netherlands Journal of Sea Research*, 20, 133-154, 1986.

List of Contributors

David G. Aubrey
Department of Geology and Geophysics
Woods Hole Oceanographic Institution
360 Woods Hole Road, MS 22
Woods Hole, MA 021543-1541

Jack Blanton
Skidaway Institute of Oceanography
10 Ocean Science Circle
Savannah, GA 31411

Bart Chadwick
Environmental Sciences Div., NRaD
53475 Strothe Road
San Diego, CA 92152

Ralph T. Cheng
Water Resource Division
U. S. Geological Survey
345 Middlefield Road, MS 496
Menlo Park, CA 94025

Cynthia N. Cudaback
University of Washington
Geophysical Program
202 ATG
Seattle, WA 98195

J. Dronkers
Directoraat-Generaal Rijkswaterstaat
National Institute for Coastal and Marine
Management/RIKZ
Central Office The Hague
P.O. Box 20907
2500 EX The Hague, The Netherlands

Carl T. Friedrichs
Virginia Institute of Marine Science
School of Marine Science
College of William & Mary
Gloucester Point, VA 23062-1346

W. Rockwell Geyer
Applied Ocean Physics and Engineering
Department
Woods Hole Oceanographic Institution
98 Water Street, Big. 106, MS #12
Woods Hole, MA 02543-1053

Xinyu Guo
Department of Civil and Ocean Engineering
Ehime University
3 Bunkyo-cho
Matsuyama 790, Japan

John M. Hamrick
2520 West Whittaker Close
Williamsburg, VA 23185

Charles Hannah
Oceandyne Environmental Consultants
373 Ridgevale Drive
Bedford, N.S. B4A 3M2, Canada

Clifford J. Hearn
Department of Geography and Oceanography
University College
Australian Defense Force Academy
University of New South Wales
Canberra 2600, Australia

Pablo Huq
College of Marine Studies
Robinson Hall
University of Delaware
Newark, DE 19702

Takashi Ishimaru
Tokyo University of Fisheries
Konan 4-5-7, Minato-ku
Tokyo 108, Japan

David A. Jay
University of Washington
Geophysical Program, AK-50
Seattle, WA 98195

David J. Kay
University of Washington
Geophysical Program
Atmospheric Sciences Bldg., Room 201
Seattle, WA 98195

Albert Y. Kuo
School of Marine Science
Virginia Institute of Marine Science
The College of William & Mary
Gloucester Point, VA 23062

John L. Largier
Scripps Institution of Oceanography
University of California, San Diego
La Jolla, CA 92093-0209
and
Department of Environmental &
Geographical Science
University of Cape Town
Rondebosch, 7700, South Africa

Jonathan O. Lewis
Brixham Environmental Laboratory
Freshwater Quarry
Brixham, Devon TQ5 8BA, United Kingdom

Roy E. Lewis
Brixham Environmental Laboratory
Freshwater Quarry
Brixham, Devon TQ5 8BA, United Kingdom

John W. Loder
Bedford Institute of Oceanography
P.O. Box 1006
1 Challenger Drive
Dartmouth, N.S. B2Y 4A2, Canada

Timothy P. Mavor
College of Marine Studies
Robinson Hall
University of Delaware
Newark, DE 19702

Jeffery D. Musiak
University of Washington
Geophysical Program, AK-50
Seattle, WA 98195

Heidi Nepf
Department of Civil and Environmental
Engineering
Massachusetts Institute of Technology
Cambridge, MA 02139

Marlene A. Noble
Branch of Pacific Marine Geology
U. S. Geological Survey
345 Middlefield Road
Mail Stop 999
Menlo Park, CA 94025

James O'Donnell
Department of Marine Sciences
University of Connecticut
1084 Shennecosset Road
Groton, CT 06340

Kyeong Park
Department of Oceanography
Inha University
Inchon, Korea 402-751

Jonathan R. Pennock
The University of Alabama
Dauphin Island Sea Lab
101 Bienville Blvd.
Dauphin Island, AL 36528

Joshua Plant
Department of Oceanography
University of Hawaii
Marine Science Building 508
1000 Pope Road
Manoa, HI 96822

John Rooney
Department of Oceanography
University of Hawaii
Marine Science Building 421
1000 Pope Road
Manoa, HI 96822

Toshiro Saino
Air and Water Research Institute
Nagoya University
Huro-cho, Chikusa-ku
Nagoya 464-01, Japan

William W. Schroeder
The University of Alabama
Dauphin Island Sea Lab
101 Bienville Blvd.
Dauphin Island, AL 36528

Jonathan Sharples
Department of Oceanography
Southampton Oceanography Centre
University of Southampton
Southampton SO15 3ZH, United Kingdom

John H. Simpson
School of Ocean Sciences
University of Wales, Bangor
Menai Bridge, Gwynedd, LL59 5EY, United Kingdom

Ned P. Smith
Harbor Branch Oceanographic Institution,
5600 U. S. Highway 1, North
Fort Pierce, FL 34946

Ronald Smith
Mathematical Sciences
Loughborough University
Loughborough, Leicestershire LE11 3TU, United Kingdom

Stephen V. Smith
Department of Oceanography
University of Hawaii
Marine Science Building 512
1000 Pope Road
Manoa, HI 96822

Alejandro J. Souza
School of Ocean Sciences
University of Wales, Bangor
Menai Bridge, Gwynedd, LL59 5EY, United Kingdom

J.A. Stephens
Plymouth Marine Laboratory
Prospect Place
Plymouth PL1 3DH, United Kingdom

R.J. Uncles
Plymouth Marine Laboratory
Prospect Place
Plymouth PL1 3DH, United Kingdom

Arnoldo Valle-Levinson
Center for Coastal Physical Oceanography
Old Dominion University
Crittenton Hall
768 52nd Street
Norfolk, VA 23529

Andre W. Visser
Institute for Marine and Atmospheric Research
Utrecht University
Princetonplein 5
3584 CC Utrecht, The Netherlands
now at
Department of Marine and Coastal Ecology
Danish Institute for Fisheries Research
Charlottenlund Castle
DK-2920 Charlottenlund, Denmark

William J. Wiseman, Jr.
Coastal Studies Institute
Louisiana State University
331 Howe-Russell Geosciences Complex
Baton Rouge, LA 70803

Kuo-Chuin Wong
College of Marine Studies
University of Delaware
Robinson Hall
Newark, DE 19716-3501

David G. Wright
Department of Fisheries and Oceans
Bedford Institute of Oceanography
P.O. Box 1006
1 Challenger Drive
Dartmouth, N.S. B2Y 4A2, Canada

Tetsuo Yanagi
Department of Civil and Ocean Engineering
Ehime University
3 Bunkyo-cho
Matsuyama 790, Japan